THE
OF BIR.

Data Driven Statistical Methods

CHAPMAN & HALL TEXTS IN STATISTICAL SCIENCE SERIES

Editors:

Dr Chris Chatfield
Reader in Statistics
School of Mathematical Sciences
University of Bath, UK

Professor Jim V. Zidek
Department of Statistics
University of British Columbia
Canada

The Analysis of Time Series – An introduction Fifth edition
C. Chatfield

Applied Bayesian Forecasting and Time Series Analysis
A. Pole, M. West and J. Harrison

Applied Non-parametric Statistical Methods Second edition
P. Sprent

Applied Statistics – A handbook of BMDP analyses
E. J. Snell

Applied Statistics – Principles and Examples
D.R. Cox and E.J. Snell

Bayesian Data Analysis
A. Gelman, J. Carlin, H. Stern and D. Rubin

Decision Analysis – A Bayesian approach
J.Q. Smith

Elementary Applications of Probability Theory Second edition
H.C. Tuckwell

Elements of Simulation
B.J.T. Morgan

Essential Statistics
Third edition
D.G. Rees

Interpreting Data
A. J. B. Anderson

An Introduction to Generalized Linear Models Second edition
A. J. Dobson

Introduction to Multivariate Analysis
C. Chatfield and A. J. Collins

Introduction to Optimization Methods and their Applications in Statistics
B. S. Everitt

Large Sample Methods in Statistics
P.K. Sen and J. da Motta Singer

Markov Chain Monte Carlo – Stochastic simulation for Bayesian inference
D. Gamerman

Modeling and Analysis of Stochastic Systems
V. Kulkarni

Modelling Binary Data
D. Collett

Modelling Survival Data in Medical Research
D. Collett

Multivariate Analysis of Variance and Repeated Measures
D. J. Hand and C. C. Taylor

Multivariate Statistics - A practical approach
B. Flury and H. Riedwyl

Practical Longitudinal Data Analysis
D.J. Hand and M. Crowder

Practical Statistics for Medical Research
D. G. Altman

Probability – Methods and measurement
A. O'Hagan

Problem Solving - A statistician's guide
Second edition
C. Chatfield

Randomization , Bootstrap and Monte Carlo Methods in Biology
Second edition
B. F. J. Manly

Readings in Decision Analysis
S. French

Statistical Analysis of Reliability Data
M. J. Crowder, A. C. Kimber, T. J. Sweeting and R. L. Smith

Statistical Methods for SPC and TQM
D. Bissell

Statistical Methods in Agriculture and Experimental Biology Second edition
R. Mead, R. N. Curnow and A. M. Hasted

Statistical Process control – Theory and practice Third edition
G. B. Wetherill and D. W. Brown

Statistical Theory Fourth edition
B.W. Lindgren

Statistics for Accountants
S. Letchford

Statistics in Engineering – A practical approach
AV. Metcalfe

The Theory of Linear Models
B. Jorgensen

Full information on the complete range of Chapman & Hall statistics books is available from the publisher.

Data Driven Statistical Methods

P. Sprent

Emeritus Professor of Statistics
University of Dundee
Scotland

CHAPMAN & HALL

London · Weinheim · New York · Tokyo · Melbourne · Madras

Published by Chapman & Hall, an imprint of Thomson Science, 2–6 Boundary Row, London SE1 8HN, UK

Thomson Science, 2–6 Boundary Row, London SE1 8HN, UK

Thomson Science, 115 Fifth Avenue, New York, NY 10003, USA

Thomson Science, Suite 750, 400 Market Street, Philadelphia, PA 19106, USA

Thomson Science, Pappelallee 3, 69469 Weinheim, Germany

First edition 1998

© 1998 Chapman & Hall

|9757 31 X

Thomson Science is a division of International Thomson Publishing I(T)P®

Printed in Great Britain by St Edmundsbury Press, Bury St Edmunds

ISBN 0 412 79540 X

A catalogue record for this book is available from the British Library

∞ Printed on acid-free text paper, manufactured in accordance with ANSI/NISO Z39.48-1992 (Permanence of Paper).

Contents

viii   Contents

Preface

This book is designed both as a textbook to highlight practical aspects of data analysis and statistical inference for students with a preliminary grounding in the subject and also as a reference manual for research workers, scientists, engineers, industrialists, technologists or managers in any field who have a basic knowledge of statistics such as that imparted in a service or introductory course, but who want to know more about possible ways to handle data analytic problems they meet in their work. I deal with methods stimulated by the data rather than those based upon abstract models.

Statistical methods are developing rapidly so a book at this level can only cover selected applications. I try to capture the spirit of the subject by considering a range of techniques in a way that brings together modern developments, especially those made possible by computing power now at our fingertips, but undreamt of 50 or even 20 years ago.

Many undergraduate courses now include computationally feasible methods such as permutation based inference, bootstrapping and diagnostic tests for outliers and influential observations and also introduce powerful tools like generalized linear models. Specialist textbooks on any one of these and other modern developments, admirable as many are, often do not, at least at the elementary level, draw attention to relationships between topics. One result is that many tools are not exploited to maximum advantage.

In this elementary treatment I emphasize breadth rather than depth. This lets me illustrate many relationships between topics and techniques. Despite giving priority to breadth, I could not in 400 pages include even brief accounts of many important methods and applications, but I hope the diversity in my approach will encourage instructors to use the book as a class text that can be supplemented with material from their own fields of interest or from areas particularly relevant to their students. Many teachers will want to develop some topics in more detail, and may regard others as worth less coverage than I have given them. I make no apology for this, for one of my aims is to encourage further exploration. I urge students, research workers, industrialists and others with statistical problems to look beyond this book when they need more information. I give numerous enthusiastic references to specialist books on the main topics (highlighting these by asterisks in the list of references). Because my treatment is elementary a reader wanting to use the methods I describe in any but simple applications should consult a relevant specialist text.

Examples are perhaps the most important aspect of a data-driven approach. It is fashionable in some statistical quarters to argue that all textbook examples and exercises should use only real data. I disagree. Some certainly should, but even for relatively small data sets the computations needed are often feasible only with sophisticated black-box computer programs. Because it is better not to run until one has learnt to walk, I often illustrate rationale without getting lost in computational clutter by using statistically trivial artificial examples that call only for pencil and paper, or require at most a pocket calculator. Then, when appropriate, I move to real data sets. These are mostly still simple because using large sets to illustrate key features often proves counter-productive since much real data has a unique mix of features, making it hard to demonstrate one point without unrealistically ignoring others. It is only when one has mastered the basics that one can tackle multi-facet problems with confidence. I strongly urge instructors to encourage students to analyse their own data, selecting those compatible with their interests and bearing in mind what statistical software is available.

In section 1.3 I outline assumptions I make about what readers will have covered in a preceding basic course in statistics. All such course cover slightly different material. A number of readers will have met previously some topics that are in this book and need only skim over material with which they are already familiar. For more advanced topics I indicate essentials needed to understand my approach, or else suggested background reading for those not familiar with classical model-driven approaches where I present data-driven alternatives. Several sections dealing with more advanced topics have a foot-note indicating they may be omitted at first reading.

Exercises at the end of chapters fall into two categories; the first ask for confirmation of results that are merely quoted or only covered briefly in the text; the second and more advanced exercises, many based on real data, are often open-ended to stimulate initiative, and many require reasonable computing facilities. Again, I encourage instructors, students or research workers using the book to supplement the exercises by applying the methods to data relevant to their particular interests

My sincere thanks go to those who have provided data for particular examples, acknowledged at appropriate points in the text, and special thanks go to Joseph Gastwirth for many useful discussions on points of presentation and in particular for providing me with a wealth of material on applications in a legal context, only a few of which I have been able to cover in this book.

P Sprent
September 1997

1

Data-driven inference

1.1 Data-driven or model-driven

Data are the raw material of applied statistics – providing a focal point or pivot in statistical methodology. I say 'focal' rather than 'starting' point, because even before data are collected the statistician has a role advising what data are needed, and how these should be collected to best answer questions posed by experimenters and others who use numerical information. Both the statistician's *before-* and *after-collection* roles are centered on data.

One impact of computers on data analysis has been to shift emphasis away from restricted probabilistic models chosen on the pragmatic grounds that hopefully they would reflect the main data characteristics while providing a method of analysis that was not too computationally demanding. The shift has been to computer-intensive analyses that let one explore more fully the characteristics of data without artificial constraints imposed by a particular preconceived mathematical model. Obvious limitations of simple probabilistic models are usually drawn to the attention of students early in their training, but some potential difficulties are far from self-evident, and even when they are the way round them may not be clear.

Example 1.1 gives simple data where the assumptions for a standard classic test break down. Example 1.2 covers a situation where a standard test seems reasonable, but it produces results contrary to intuition and further analysis is needed to see why this happens.

Example 1.1 Look at any statistical journal, or indeed at any scientific journal published several decades ago, and at its modern counterpart and you are likely to see changes in the type of material and often in the way it is presented. This may be due in part to changes in editorial or production policy but it also reflects rapid developments in science and increasing pressure on scientists to publish their results in reputable journals. Recently I looked at the statistical aspects of such changes in several journals. I wanted to find out for each whether annual volumes were becoming larger, if the proportion of space devoted to particular topics had changed, whether papers now tended to be shorter or longer, if the geographical distribution of authors had changed, whether joint authorship was becoming more common than individual presentation, and so on. There were supplementary questions. For example, did authors of statistical papers in the 1990s tend to list more references than those writing in the 1950s? *A priori*, this seemed likely in the light of the ever-increasing store of theory and knowledge. As part of my larger study I took a random sample of papers published in the journal *Biometrics* in 1956–7 and another sample from that journal in 1990. The numbers of references in papers in the samples were:

1956–7	2	3	6	6	6	9	9	10	14	15	16	72	
1990	6	8	9	10	16	18	20	22	22	23	23	26	28	59

Do these data support the hypothesis of a higher mean number of references in the later period? The classic test for a difference between population means is the two-sample *t*-test which is strictly relevant to random samples from normal distributions that differ, if at all, only in mean. The samples here are not from normal distributions, but from finite populations of, in each case, less than 100 papers. Also the data are counts, not the continuous variables implicit in a normal distribution. However, a wealth of empirical experience plus certain theoretical consequences of the often-quoted **central limit theorem** indicate that these factors may not matter greatly and that the *t*-test may still be useful. But that test is not satisfactory here because the observations 72 and 59 are isolated from the rest of the data and strongly suggest that the underlying distributions are skew, or that something is peculiar about these two observations. Some people would call them outliers – a concept I describe in section 3.2. The data reflect the fact that *Biometrics* publishes many papers presenting new theory and methods with typically a small to moderate number of references and also a few expository or review papers covering broad subject areas that usually include extensive bibliographies. This implies we are sampling from a mixture of at least two populations, one of which (theory and method papers) is the larger and thus more strongly represented in the random samples. This mixture takes away the near-symmetry expected in samples from a normal distribution even though there is no evidence here that the samples come from populations with different variances. This lack of symmetry reduces the ability (*power* is the technical term described in section 1.7) of the *t*-test to detect shifts in the mean. Indeed, for these data the *t*-test supports the hypothesis that the population means do not differ. Yet over half the papers in the 1990 sample give 20 or more references while only one paper in the 1956–7 sample does. In section 4.1, 4.2 and 4.6 I discuss tests that are in these circumstances more appropriate than the *t*-test for detecting evidence of any location difference.

Example 1.2   Statisticians often meet data that are counts of individuals with various characteristics. A simple case is that where items are allocated to one or other of two categories for each of two criteria. This gives rise to the familiar 2×2 contingency table. We often want to know if there is evidence for association between criteria. Gastwirth and Greenhouse (1995) give data for promotions (first criterion) among 100 employees in a firm where each employee belongs to either a majority or a minority group (second criterion). Data like these are of interest when there are charges of discrimination against some group on the grounds of race, sex, age, religion, etc. The data given for firm A were:

	Promoted	*Not Promoted*	*Total*
Minority group	1	31	32
Majority group	10	58	68

If the data were presented in a court case alleging discrimination on, say, ethnic grounds, Gastwirth and Greenhouse indicate that US courts may seek an explanation from a company as to why its policy is not discriminatory if the appropriate statistical test rejects at the 5% significance level a hypothesis that promotion is independent of ethnic grouping. The company must then try to explain the discrepancy on grounds other than ethnic considerations. It might argue, for example, that promotion required either a prior agreement or a willingness to work unsocial hours or spend long periods away from home and that relatively few in the minority group met that requirement. One appropriate test for independence is Fisher's exact test. If you are not familiar with this test it is described in section 4.6, but it suffices here to say that although the proportion promoted in the minority group is only 1/32, or 3.125% compared to the overall rate of 11% the hypothesis of independence is not rejected at the 5% level by the Fisher test. Gastwirth and Greenhouse give a similar table for a comparable group of 100 employees recruited from the same population for a second firm, B. For this firm the data are

	Promoted	Not Promoted	Total
Minority group	2	46	48
Majority group	9	43	52

The promotion rate is 4.167% in the minority group, greater than the corresponding proportion for firm A, while the overall rate is again 11%, so clearly there is not going to be significant evidence of association.

I made that statement with tongue in cheek, for if we apply the Fisher exact test to confirm lack of association the result is significant at the 5% level, i.e. there is an indication of association that we did not get for firm A! Unlike the situation in Example 1.1, however, it is not the test that is inappropriate. I leave you in suspense until I return to this problem in section 13.6, telling you only that the anomaly arises because we have ignored information in the data that affects the relative performance of the test in the two cases.

In modern statistical inferences two complementary but often overlapping approaches are:

• *Model-driven analyses.* These are based on probabilistic or mathematical models, simple or sophisticated, to encapsulate the main features of data. Once the model is specified, analyses tend to be driven by that model. A well known example is the general linear model that covers analysis of variance and linear regression; validity of these analyses in, for example, basic analysis of variance depends upon assumptions like additivity of effects and homogeneity of error variance. Many inferences are strictly valid only under further assumptions of normality. Transformation of data sometimes induces such conditions, but the price to pay may be increased difficulty of interpretation.

• *Data-driven analyses* There are three subcategories. The first comprises methods that work over a range of potential models and includes exploratory and robust methods. These are useful if there is not enough data-based or other information to select any one probabilistic model. *Robust methods* are ones that perform well for several models, even if optimal for none or only some.

The second kind of data-driven analyses use the data to squeeze out information with only limited assumptions about potential models. These include permutation tests and also the bootstrap and jackknife described in Chapter 2.

The third type of data-driven analyses use diagnostic tools. These are often follow-ups to analyses based on a specific model and show how relevant or reliable that model-based analysis is in coping with the data. Diagnostic tests are widely used in regression. Why use a model-based analysis if one doubts the adequacy of that model? We may do so because there is no readily available data-driven alternative or because it is not clear which of several data-driven alternatives is appropriate. Then it is often best to use a model-based analysis and follow it with diagnostic tests to see whether the analysis is adequate, and if not, what causes any difficulty. Diagnostics may indicate that some particular data-driven, or perhaps an amended model-driven analysis is more appropriate.

Data-driven analyses nearly always use models, but these are usually, at least at the early stages, less specific than the probabilistic ones at the core of the model-driven approach. In data-driven inference the model might be that the data are a random sample from a population with a symmetric distribution (without further specifying that distribution), or that two sets of data are random samples from populations with unspecified distributions that might differ in location (mean or median), but are otherwise identical.

Many methods used in data-driven analyses are computer intensive and may involve anything from small univariate to large multivariate data sets.

There is no dramatic distinction between, or clear boundary to separate, data- and model-driven analyses, but rather a difference in emphasis. In many scientific investigations progress often calls for a data-based analysis that suggests a model which is then tested against further data and that test may show a need to abandon or modify the model, the process often being repeated many times before a satisfactory model is obtained. In Chapter 14 I describe some aspects of a merging of the data- and model-driven approaches.

1.2   Some data-driven methods

This book is about procedures mainly in the data-driven category. It is often illuminating to try both data- and model-driven approaches, or to combine them, as when diagnostic tests are used after a conventional regression analysis.

My cover is far from exhaustive. For example, I do not deal specifically with the data-driven approach widely described as *exploratory data analysis* or *EDA*.

That is an important and useful concept but at the elementary level it is more concerned with highlighting characteristics of data than with making inferences. At a more sophisticated level I have omitted empirical Bayesian methods which many would regard as data-driven. The main themes I cover are:

- randomization and permutation tests and related estimation procedures;
- bootstrapping and related methods;
- robust estimation and treatment of outliers;
- diagnostics, especially for regression;
- inter-relationships between inference methods.

Many data-driven procedures are based on **nonparametric** or **distribution-free** methods, closely related but not synonymous concepts. However, each implies only limited assumptions about populations from which data may have been obtained in the form of samples. Serious use of such procedures began about 1900 when Karl Pearson introduced the chi-squared test for goodness of fit and a few years later C. Spearman proposed a rank correlation coefficient. R.A. Fisher and E.J.G. Pitman independently proposed randomization or permutation tests as widely applicable minimum-assumption exact tests in the 1930s, but these tests were then computationally impracticable. Development of permutation tests using ranks in the late 1940s stimulated the use of distribution-free methods when assumptions for tests based on normal distribution theory broke down or were of doubtful validity. Wide use of more general nonparametric and distribution-free methods based on permutation theory came in the 1980s with the advent of relevant computer software. These developments led to a greater awareness of relationships between ways for tackling apparently diverse problems.

The bootstrap came into prominence in the late 1970s and was soon seen to have links with permutation or randomization tests. The name *distribution-free* carries an implication of robustness – the property of not depending too critically on distribution characteristics. Other robust methods were stimulated by studies of what is termed the **outlier** problem. One group of robust estimators, called *M-estimators*, derive their name from the property that they behave like maximum-likelihood estimators when these are appropriate, but are not too greatly influenced by a few observations clearly contrary to the pattern in the bulk of the data, or to mild perturbations in many observations.

Diagnostic techniques are useful to detect data values that exert, or have the potential to exert, a strong influence on the outcome of an analysis or that are in some way 'out of tune' with the bulk of the data.

In this book I introduce and discuss relationships between these and other topics. I stress the word *introduce* because all these topics have produced at least one well-received student text or reference book for research workers. Excellent as these key expositions are in their own field, and I often refer enthusiastically to them, their concentration on one theme often means they make only passing mention of inter-relationships between topics.

I show – at an elementary level – how these approaches, all important on their own, provide a broad base for many inter-related data-driven methods. Because there is no clear divide I describe some analyses which are essentially model-driven, but which link closely to data-driven methods.   Many diagnostic procedures are of this kind.

1.3 The approach

This book is about interpreting and analysing data and I keep mathematics and formal proofs to the minimum needed to understand the rationale. Mathematical arguments are vital for the theoretical framework that justifies or validates techniques, but an applied statistician needs to understand the implications of this theory rather than the nuts and bolts. Where possible I illustrate points by numerical examples, but I use algebra when it is needed for a proper under-standing or leads to a useful generalization.

I assume a knowledge of basic statistics – familiarity with the normal, binomial and Poisson distributions and an understanding of concepts such as mean, median, variance, standard deviation, symmetric and skew distributions, together with the fundamentals of hypothesis testing and, more important in practice, of interval estimation. I also suppose that the concept of a 'random' sample from a population where the observations $x_1, x_2, \ldots, x_n$ in the sample are independently and identically distributed (i.i.d) is understood.   Some of the conceptually simplest inference problems are about whether or not two or more samples come from populations having the same distribution and their validity often depends upon these samples being random and often on their also being independent of one another. Most one-semester introductory courses in statistics cover these matters in enough depth for present purposes, but please refer to a standard undergraduate statistical text like Rees (1994) or the more advanced Chatfield (1983) if you have difficulty interpreting arguments where I use these or other basic statistical concepts. When more advanced statistical ideas are needed I give either appropriate references or a brief description in the text.

Sections 1.5 and 1.6 give an informal introduction to permutation tests and I use these as a peg for discussing points about inference that are especially relevant to data-driven analyses. Permutation tests feature prominently in the book and are developed in detail from Chapter 4 onward. I introduce the bootstrap in Chapter 2, both as an alternative to permutation tests and more importantly as a tool in its own right when permutation tests may not be available. Bootstrapping is further developed in later chapters when appropriate. Chapter 3 is an overview of key ideas about outliers, influential observations and robust estimation; these too are developed in differing contexts in later chapters. Apart from simple examples in section 3.3 I leave diagnostics to Chapter 11 which is about regression, a topic that brings together four main types of data-driven analyses – permutation tests, bootstrapping, robust methods and diagnostics.

I use small artificial data sets to explain rationale, avoiding a clutter of computational detail inseparable from larger real data sets. Then, where appropriate I show, using real data, how common procedures are implemented by practising statisticians and by research workers to help interpret data and make inferences, highlighting interconnections between methods. For example, I show, particularly in Chapters 9 and 12 that many, though not all, permutation tests are basically about correlation measures that are either generalizations of, or special cases of, the Pearson product-moment coefficient. Thus tests as diverse as the well-known Wilcoxon– Mann–Whitney test (sections 4.2) and an *ad hoc* test for deciding whether side-effect incidence for a drug depends upon dose levels (section 12.4) are both based on the same underlying model. Because many elementary tests are special cases of more advanced procedures I often, as I do here, give cross references to later material. At a first reading you may prefer, indeed be wise, not to take up these references until after you have met that later material.

End-of-chapter exercises either ask you to verify results that are only stated in the text or to analyse some data. The range of potential applications of most techniques in this book is wide and although I cover in the text or as exercises applications in many fields some of these may not be of direct interest to individual readers. I urge readers to apply methods to data sets from their own field of interest, even collecting such data for themselves if that is appropriate.

1.4 Different kinds and sources of data

Data commonly consist of measurements, counts or scores. We measure the thickness of metal sheets produced by one or more machines, record the times of survival of patients subjected to various treatments, count the number of defects in each square metre of glass, assign to each of several varieties of canned soup a score on a scale 1 to 5 to indicate a 'satisfaction' rating for each. In data-driven analyses we often replace or 'score' full data such as measurements by ranks that reflect order. For example, we might assign ranks 1, 2, 3, . . . to an ordered set of weights, the smallest being ranked 1, the next smallest 2, and so on.

The validity, range and usefulness of statistical analyses and inferences depend critically upon how data are obtained. The concepts of **population** and **random sample** are often regarded as basic in statistics yet many valid experiments are carried out on suitable material that just happens to be available. For example, the first clinical test of a new drug may be one where responses to that drug are compared to those of a standard drug or other alternative treatment using a small group of patients in one district or even in one hospital. In no way are these patients a random sample from all patients with the condition being treated. However, if the new drug and the standard treatment are allocated to patients in that small group **at random** we may make valid inferences about the performance of the new drug relative to the standard treatment for the particular patients in the experiment, even though **not all** those patients have been treated with the new drug. It is often reasonable to believe conclusions from such an experiment apply to all patients who are in a broad sense similar to those in the experiment. It is another matter whether the treatment would have the same effect for other types of patient. If patients in the experiment were all male, the result might be different for females; if they were all non-smokers, the outcome might not be the same for smokers. In practice clinicians are often confident, on the basis of past experience, that inferences carry over to larger populations including individuals in different circumstances. Such 'hunch-based' inferences, correct as they often turn out to be, are not statistical inferences in the accepted sense of that term.

At the opposite extreme to an experiment with subjects chosen on the basis of availability or convenience but then allocated at random to treatments, is one where genuinely random samples are taken from large populations. Then a range of statistical techniques can be used to make inferences about populations from which samples are taken. There are also situations, as for the data in Example 1.1, where random samples are taken from small to moderate-sized populations and here valid statistical inferences are also possible. Sampling theory deals specifically with inferences in these circumstances.

Another common type is **observational** data, often data collected as they come to hand or culled from existing records. Inferences from observational data are often made using dubious, even pious, assumptions. It is often assumed that such data exhibit characteristics similar to those of a random sample from a larger population, or if two sets of observational data are being compared, that they are behaving as though the units had been allocated to the sets at random. The pitfall with such assumptions is that differences may be due to factors other than those under study. For example, in comparing the merits of two methods of teaching children to read, if the data are performance measures for a class in one school taught by one method and for a class of pupils of similar age in another school taught by the other method, an apparent difference might not be due to the superiority of one teaching method, but to the fact that one school had an intake of more able pupils, employed more competent teachers, devoted more time to teaching reading, or used different criteria for measuring achievement.

Matters touched upon in this section are well-covered in the literature on experimental design, sampling theory, and fundamentals of statistical inference. An excellent general review of the practical aspects of data collection is given by Altman (1991, Chapters 1–5)

1.5 Randomization and permutation tests

The central role of a randomization test, a specific form of permutation test, in statistical inference from designed experiments was recognised by Fisher (1935) and extended and developed in a trilogy of papers by Pitman (1937a,1937b,1938) who described these tests as *significance tests that may be applied to samples from any population.* This is true in the sense that the tests are conditional only upon the data, and make no assumption about the samples coming from any specific family of distributions, or that they are random samples. Occasionally the samples themselves constitute the entire populations of interest. However, the **domain of validity** of inferences based upon the tests depends strongly on how the samples were obtained. This is explained in section 4.3 and after Example 4.10. Unfortunately, Pitman's wording is sometimes taken to imply that such tests may be used to make inferences without **any** distributional assumptions. Although no definite family of distributions need be specified for populations, a meaningful test often requires assumptions about continuity or symmetry, or about populations differing, if at all, only in location or only in scale.

Both Fisher and Pitman specifically recognized the relevance of such tests in designed experiments where the units used often consist of material chosen only because it is available and suitable for experimental purposes. Here randomization plays a key role; valid inferences about treatment effects can only be made if treatments are allocated **at random** to the units. If this is done, exact test and estimation procedures are based on comparisons between all possible randomizations of the data subject to any constraints imposed by the experimental design. Such nonrandom selection of experimental material, followed by random allocation of treatments to units, is so common in practice that it is virtually the norm. Exact randomization tests for designed experiments belong to a wider class of tests called permutation tests, but many authors, including Edgington (1995), distinguish between tests with random allocation of units to treatments, calling these randomization tests, and more general permutation tests, whereas other writers use the terms *permutation test* and *randomization test* interchangeably.

Fisher (1935) realized implicitly that in a designed experiment using available, rather than randomly selected material, a randomization test is exact in a sense I explain in section 1.7, while the more familiar tests based on the fiction of experimental units being samples from a normal population are only approximate, a theme that has been pursued by many, including Kempthorne (1952, 1955 and elsewhere) and Edgington (1995). Normal theory approximations are valuable, and may not be seriously misleading, thanks to the central limit theorem, when some implicit assumptions do not hold. This justifies their continuing use even though exact randomization tests are now, for small samples at any rate, often as easy, or easier, to carry out. There are also situations where there is no randomization test corresponding to one readily available in, for example, parametric analysis of variance. An example is given in section 7.6. Bailey (1991) explores some parallels and differences between the two approaches.

Additional support for wide use of normal theory parametric tests came when Pitman (1937a) showed that the t-test, which is optimal when testing for a location shift under appropriate assumptions of normality and homogeneity of variance, was often a close approximation to a randomization test applied to the same data even if the relevant assumptions for the t-test clearly did not hold. Pitman's finding, and a less general one by Fisher (1935, section 21) in a different context, were taken to justify using the then easier-to-compute t-test or F-tests as practical tools when normality or homogeneity assumptions were suspect, or there was insufficient evidence to decide whether these assumptions were reasonable. This argument (sections 4.1, 4.2) overlooks questions of robustness.

Pitman applied randomization tests, which I refer to as Fisher–Pitman tests, to measurement data. Replacing measurements by ranks gives robustness against outlying, or rogue, observations and their advantages for computation and testing in the pre-computer era were recognised by Wilcoxon (1945), while Pitman (1948) showed that replacing data by ranks led to only a small loss of efficiency.

Computing problems that inhibited widespread use of permutation tests other than those based on ranks or on counts have been eliminated by modern statistical software. Such software also allows use of robust methods that reduce the influence of contaminated observations or failure of assumptions, and also make possible resampling methods including, in particular, the bootstrap.

1.6 A basic permutation test

Modern statistical software performs in seconds permutation tests that were once impractical for all but unrealistically small data sets. Some software packages allow analysis of the same data in different ways; this is valuable, but a recipe for trouble unless care is taken to make appropriate choices. Many algorithms depend upon inter-relationships between techniques of a kind described by Graubard and Korn (1987), Kendall and Gibbons (1990), Sprent (1993) and others. In particular, expressing many techniques in a categorical data format as two or higher dimensional tables of counts show relationships that help develop *ad hoc* permutation tests for specific problems and data. Such relationships are demonstrated throughout this book.

A classic permutation test is the two-sample Fisher–Pitman test. I apply it here to data from an experiment with 8 units where each unit is allocated at random to one of two treatments and a characteristic is measured on each unit after treatment. Typically in such situations the null hypothesis, denoted by H_0, is

H_0: *There is no differential response in the measured characteristic between treatments*

and a simple alternative hypothesis H_1 is

H_1: *The measured characteristic is greater for one specified treatment.*

These hypotheses make no reference to populations or to parameters because any inferences are only strictly valid to the material used and without further assumptions cannot be made too specific at this stage, although H_1 could be made more precise by saying, for instance, that the effect of the beneficial treatment has been to increase the response by a constant amount for each unit relative to what it would have been if the other treatment had been applied to that same unit, if indeed we had theoretical grounds to believe, or intuition suggested, that is what might happen. This stronger assertion does not change the test procedure but it has implications for the estimation problems I discuss in section 1.8.

Example 1.3 Eight batches of alloy are available to measure the amount of an impurity in metal bars cast at two different temperatures. The experiment is conducted to see if the impurity tends to burn-off at the higher temperature. Batches are allocated at random to each casting temperature and the parts per thousand of impurity measured after casting. I describe the Fisher-Pitman permutation test for H_0: *no difference in impurity level* against H_1: *higher temperature reduces the impurity level.*

The results were:

Higher casting temperature (Group I)	0	3.5	4.2		
Lower casting temperature (Group II)	4	4.6	7	9	10.5

Pitman saw the kernel of the problem this way:

If Group C (C for combined) is the set of all values in Groups I and II, then Groups I and II are discordant if Group I is an unlikely group among all groups of that size that could be obtained by random sampling from Group C. This implies that if Group I is discordant then Group II is also discordant. Discordance suggests rejection of H_0 is appropriate.

This simple statement needs elaboration. What is meant by *unlikely*? Under randomization all possible permutations of 3 from 8 units are equally likely, but *an unlikely group* is one more likely under H_1 than under H_0. To carry out the permutation test we compare the observed group I with all possible groups of 3 that can be formed from the combined group of 8, i.e. from

$$0 \quad 3.5 \quad 4.2 \quad 4 \quad 4.6 \quad 7 \quad 9 \quad 10.5$$

Standard permutation theory shows there are $^8C_3 = 56$ distinct sets of 3 items (here observations) that may be selected from 8 different items. Each set is equal likely under random selection. We could write down all possible selections, but need not do so. If the alternative hypothesis H_1 is true, then Group I values will tend to be lower than Group II values. A reasonable measure of the *smallness* of values in a group of fixed size is the group mean, or even simpler, the sum of the group values. The following procedure is intuitively reasonable:

- Arrange all groups of 3 from the 8 observations in increasing order of their sum, S_1.
- Reject H_0 at the $100\alpha\%$ significance level if the observed group is among the $100\alpha\%$ of groups with the lowest sums. This one-tail test is appropriate because we have specified a lower impurity level at higher casting temperature under H_1.

In classic hypothesis testing common choices of α are $\alpha = 0.05$ or $\alpha = 0.01$ where these are the probabilities, when H_0 is true, of getting a sum S_1, as low or lower than that observed. In this example because there are 56 different groups of three, 5% of these would be 2.8 groups. Clearly in practice we must choose a critical region, a term explained in the next section if you are not already familiar with it, containing either 2, 3 or some other integer number of 'extreme' groups and it is easily seen that choosing 2 and 3 corresponds to tests at significance levels $(2/56) \times 100 = 3.57\%$ and $(3/56) \times 100 = 5.36\%$. For illustrative purposes, I choose the smaller number and test at the 3.57% level. We describe this, or any other realizable level, as an exact level, and call the associated test **exact** for reasons I discuss in Section 1.7, where I consider possible significance levels for permutation tests more fully. We reject H_0 only if the observed Group I has either the smallest sum or the second smallest sum among all possible groups of 3; for then, and only then, is our result in the critical region. Eye inspection of the data shows that the groups with the three smallest sums are:

$$\begin{array}{cccl} 0 & 3.5 & 4 & (\text{sum } 7.5); \\ 0 & 3.5 & 4.2 & (\text{sum } 7.7), \text{ our observed group}; \\ 0 & 3.5 & 4.6 & (\text{sum } 8.1). \end{array}$$

Thus, we reject H_0 at the 3.57% significance level. We only need to calculate these 3 from the 56 possible sums to justify this conclusion.

Other test statistics may be used. I could have selected the groups of 5 with the largest sums and rejected the hypothesis if the observed group of 5 had the largest or second largest sum. This is because the sum for the combined group of 8 is fixed, so the group of 5 with the largest sum matches the group of 3 with the smallest sum, and so on. Other different, but equivalent, test statistics include the mean of the smaller group or the observed smaller group sum minus the mean of that sum for all possible groups of three. Another alternative is the difference between the smaller and larger group means. In this and some other permutation tests, if S is an appropriate test statistic so is $T = a + bS$ where a and b are constants, although of course the numerical values of T and S that indicate significance at a specified level will be different. Confusion often arises because different statisticians, or different programs, may use any of these, or other **equivalent** statistics. For example, it can be shown (see Exercise 1.1) that for the Pitman–Fisher test we may use the two-independent sample t-statistic, because if we compute t and the S_1 used above for all permutations they will be in the same order. However, the value of t required here for significance is no longer given by conventional normal theory t-tables. This should be no surprise, because we have not assumed our data are a random sample from a normal, or any other large population. The arguments used here for equivalence of tests based on statistics that are the sum of values for one group or sample, that group mean, the difference between group means or the t-statistic apply to this particular test and some others. This equivalence (Example 4.2) does not hold if we based our test on group medians rather than means, so care is needed in selecting an appropriate statistic.

The above test is simple only because the artificial data in Example 1.3 is a small set. If, more realistically, we had 10 observations in the first and 12 in the second group, and all were different, then there are $^{22}C_{10} = 646\,646$ different groups of 10, and we easily verify that for a test at an exact significance level as close as possible to, but not exceeding 5%, (exact level $16\,166/323\,323$) we must decide whether our observed group of 10 is among the 32 332 groups having the smallest sums. To see the magnitude of the task without a computer try to write down the five groups of 10 with the smallest sums selected from a combined sample associated with the following data:

Group I	2.3 2.7 3.1 4.7 4.9 5.4 5.7 6.1 6.3 7.2
Group II	2.8 2.9 3.2 4.8 5.2 6.2 7.1 7.9 8.1 8.4 8.7 9.3

Do not waste much time on this. Using the software package StatXact 3 it took about one second to show that this group is not among the 32 332 groups with the lowest sums. The output did not tell me exactly how many groups had a lower sum, but it was easy to deduce that the number was about 45 200.

Extension from a one-tail to a two-tail test with H_0 as in Example 1.3 but the alternative H_1:*one temperature gives a different (i.e. either higher or lower) impurity level* is discussed in another context in section 1.7.

We are often interested in more specific hypotheses that imply some model; e.g. the model may postulate firstly that under H_1 the level of impurity for one treatment relative to the other is reduced by some, usually unknown, constant amount $\theta > 0$. This is called a *fixed effect* model. A slightly more complicated model known as the *random effect* model replaces θ by a random variable, τ, say, with (again usually unknown) mean, $E(\tau) = \theta > 0$. For validity of Student's t-test there are further assumptions about normality and homogeneity of variance and identity of distribution of all observations under H_0. An even more general model might simply specify that reduction for unit i is $\tau_i \geq 0$ with strict inequality for at least one i without specifying a distribution for the τ_i. I consider models like these in a distribution-free context in Chapter 4.

The problem in Example 1.3 may be put in a contingency table format. I assume familiarity with the concept of an $r \times c$ contingency table of r rows and c columns, where each row corresponds to a different attribute, or measure, of one characteristic and each column to a different attribute, or measure, of another characteristic. The number at the intersection of row i and column j [called cell (i, j)] is the number of items that have the attributes specified by that row and column combination. Example 1.3 needs a 2×8 table .Each row corresponds to a group (I or II) and each column to one of the 8 observed values. It is convenient (but not essential) to arrange columns in ascending order of observed values.

Since each observed value occurs only once, in any column the two cell entries are either 0 (for absence) or 1 (for presence), the positions of the 0 and 1 depending upon whether an observed value is in Group I or Group II. Row totals represent the number of observations in each group, and the column totals are 1 for each column. Table 1.1 shows the data for Example 1.3 in this format.

Each possible arrangement of zeros and ones in the cells, subject to the constraints that row and column totals are fixed at the values in Table 1.1, corresponds to a permutation of the data. The sums used for testing in Example 1.3 were those of the 3 Group I values for the permutations of interest. We recover these and other equivalent statistics from Table 1.1 by allocating certain scores to each row and column. Specifically, we take the observed values (as in the column headings) as column scores. We allocate the scores 1, 0 respectively to the rows. Denote, more generally, the score for the ith row by u_i and that for the jth column by v_j and the entry (here either 1 or 0) in cell (i, j) by n_{ij}. It is easily verified for the set-up above that the first group sum S_1 is given by

$$S_1 = \sum u_i v_j n_{ij}$$

where $\sum$ refers to summation over all cells in the table.

Table 1.1 A contingency table format for the data in Example 1.3

Obs. values	0	3.5	4	4.2	4.6	7	9	10.5	Row total
Group I	1	1	0	1	0	0	0	0	3
Group II	0	0	1	0	1	1	1	1	5
Column total	1	1	1	1	1	1	1	1	8

For each permutation there is a different pattern of 0s and 1s constrained only to give the same row and column totals. With the scores unchanged, S_1 again gives the sum of the 3 Group I values for the permutation being considered. To base the test on sums of values in Group II, the row scores 1,0 are changed to 0,1.

This contingency table approach may look like using a sledge-hammer to crack a nut, but it is important because

- it helps to show how this particular test is related to several I describe later; some of these are alternative two-sample tests; others are more general;
- it provides a useful framework for dealing with situations where there are observations with tied values or with grouped data (e.g. Example 4.8 in section 4.4) or other special features;
- it is basic to some fast and accurate computer algorithms.

In this book the first and second points are illustrated in a number of cases. The last one is important for computation, but primarily concerns software developers.

1.7 More about tests

The Fisher–Pitman test described in section 1.6 is a **conditional** test because the value of the test statistic needed for significance at any level is *conditional* upon the data values. This differs from Student's t-test, where for samples of fixed sizes (or more precisely for given degrees of freedom), the value of t for significance at a level $100\alpha\%$ is fixed irrespective of the sample values.

Modern computing methods have influenced our approach to hypothesis testing and estimation and also some aspects of classic hypothesis testing need modification when used in permutation tests and other distribution-free or nonparametric contexts. I discuss some of these changes informally, assuming familiarity with basic classic statistical inference.

Significance levels. Historically, tests like the t-test were carried out at a **preassigned** significance level expressed either as a probability, usually denoted by α, or an equivalent percentage, $100\alpha\%$. In this book I refer to percentage significance levels, but use probability in relation to sizes of critical regions, etc. Typical preassigned values were $\alpha = 0.05$ or 0.01, values partly dictated by constraints imposed by published tables.

Modern computing methods let us do better. Most statistical software for standard tests like the t- or F-tests give the exact probability that the observed or greater (or the observed or less, if appropriate) value of a calculated test statistic will occur under H_0 (i.e. when H_0 holds). This value is sometimes referred to as the significance probability, but more commonly as an exact **p-value**, and if we reject H_0 in a valid one tail-test on the basis of this evidence, p is the exact probability of making an **error of the first kind**, or of rejecting H_0 when true.

Suitably **small** p-values are effectively exact significant levels. This attitude and terminology may worry those who believe that a significance level should be chosen *a priori* at some fixed value, e.g. 5%, and significance is declared if and only if we obtain a lower p-value, but I do not see this as a practical difficulty. If a distinction is thought necessary, the term p-value provides an alternative to the term *exact significance level*. However, in practice p-values may be large, e.g. $p = 0.23$, and this cannot imply significance in any meaningful sense; only suitably small p-values imply significance. In this context an exact p-value calculated, for example, on the basis of the t-distribution is, strictly speaking, only exact if the conditions for validity of that test hold. However, for a permutation test like that in Example 1.3, subject to appropriate random allocation to treatments, the calculated p-value is the exact probability of rejecting H_0 when it is indeed true, conditional only upon our observed data, and no further assumptions are involved in calling this an exact level. It is in this sense that permutation tests are exact.

Discontinuities in possible p-values in permutation tests as in Example 1.3 cause some logical difficulties in the Neyman–Pearson theory of inference, in particular in relation to power comparisons (a concept described below), but these are usually unimportant in practice.

A decision whether to reject at a significance level corresponding to an observed *p-value* is sometimes made in the light of consequences of a wrong decision. The conventional 5% and 1% significance levels are good yardsticks but should not be used as sharp cut-off points. If an exact test gives $p = 0.053$ report this. More importantly, if an exact p-value (e.g. $p = 0.013$) is less than some preselected $\alpha = 0.05$, say, it is more informative to say a result is significant at an exact 1.3% level than merely to state that it is significant at a (nominal) 5% level but not significant at the 1% level.

Since p-values for tests like the t-test are only exact, and the corresponding test is only exact, if the assumptions needed for validity hold, in cases of doubt an appropriate permutation test may be preferable because it tells us with only minimal assumptions the exact probability of making an error of the first kind, i.e. of rejecting the null hypothesis when it is true.

As in Example 1.3, exact permutation tests may be possible only at certain discrete values of p, but as I indicated in section 1.6 the discontinuity problem may be trivial for large or even moderate sized data sets.

If large (small) values of a statistic indicate significance the usual practice, and the one I follow, is to include all values of the statistic greater (less) than or equal to that observed in the region, known as the **critical region**, that lead to rejection. The sum of the associated probabilities under H_0 is the corresponding p-value. Values of the statistic outside the critical region lie in a region called the **acceptance region.** An alternative is to split the probability corresponding to the **observed** value of a statistic equally between the critical region and the acceptance region. Programs, such as those in StatXact or TESTIMATE described in section 1.10, allow a choice. When relevant, both programs quote the probability under H_0 that a test statistic S will be greater than or equal to its observed value (or be less than or equal to, if appropriate), and also give the probability associated with the observed value itself, enabling one to adjust the exact significance level if this probability is split between the regions.   The second approach is described as using **mid-p values**; if you prefer this approach use it consistently. If there is doubt, say which approach you are adopting. The difference is usually only of practical importance for small data sets, or sometimes when there is a mixture of small and large samples.

Example 1.4 In the Fisher–Pitman test if we have samples of 4 and 5 that are random samples from some populations and wish to test the hypothesis

H_0: *Samples are from identical populations*

against

H_1: *First sample is from a population with a greater median*

then, as in Example 1.3, an appropriate critical region is determined by selecting from among the first sample sums for all permutations, a set that have the largest values. Arguing as in Example 1.3 there are $^9C_4=126$ permutations.   Under H_0 the probability of our data giving a statistic corresponding to any sum (e.g. the third largest) is 1/126, and the overall probability associated with that sum for all permutations will be $r/126$ if r permutations give that same sum. If the observations correspond to the permutation with the third largest sum, and no other permutation gives that sum, using the 'greater than or equal to' convention the exact test significance level is $(3/126)\times100 = 2.38\%$. If we adopt the mid-$p$ value convention we would claim an exact significance level of $(2.5/126)\times100 = 1.98\%$.

If more than one permutation gives the same sum as the observed statistic, S_1, these levels are modified. For example, if the data are

| First sample | 2.2 | 2.3 | 2.5 | 3.1 | |
| Second sample | 1.3 | 1.6 | 1.8 | 2.1 | 2.4 |

then the permutation with the largest first sample sum is 2.3, 2.4, 2.5, 3.1 with $S_1 = 10.3$, while the sample 2.2, 2.4, 2..5, 3.1 give $S_1 = 10.2$ and the samples 2.2, 2.3, 2.5 and 3.1 (that observed) and 2.1, 2.4, 2.5 and 3.1 both give $S_1 = 10.1$. Thus the value $S_1 = 10.1$ arises with two permutations and has associated probability 2/126. Our exact test significance level under the 'greater than or equal to rule' is $(4/126)\times100 = 3.17\%$ and using a mid-p value it is $(3/126)\times100 = 2.38\%$.

The effect of discontinuities in exact p-values may be substantial for small samples, but rapidly diminishes as sample size increases providing there are not many tied values in the data. With samples of size 9 and 14 there are $^{23}C_9 = 817\ 190$ permutations. The probability associated with an S that is given by only one permutation is thus $1/817\ 190$. Even if 10 permutations give the same S the probability associated with that sum is only 0.000 012.

One- and two-tail tests. The appropriate test for a possible location shift, e.g.

$$median\ for\ population\ II\ -\ median\ for\ population\ I = \theta$$

of the form $H_0:\theta = 0$ against the alternative $H_1:\theta > 0$ is a one-tail test. This formulation for a one-tail test is not always appropriate, because although we may only be interested in changes represented by a positive θ, it excludes any possibility of a negative θ. If, as is often the case, this is a possibility, the null-hypothesis should be stated as $H_0:\theta \leq 0$ but the test procedure is unaltered by this often more realistic specification.

A one-tail test is so-called because the critical region is located in one tail of the distribution of the statistic. It is important to select the tail appropriate to the alternative hypothesis, H_1. If H_1 specifies a higher median for population II and S_1 is the sum of the first sample values, then clearly the appropriate critical region lies in the lower tail, for low values in the first sample, relative to those in the second, indicate a lower population I median and supports H_1.

A test of $H_0:\theta = 0$ against the alternative $H_1:\theta \neq 0$ is a two-tail test because either very large or very small values of S_1 indicate significance. In classic parametric tests a common practice for two-tail tests is to double the probability appropriate to a one-tail test. This is ideal for an exact permutation test if the permutation distribution of the test statistic is symmetric. If it is not, an alternative procedure is to form a critical region based upon those permutations that give S values that differ from that statistics' mean over all permutations by a magnitude equal to or greater than that of the observed value. The size of the resulting critical region is the sum of the probabilities associated with all such S values included in either tail. Programs such as StatXact give information that allows a choice between these approaches when a two-tail test is appropriate.

Example 1.5 In Example 1.4 I considered the data

First sample	2.2	2.3	2.5	3.1	
Second sample	1.3	1.6	1.8	2.1	2.4

If we are interested in a median shift, θ, and wish to test $H_0:\theta = 0$ against $H_1:\theta \neq 0$ a two-tail test is appropriate. The sum of the first sample values is $S_1 = 10.1$ and this corresponds to a one-tail significance level of $(4/126)\times100 = 3.175\%$ (because one other permutation gives the same sum and two give greater sums). Doubling this figure for a two tail-test gives a 6.35% level.

Effectively this means we include also the four permutations corresponding to the lowest S_1. However, the S_1 values are not symmetrically distributed about their mean. A standard result in sampling theory (see, e.g. Cochran (1953, p.14) tells us that the mean value of S_1, the sum of the values for all possible samples of size m (here these are the samples corresponding to all permutations of 4 from 9) from a combined sample of size $N = m + n$ is obtained by multiplying the sum of all observations in the two samples by $m/(m + n) = m/N$. Here $m = 4$, $n = 5$ and the sum of all the observed values is 19.3, whence the mean or expectation of S_1 is $E(S_1) = (4/9) \times 19.3 = 8.58$. Since the observed $S_1 = 10.1$ exceeds $E(S_1)$ by 1.52 the approach that includes values of S_1 at the same or a greater distance below the mean leads to inclusion also of all $S_1 \leq 8.58 - 1.52 = 7.06$. Because the data are given to only one decimal place this implies that we include all permutations giving $S_1 \leq 7.0$ in the critical region. Inspecting the data verifies that the relevant permutations of 4 and their sums are:

Permutation	S
1.3 1.6 1.8 2.1	6.8
1.3 1.6 1.8 2.2	6.9
1.3 1.6 1.8 2.3	7.0

contributing a probability 3/126 to the critical region size in addition to the probability 4/126 from the upper tail, giving an exact significance level for a two-tail test of $(7/126) \times 100 = 5.56\%$ compared to 6.35% obtained by doubling the one-tail exact probability. Sometimes the difference is more marked and it is even possible that there may be no values in the opposite tail differing from $E(S_1)$ by more than the difference for the observed S_1. In that case, if the latter method is used, one gets the same probability for a one- or a two-tail test. Arguments can be made for and against either approach. I prefer to double the one-tail probability, but as in the situation where one chooses between allocating the whole, or only half, the probability associated with the observed S_1 to the critical region, my advice is to say what you are doing and to do it consistently, or do what StatXact or TESTIMATE do, and provide sufficient information for users to choose their preferred option. These, and other possible choices, are considered in particular contexts by Yates (1984) and in the published discussion thereon.

For a given data set the mean or expectation of S_1 is a constant calculated in the way given in Example 1.5. Subtracting that constant from S_1 gives an alternative statistic that is valid for performing tests. Many writers do this, and computer programs may give that alternative statistic, but again be consistent in what you do. Some programs calculate the usual t-statistic for each permutation and as already indicated (see also Exercise 1.1) this is also a valid statistic for an exact test that is based on the permutation distribution of t for the observed data rather than upon the normal-theory Student's t-distribution.

Power and efficiency. Two kinds of 'wrong' decision may result from a hypothesis test. Rejecting H_0 when it is true is *an error of the first kind* considered above. In an exact test at the $100\alpha\%$ significance level the probability of doing so is α. Accepting H_0 when H_1 is true is an *error of the second kind*. The probability of making an error of the second kind is often denoted by β. Even if we choose α and the test is exact, we generally do not know β and indeed in a test for location shift, for instance, its value depends upon how far the true θ departs from the value specified in H_0 and also upon the choice of α and the sample sizes. To test $H_0:\theta = 0$ against $H_1:\theta \neq 0$, we instinctively seek a test for which β is small when θ departs markedly from zero. For good tests with a given α, the value of β decreases as $|\theta|$, or the sample sizes or both, increase. The quantity $1 - \beta$ is called the *power* of a test, and a good test is one that has high power for detecting departures of interest. The power equals the probability of getting a result in the chosen critical region (thus rejecting H_0) when H_1 is true. This is what we want to do. The goal is a uniformly most powerful test which is a test that is more powerful than any other test for all alternative hypotheses of interest. Such tests do not always exist and even when they do, they may not perform well when relevant assumptions break down. A locally most powerful test, or one that is most powerful for determining small departures from H_0, is a desirable goal when there is no uniformly most powerful test, for such tests will still usually be quite powerful for detecting larger departures, even though in that situation a more powerful alternative test may exist. The power of a test against a specific alternative is a useful concept when deciding how large a sample is needed to give a reasonable prospect of rejecting H_0 when the alternative is in fact true. There are sometimes practical difficulties in determining power in real-life applications.

Allied to the concept of power is that of efficiency. The efficiency of one test relative to another is measured by the ratio of the sample sizes needed to attain the same power for detecting a given true specified θ included in H_1 for a chosen α. In general, attaining a given efficiency depends on sample size (which we may be able to control), our choice of an *a priori* α (which is in our hands), the power of a test (outwith our control except insofar as we aim to select a test with good power). A widely used general measure of efficiency for permutation tests relative to one another or relative to parametric tests was introduced by Pitman (1948). Based on earlier work on similar concepts he called it the **asymptotic relative efficiency**. Alternative asymptotic measures of efficiency have been proposed, but in this book quoted asymptotic efficiencies are Pitman ones and to make this clear I refer to them as **Pitman efficiencies**. Examples of the calculation of these efficiencies for many tests are given by Gibbons and Chakraborti (1992). Pitman efficiency is independent of choice of α or β, applies strictly to large samples (hence the name asymptotic) and is relevant to the detection of even small departures from H_0. A test having Pitman efficiency of 1 relative to some other test means that for large samples of a fixed size both tests should perform equally well. A Pitman efficiency of one test relative to another greater than 1 indicates that, for large samples at any rate, the first is superior. The actual efficiency in small samples is often close to, or sometimes greater than, the Pitman efficiency. When one of two competing tests has a Pitman efficiency less than 1 relative to the other, we may sometimes prefer the one with lower efficiency. For instance, if that test requires only ranked data while the alternative test needs exact measurements, the ranked data might be easier and cheaper to obtain, making it economic to get the same power by taking a larger sample and making only the simpler measurement. Comparing Pitman efficiencies must be related to available information. For example, I point out in section 4.5 that there is a test with Pitman efficiency 3 relative to the t-test for data sampled from an exponential distribution, yet when sampling from the normal distribution the efficiency of this test relative to the t-test is less than 1.

Care is needed in interpreting p-values. While a high p-value (e.g. $p > 0.10$) provides no evidence for rejecting H_0 it may arise because the test applied lacks power. Many examples in this book demonstrate the desirability of selecting a powerful test. If one has sufficient data and uses an appropriate test it is nearly always possible to obtain a small p-value (e.g. $p < 0.05$) thus providing strong evidence against H_0, but such statistical significance need not imply practical importance. For example, there are a number of drugs that will reduce blood pressure by between 15 and 20 mm Hg, and because of the natural variability of blood pressure only differences of this, or sometimes greater, magnitudes are clinically important. If a new drug for reducing hypertension is being tested with hypotheses

$$H_0 : \textit{Average reduction, } \theta = 0$$

against

$$H_1 : \theta > 0$$

we want in practice to be reasonably sure of detecting values of $\theta \geq 20$. We will be even happier if, when there is a significant difference, we can be fairly certain whether that difference is less than 20 or is 20 or more. That idealised state of affairs is generally not attainable, but studies of precision enable us to design experiments that have a good chance of achieving this first objective and will also give a good estimate of the true difference in the sense I describe below by making use of a confidence interval.

1.8 Confidence intervals

For location differences, e.g. a difference between population medians, we may wish to test $H_0 : \theta = \theta_0$, where θ_0 is some specified non-zero value against either a one- or two-sided alternative specifying different values of θ. If the data are of the form

$$
\begin{array}{ll}
\textit{Sample I} & x_1\ x_2\ x_3\ .\ .\ .\ x_m \\
\textit{Sample II} & y_1\ y_2\ y_3\ .\ .\ .\ y_n
\end{array}
$$

and we claim under H_0 that

$$\theta_0 = (\textit{median population II}) - (\textit{median population I})$$

where θ_0 is specified, then the problem of testing $H_0{:}\theta = \theta_0$ against $H_1{:}\theta \neq \theta_0$ may be reduced to one of testing for a zero population difference under H_0 if we first replace each y_i by $v_i = y_i - \theta_0$. This transformation is important in understanding the rationale of forming confidence intervals.

I assume familiarity with the way to find a 95% confidence interval for a mean difference based on the distribution of the statistic t used in a t-test.

We often want a confidence interval associated, for example, with either a difference, θ, between the medians, means or other indicators of location where that interval is based solely on information derived from two samples. What is measured may be regarded as an estimate of a parameter that indicates location difference between otherwise identical distributions, however vaguely these are specified. In that sense confidence intervals based on permutation tests are better termed distribution-free rather than nonparametric, for we are looking upon a location difference as a parameter difference, where 'parameter' is used in a broad sense for some measurable population characteristic rather than in the restricted sense of a constant appearing in the mathematical expression for a distribution function. A $100(1 - \alpha)\%$ confidence interval for a single parameter may be interpreted as an interval that specifies all possible values θ_0 of θ for which we would accept $H_0{:}\theta = \theta_0$ in a test at the $100\alpha\%$ significance level. Confidence intervals may be associated with a one-tail or a two-tail test. The latter are usually of prime interest, so I consider them first.

In cases like that in Example 1.3 where the data are the entire population and the permutation test is justified by random allocation of treatments it is still possible to establish a confidence interval for the mean treatment effect implied by a decision to reject H_0 at some specific significance level, but the result would be applicable strictly only to those units (which would be allocated to treatments differently under other randomizations).

For the Fisher–Pitman test the rationale for determining the set of parameter values for which we would not reject a given null hypothesis is almost self-evident in principle and is easily described, but the technique for determining the interval for a particular data set requires subtle though simple reasoning. The difficulty stems from the conditional nature of the test. For each value θ_0 specified in H_0 the transformation $v_i = y_i - \theta_0$ needed to reduce this to a test of zero difference leads to a permutation test on a new data set because different v_i

replace the y_i. Thus different numerical values of S_1 will indicate significance for different values of θ_0. An example shows how to deal with this problem.

Example 1.6. Consider the data

Sample I	1.9	2.3	2.5	3.1	
Sample II	1.4	1.6	1.8	2.1	2.4

which may be either random samples from large populations or groupings based on random treatment allocations to units that happen to be available and suitable for experimentation..

As indicated at the start of this section, if $\theta = $ *median population II – median population I* and we want to test $H_0:\theta = \theta_0$ against $H_1:\theta \neq \theta_0$ we replace the second sample observations by $1.4 - \theta_0,\ 1.6 - \theta_0,\ 1.8 - \theta_0,\ 2.1 - \theta_0,\ 2.4 - \theta_0$. and then test $H_0:\theta = 0$. To obtain a $100(1 - \alpha)\%$ confidence interval for the location difference we need to determine **all** θ_0 for which we would accept the original H_0 when testing at the $100\alpha\%$ significance level. In this example, since there are $^9C_4 = 126$ permutations, we might perform for any θ_0 a two tail test at an exact $(6/126) \times 100 = 4.76\%$ significance level, implying a critical region consisting of the three highest and the three lowest values of the statistic S_1, the sum of permutations of four. In theory we may consider all θ_0 between $-\infty$ and ∞ but clearly if $\theta_0 = -\infty$, or indeed takes any value less than -1.7, our observed S_1 would be smaller than that for any other permutation because our amended sample II values would all be greater than any in the observed sample I, so any other permutation would give a greater S_1. Now consider what happens as θ_0 increases. As it increases through -1.7 it takes a value at which one of the permutations, other than that corresponding to our first sample, gives the same value as S_1, i.e. at that stage one of the original first sample observations x_i could be replaced by one of the $v_j = y_j - (-1.7)$ where y_j is a second sample observation, to give a sum equal to the existing S_1, and as soon as θ_0 increases a little further the original S_1 is no longer the smallest permutation sum. A moment's reflection shows that if we continue increasing θ_0 then at some later stage once again a different permutation will give the same S_1 as that for our data and that as θ_0 continues to increase, eventually our original S_1 will be a sum larger than that corresponding to any other permutation, because when θ_0 becomes large and positive the $y_j - \theta_0$ will all become negative and clearly any permutations involving one or more $y_j - \theta_0$ in our statistic will have a lesser sum than the S_1 for our data. Elementary algebra shows how we use this information and the way it generalises. Call the four Sample I values $x_1,\ x_2,\ x_3,\ x_4$ and the modified sample II values $y_1 - \theta_0, y_2 - \theta_0, y_3 - \theta_0, y_4 - \theta_0, y_5 - \theta_0$. If, say, for some θ_0, we find $x_1 + (y_3 - \theta_0) + x_3 + x_4 = x_1 + x_2 + x_3 + x_4 = S_1$, then S_1 changes its order among the first-sample sums for all permutations as θ_0 increases through this value. This equality implies $y_3 - \theta_0 = x_2$ when the former replaces the latter. This in turn implies that the change has occurred when $\theta_0 = y_3 - x_2$. The argument generalizes and a change in the ordering occurs for any $\theta_0 = y_i - x_j$. Clearly there are $4 \times 5 = 20$ such pairings of y_i and x_j. Changes at certain values of θ_0 also occur if two, three or all four of the x_i are replaced by the same number of $y_j - \theta_0$. For example, if

$$x_1 + (y_3 - \theta_0) + (y_1 - \theta_0) + x_4 = x_1 + x_2 + x_3 + x_4 = S_1,$$

this implies $\theta_0 = (y_1 + y_3)/2 - (x_2 + x_3)/2$. I ask you to verify in Exercise 1.2 that this argument generalises to all possible pairings of the x and the y. Clearly also we may get order changes of S_1 where 3 or all 4 of the x_i are replaced by different $y_j - \theta_0$. In these cases the value of θ_0 at which this occurs is the difference between the mean of the y values that enter S_1 and the mean of the x values that are replaced (Exercise 1.2). Straightforward combinatorial algebra shows that there are ${}^5C_2 \times {}^4C_2 = 60$ two-replacement changes, ${}^5C_3 \times {}^4C_3 = 40$ three-replacement changes and ${}^5C_4 \times {}^4C_4 = 5$ four-replacement change, giving a total of $20 + 60 + 40 + 5 = 125$ changes in the ordering of the observed S_1 among those corresponding to all 126 permutations, sufficient to change it from the smallest to the largest possible S_1. To get an appropriate confidence interval we arrange the θ_0 at which these reorderings occur in ascending order, and the end points of our required $(1-4.76)100\%{=}95.24\%$ interval are given by the third smallest and the third largest θ_0. Ordering all 125 relevant change-values of θ_0 would be a formidable task, but it is not hard to find the three largest and the three smallest for our small samples. Among the differences $y_j - x_i$ only the three largest and three smallest can possibly be relevant and these are easily obtained by inspecting the data. We need only look at paired differences starting with the largest and smallest until we find one that is not included in the extremes already computed and so on. The relevant cases are listed in Table 1.2. The entries in the first line are the difference between the largest y and the smallest x, the means of the two largest y and the two smallest x, and the means of the three largest y and the three smallest x. Check that you see how the remaining entries are obtained, e.g. that for the third largest change point with one x changed is the difference between the largest x and the third largest y. Verify (Exercise 1.3) that no changes including four second-sample entries are relevant.

The exact $100(1 - 6/126) = 95.24\%$ confidence interval is the interval between the third smallest and the third largest entry in Table 1.2, i.e. $(-1.3, 0.15)$. The end points of this interval are the confidence limits. Because of discontinuity in p-values and the fact that the end points of the interval correspond to passing through tied values that correspond to change points in the ordering of S there are discontinuities in the confidence level at these points. This is discussed further in Exercise 1.4.

The arguments in Example 1.6 extend in an obvious way to samples of m, n observations, but computation becomes formidable for medium to large m, n.

A two-tail confidence interval is usually more appropriate than a one-tail interval where we exclude values of θ that are either *too large* or *too small* but not both. In the one-tail case, if a lower tail is our critical region a confidence interval take the form (a, ∞) while if an upper tail critical region is appropriate the interval takes the form $(-\infty, b)$ where a, b are finite. If, in Example 1.6 we wanted a one-tail confidence interval at the 95.24% level we would exclude only the six largest

Table 1.2 Largest and smallest θ at which ranking of S changes among permutations

	One x replaced	Two x replaced	Three x replaced
Largest	2.4–1.9=0.5 2.1–1.9=0.2 2.4–2.3=0.1	2.25–2.1=0.15	2.1–2.23= –0.13
Smallest	1.4–3.1= –1.7 1.6–3.1= –1.5 1.8–3.1= –1.3	1.5–2.8= –1.3	1.6–2.63= –1.03

changeover values of θ.   Berry and Armitage (1995) point out that exact confidence intervals may also be based on mid-p values.

Intuitively reasonable point estimates of θ are either the difference between the sample medians or the difference between the sample means or perhaps the mean or median of the 125 values of θ corresponding to the order change points for S_1. The difference between the sample means is easy to calculate and for small samples especially, is preferred to the sample median difference which tends to be erratic if based on few data.   Use of the mean may further be justified because, as indicated in Example 1.3, the mean value of the statistic S_1 for a given θ is

$$E(S_1,\theta) = \frac{m}{m+n} \ (\Sigma x_i + \Sigma y_j - n\theta) \tag{1.1}$$

Also, in Example 1.3 I pointed out that we could replace S_1 by

$$S_0 = S_1 - E(S_1,\theta) \tag{1.2}$$

It is reasonable to choose θ so that the observed S_1 is equal to its mean value, i.e. so that $S_0 = 0$.   From (1.1) and (1.2) we form an estimating equation for θ, i.e.

$$S_1 - E(S_1,\theta) = 0 \tag{1.3}$$

where $S_1 = \Sigma \ x_i = m\overline{x}$ and $E(S_1,\theta)$ is given by (1.1). If we write $\overline{x} = (\Sigma x_i)/m$ and $\overline{y} = (\Sigma \ y_j)/n$ using elementary algebra, (1.3) may be rewritten

$$\frac{mn}{m+n} \ (\overline{x} - \overline{y} + \theta) = 0,$$

giving the estimate $\theta_0 = \overline{y} - \overline{x}$.

The concept of an estimating equation like (1.3) is common (e.g. in least squares or maximum likelihood estimation) and I use it elsewhere in this book.

1.9 Asymptotic results

When m, n are large exact permutation theory hypothesis tests or estimation of confidence intervals call for suitable computer programs and even highly efficient specialist programs like StatXact or TESTIMATE (section 1.10) may take a long time to, or be unable to, produce exact results. An alternative implemented by StatXact is to obtain Monte Carlo approximations and both StatXact and TESTIMATE and many general statistical programs give asymptotic results. Some general statistical programs that give asymptotic but not exact results report these asymptotic results for small as well as large samples. It is often instructive, and sometimes alarming, to compare asymptotic p-values with exact results based on permutation theory. Maritz (1995, p.84) discusses the asymptotic distribution theory associated with the statistic S_0 defined by (1.2) and shows how this may be used to test hypothesis about and obtain approximate confidence limits for a location difference measured by θ. Such tests assume that the distribution of the statistic involved is approximately normal, and for large samples this is so for the distribution of S_0 defined above.   As an alternative equivalent to Maritz's approach I consider S_1, the sum of the values in the sample of m rather than S_0. I denote a typical value in the first sample by x_i and in the second sample by y_j. The permutation test procedure can be looked upon as sampling without replacement from a finite population (here the combined sample values) which for notional convenience I denote by $z_1, z_2, \ldots z_m, z_{m+1}, \ldots z_{m+n}$.

Standard sampling theory results for sampling from finite populations [see, e.g. Cochran (1953, pp.14-16)] give

$$E(S_1) = m(\textstyle\sum z_i)/(m + n) \tag{1.4}$$

and

$$\mathrm{Var}(S_1) = \frac{mn}{(m + n)(m + n - 1)} [\textstyle\sum z_i^2 - \frac{1}{m + n}(\textstyle\sum z_i)^2] \tag{1.5}$$

where the summations are over all $N = m + n$ values of i.

Assuming normality for the large sample distribution of S_1, then

$$Z = \frac{S_1 - E(S_1)}{\sqrt{[\mathrm{Var}(S_1)]}} \tag{1.6}$$

approaches a standard normal distribution as $N \to \infty$ providing both m and n are also reasonably large. Precise conditions are given by Maritz (1995, section 1.9.3).   If S_0 is used as the test statistic it has mean zero, and if S_2, the second sample sum is used, it has a mean given by interchanging m and n in (1.4) and for both S_0 and S_2 the variance is given by (1.5).

The general formulae for mean and variance, suitably modified, may be used with ranks or other scores in place of the original data, and I make appreciable use of them in Chapter 4.

1.10   Computer software

Newer releases of several general statistical packages include more nonparametric or distribution-free methods than earlier versions. Exact methods based on permutation tests are sometimes given, but there is a tendency to rely on large-sample approximations, i.e. asymptotic results, sometimes without any warning about sample sizes needed to ensure reasonable approximations to exact p-values or exact confidence intervals. Specialist packages for exact inference based on permutation tests include two already mentioned, StatXact developed and distributed by Cytel Software Corporation, Cambridge, Ma, USA, and TESTIMATE from IDV, Munich, Germany. There is considerable overlap in the topics covered but some important methods are included in one but not the other. Output for procedures included in both packages is generally similar but small differences may enhance the appeal of one or the other for users with special requirements. There are one or two more specialized packages for particular types of analyses, and several textbooks include program listings for specific problems. EGRET and LogXact are packages especially relevant to logistic regression that I discuss briefly in Chapter 14, while GLIM is particularly useful for these and other generalized linear models also discussed in that chapter. New software appears regularly and readers should check advertisements in relevant journals or the reviews of statistical software featured in publications such as *The American Statistician* to find relevant new products. It is also worth checking whether there are additional relevant tests included in up-dates of general statistical packages. For example, some exact permutation tests have been added as a module in recent releases of SPSS. When possible, always use the latest release of software. Some textbooks carry listings of specialist programs, e.g. Neave and Worthington (1988) and Edgington (1995). Most statistical packages have facilities to generate random samples of m from $m + n$ observations and these may be used when writing an appropriate macro to form a large number of Monte Carlo samples and to approximate, for example, the distribution of the sum S_1 for the Fisher–Pitman test. By determining the number of S_1 in such samples that are less than the observed sum for the data one obtains an estimate of the exact p-value. Programs that specifically provide Monte Carlo approximations to many exact tests are listed in the appendix to Manly (1997).

I have not in this chapter dealt with techniques such as bootstrapping (Chapter 2) or robust estimation (Chapter 3) or with regression diagnostics (Chapter 11) but these are all computer intensive. General statistical packages cover these topics at varying depths and it is possible to enhance their usefulness in these fields by writing simple macros. For example, in applying the bootstrap any general purpose program that has a facility for generating random samples with replacement and carrying out Monte Carlo simulations of a range of statistical procedures and then producing good summary statistics can easily be adapted to bootstrapping, although one with special language facilities like S or S-Plus is desirable for sophisticated applications. However, for most of the examples I give for bootstrapping a package like the widely-available MINITAB may be used, tailoring output to personal needs with easily written macros.

Throughout this book I assume that the reader has access to at least one major statistical software package allowing at least Monte Carlo approximations to permutation tests.

Exercises

1.1   To establish that the conventional t-statistic (but not the conventional tables) may be used in place of the statistic S_1 used in Example 1.3 one must show that t is a monotonic 1:1 transformation of S_1. This is not difficult to establish by algebraic manipulations that enable one to recognise components of t that are invariant under permutation. To do this you may find it helpful to recall that t^2 has an F-distribution with 1 and $m + n - 2$ degrees of freedom and that the total sum of squares in an analysis of variance is invariant under permutation.

1.2   In the generalization of the two independent sample confidence interval problem considered in Example 1.6 show that if we take samples of m, n distinct values and decrease all second sample values by θ then the ordering of the sum S_1 among sums for all possible permutations of m will change when θ passes through the difference between the means of all observations entering and all observations leaving the first sample at that change.

1.3   Determine the largest and smallest values of θ at which four y values replace all four x values in Example 1.6 and hence verify that all four-value changes are irrelevant to determining the 95.24% confidence limits in that example.

1.4   To confirm the confidence intervals obtained in Example 1.6 one may subtract the limits -1.3 and 0.15 obtained there from all second sample values and then determine the relevant one-tail p-values which should be such that they are equal to or are slightly greater than $3/126$, whereas if we subtract , say -1.31 or 0.16 or any other numbers lying just outside the limits each one tail p-value does not exceed $3/126$. You may check these results numerically using the method outlined in Example 1.3 for these small data sets, but more generally the results may be checked using a program such as that in StatXact or TESTIMATE using the given first sample values and the second sample values after the adjustments indicated above.

1.5   Compare the p-values appropriate to a two-tail test (i) formed by doubling the single tail p and (ii) by including only opposite tail probabilities for values of the statistic at the same or at a greater distance from the mean value of that statistic using the data in Example 1.5 after modifying them by replacing the observation 3.1 by 4.1.

1.6   Ten units are randomly allocated each to one of two treatments. Six receive treatment A and the remaining four receive treatment B. A Fisher–Pitman test is carried out and all the treatment B response variable sums under permutation are unequal. What are the possible exact p-values less than 0.05 relevant to (i) a one-tail and (ii) a two-tail test, if the latter is performed by doubling a relevant one-tail probability? What would be the corresponding sets of relevant p-values if the ten units had been allocated five to each treatment?

1.7   A Fisher–Pitman test is performed on measurements on samples of sizes 7 and 9, each subjected to different treatments. It is found that 523 permutations in the relevant reference set give values of S, the sum of permutations of 7 observations, as small or smaller than that observed. Would this indicate significance in a one-tail test, and if so at what exact level?

2

The Bootstrap

2.1 Basic concepts

This chapter is about inference methods that stem from a different philosophy to that leading to permutation tests, but that often gives rise to closely related methods and results. The approach is motivated by a notion basic to most statistical inference – that random samples broadly reflect characteristics of the populations being sampled. Thus the sample mean is generally a good, often the best, estimate of the population mean; the sample variance is *almost* the best estimate of the population variance. I say almost because it is known to show bias which may be important with small samples, but which tends to zero as the sample size increases. For samples from a bivariate distribution the sample correlation coefficient is often a good estimator of the population correlation coefficient. The new feature of the main method described in this chapter, called **the bootstrap**, is that it carries this notion a stage further by using samples of samples to further explore characteristics that reflect those in the original sample and so in turn may reflect those in the population

The key to the way population characteristics are reflected in samples is the fact that for all but very small random samples the sample cumulative distribution function (cdf) defined below is usually close to the population cdf.

The sample cdf $S(x)$ is a step function with a step $1/n$ at each sample value x_i. If the n sample values are all distinct and arranged in ascending order these ordered values, called the **order statistics**, are denoted by $x_{(1)}, x_{(2)}, \ldots x_{(n)}$ where $x_{(1)} < x_{(2)} < \ldots < x_{(n)}$. A full description of order statistics and their properties is given by Gibbons and Chakraborti (1992, Chapter 2). The sample cdf $S(x)$ defined in terms of the order statistics is

$$S(x) = 0, \ x < x_{(1)},$$
$$S(x) = 1/n, \ x_{(1)} \leq x < x_{(2)},$$
$$S(x) = 2/n, \ x_{(2)} \leq x < x_{(3)},$$

$$\cdots \cdots \cdots \cdots$$
$$S(x) = (n-1)/n, \ x_{(n-1)} \leq x < x_{(n)},$$
$$S(x) = 1, \ x \geq x_{(n)}.$$

The sample distribution for a sample of n observations is discrete with probability mass function $p_i = \Pr(X = x_{(i)}) = 1/n$ for all $x_{(i)}$. Modifications are needed if some of the x_i are equal, the effect of equality being to replace a step $1/n$ by a step r/n if there are r tied values each equal to x_i.

For random samples from a continuous distribution with cdf $F(x)$ if n is large $S(x)$ will usually be close to $F(x)$ for all x. Even for n about 20 some, though not always all, key population characteristics are usually reflected quite well in the sample, e.g. the sample mean is usually tolerably close to the population mean, or if the population has a markedly skew distribution this should be evident in the sample, and so on. The differences $F(x_{(i)}) - S(x_{(i)})$ will tend to be small. For samples of about 20 the maximum $|F(x_{(i)}) - S(x_{(i)})|$ will typically be less than 0.2, and often of the order of 0.1.

Example 2.1 Random samples of 20 were taken from each of 3 known populations. The samples, denoted by A, B and C, with the data ordered, were

A 0.20 0.46 0.58 1.13 1.51 1.67 1.78 1.87 2.01 2.08 2.18 2.19 2.23 2.34 2.44 2.70 3.12 3.31 3.48 3.57
B 0.29 0.50 0.59 0.99 1.23 1.57 1.75 1.84 1.88 1.92 2.11 2.12 2.88 2.97 2.98 3.02 3.08 3.12 3.33 3.62
C 0.11 0.12 0.17 0.18 0.22 0.32 0.34 0.37 0.40 0.50 0.51 0.55 1.23 1.84 1.94 2.50 2.86 3.38 3.81 3.86

Eye examination of these data, without knowing how they were generated, suggests that samples A and B may be from the same populations but that C is almost certainly from a different population because, for instance, there are appreciably more values less than 1 in that sample. Thus, despite the range of observations in all three samples being broadly similar, C, shows clear signs of skewness.

Summary statistics in Table 2.1 back these eye observations. These statistics are estimates of the population equivalents based on the sample data.

Figure 2.1 shows histograms for each of the 3 data sets with unit class intervals. These are only moderately informative for these small data sets, but that for sample C highlights the skewness. There is a hint that the data for sample B has a more uniform spread than that for sample A, but those familiar with sampling variation would not put much weight on this as an indicator of samples being from different populations. In Exercise 2.1 I suggest confirming these tendencies using a box and whisker plot.

Sample A was from a N(2, 1) distribution and the sample mean and standard deviation are close to their population counterparts. Sample B was from a continuous uniform distribution over (0, 4), which has the same mean as the normal distribution giving Sample A, and the population standard deviation of 1.15 is only slightly greater so it is not surprising that the sample values of these statistics are close in A and in B. The slight differences in the histograms are in line with the different population distributions. Sample C came from an exponential distribution with unit mean and variance.

The Kolmogorov test described in section 6.7 indicates that the samples are all consistent with their known parent distributions. The differences $F(x_{(i)}) - S(x_{(i)})$

Table 2.1 Summary statistics for samples $A - C$ in Example 2.1

Sample	Mean	median	Standard deviation	Std. error of mean
A	2.04	2.13	0.96	0.21
B	2.09	2.02	1.01	0.23
C	1.26	0.51	1.33	0.30

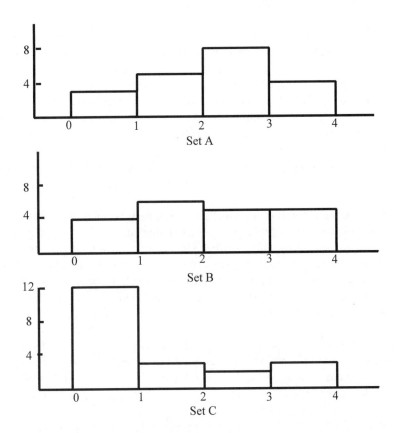

Figure 2.1 Histograms for data in samples A, B, C in Example 2.1.

of greatest magnitude, which are closely related to, but not quite the same as, the statistic used in the Kolmogorov test, were respectively –0.08, 0.12, 0.18 for samples A, B, C; all within the acceptable limit (i.e. < 0.2) suggested above. The samples were computer generated using MINITAB. When using software to generate samples it is wise to test for conformity with a true random sample from the alleged population distribution, because not all packages have satisfactory random number generators, although better known and longer established packages usually have adequate generators.

2.2   The bootstrap concept

In practice we seldom know from what distribution a sample comes except in broad and sometimes vague terms reflecting past experience, reasonable theory or inspection of large data sets. Indeed, if we knew precise source distributions with known parameter values, there would be no inference problem! The more vague the suppositions about population distributions, the more useful the bootstrap becomes. Repeated resampling of data to tell us more about characteristics of populations is basic to the bootstrap. Resampling, as we saw in sections 1.5 and 1.6, is also at the heart of randomization tests. Bootstrap sampling is done in a different way and is applicable also in some situations where no permutation test is available or appropriate. Successful use generally requires careful evaluation of properties of data generated by the bootstrap. Although strongly data-driven in the sense that the same procedure may be relevant without any *a priori* specification of a probabilistic model, the method can itself often be justified only by complicated mathematics. For its application the main requirement is suitable computing facilities. Unlike permutation tests which lead to exact inferences, bootstrapping usually only leads to approximate results. These results, however, are often more accurate and informative than those obtained by fitting a wrong model to data (e.g. assuming normality when that is not valid).

There is seldom a unique or best bootstrap solution to a problem. In practice the method is used mainly either when there is no analytic solution available, when a permutation test, or the facility to carry it out, is unavailable, or when there is doubt whether conditions needed for a particular analytic solution or permutation test actually hold.

Efron and Tibshirani (1993, Chapter 2) suggest that the computational burden of the bootstrap is offset by

- an increased range of problems that can be analysed;
- a reduction in the assumptions needed to validate the analysis;
- elimination of many routine but tedious theoretical calculations, especially those associated with assessing accuracy.

The approach is intuitively reasonable in many applications and confidence in the method is enhanced because when correctly used the technique leads to inferences similar to those given by analytic or permutation solutions when these exist and are relevant.   Important applications are either to complex data structures or ones that involve inferences about concepts such as correlation, ratios of variables or eigen-values of matrices where analytic results are not readily available except under restrictive distributional assumptions.

In this chapter I consider mainly applications involving a random sample of n observations $x_1, x_2, \ldots, x_n$ from some unspecified population, turning briefly to the two-sample problem in section 2.7.

Given sample data $x_1, x_2, \ldots, x_n$ a bootstrap sample is a random sample of size n obtained from these data by sampling **with replacement**. Thus, some of the x_i may occur more than once and others not at all in a bootstrap sample.   It is notionally possible to determine the distribution of all possible bootstrap samples and the distribution for many associated statistics such as that of the means or medians of the bootstrap samples.   Such distributions are called **true** bootstrap distributions to distinguish them from estimates based on a random sub-set of all possible bootstrap samples.   The latter play an important role in practical applications because determining true bootstrap distributions analytically is a formidable task for all but small n except in the case of a few statistics where some general analytic results hold.   Further, such general results usually arise in cases where inferences can be made easily without bootstrapping, and the main interest then is to confirm that the bootstrap, if it were used, would lead to similar conclusions to established theory.   In practice therefore bootstrap inferences are usually made by taking a sample of all possible bootstrap samples using Monte Carlo sampling not unlike that sometimes used, for example, to estimate p-values in permutation tests when the data sets are too large for practical computation of exact p-values.   The main difference is that sampling is **without** replacement for permutation tests and **with** replacement for the bootstrap.

Use of the bootstrap stems largely from work by Efron (1979) who realized that bootstrapping gives estimators similar to those arising from classical theoretical methods when these are appropriate.   A simple example is the estimation of the standard error of a sample mean.   It is well-known that the appropriate estimator of the standard error (se) of the sample mean $\bar{x}$ as an estimator of a population mean, μ, is $\mathrm{se}(\bar{x}) = s/\sqrt{n}$ where $s^2 = \sum(x_i - \bar{x})^2/(n-1)$. For a normal distribution $\bar{x}$ is distributed $N(\mu, \sigma^2/n)$ and if σ^2 is not known, we base inference on the fact that $t = (\bar{x} - \mu)/(s/\sqrt{n})$ has a t-distribution with $n-1$ degrees of freedom.   For large n justification for still basing inference about means on t for nonnormal distributions rests on the central limit theorem.   Although that is an asymptotic result, the approximation to normality works reasonably well for moderate values of n, sometimes as low as $n = 10$. However, such approximations are almost unique to the mean and analytic solutions for other simple statistics like the median or correlation coefficient require fairly rigid distributional assumptions.

I show in section 2.3 that if we consider the true bootstrap distribution for samples of size n, this distribution has mean $\bar{x}$ and that the bootstrap mean has a standard error $[\sum(x_i - \bar{x})^2/n^2]^{1/2}$. This differs from $s/\sqrt{n}$ by a factor $\sqrt{[n/(n-1)]}$, which is close to 1 for large or even moderate n. The bootstrap is important because it provides estimates of quantities like the standard error of, and approximate confidence intervals for, other statistics such as the median and the correlation coefficient or more complicated statistics without the need for any population distributional assumptions. Before showing how this is done I discuss in the next section the exact distribution of bootstrap samples. As already indicated except for small n, it is virtually impossible to make use of the exact bootstrap distribution in practice, but good approximations are obtained by Monte Carlo sampling generating relatively small numbers (often between 20 and 200) of all possible bootstrap samples and estimating bootstrap statistics like a standard error from these. Much larger samples (often between 1000 and 2000 or more) are usually needed for confidence intervals.

2.3 The bootstrap sampling distribution

Consider a sample of n formed with replacement from n items. For bootstrapping these items are the observations $x_1, x_2, \ldots, x_n$ that form a random sample from some, often vaguely specified, population. Because of replacement, the probability of any x_i being selected remains $1/n$ throughout and in a bootstrap sample each x_i may appear r_i times, $r_i = 0, 1, 2, \ldots n$ subject to the constraint $\sum_i r_i = n$, where the probability associated with each possible sample is given by

$$\Pr[n(x_1) = r_1, n(x_2) = r_2, \ldots, n(x_n) = r_n] = \frac{n!}{r_1! r_2! \ldots r_n!} (1/n)^n \qquad (2.1)$$

where $n(x_i)$ is the number of times x_i occurs and $r_1 + r_2 + \ldots + r_n = n$. This is the probability mass function of a multinomial distribution.

Example 2.2 For $n = 3$ it is easy to record all possible bootstrap samples and the associated probabilities given by (2.1). Table 2.2 lists the 10 possible samples and the associated probabilities, and also the mean and median for each sample, assuming, without loss of generality, that $x_1 < x_2 < x_3$.

The values in Table 2.2 for the sample means and medians are called bootstrap sample means and medians. It is clear from the third column of Table 2.2 that each bootstrap sample has a different mean (although for particular values of the x_i some may be numerically equal).

Table 2.2 Bootstrap sample distribution for $n = 3$ and sample means and medians

Sample	Probability	Sample mean	Sample median
x_1, x_1, x_1	1/27	x_1	x_1
x_1, x_1, x_2	3/27	$(2x_1 + x_2)/3$	x_1
x_1, x_1, x_3	3/27	$(2x_1 + x_3)/3$	x_1
x_1, x_2, x_2	3/27	$(x_1 + 2x_2)/3$	x_2
x_1, x_2, x_3	6/27	$(x_1 + x_2 + x_3)/3$	x_2
x_1, x_3, x_3	3/27	$(x_1 + 2x_3)/3$	x_3
x_2, x_2, x_2	1/27	x_2	x_2
x_2, x_2, x_3	3/27	$(2x_2 + x_3)/3$	x_2
x_2, x_3, x_3	3/27	$(x_2 + 2x_3)/3$	x_3
x_3, x_3, x_3	1/27	x_3	x_3

Since, for each sample, there is an associated probability given in column 2 of Table 2.2 we know the probability distribution not only of the bootstrap samples, but in this case also the bootstrap sample means have the same associated probabilities. In particular, applying the usual formulae for mean and variance of a discrete distribution, i.e.

$$E(Y) = \sum p_i y_i, \quad Var(Y) = E(Y^2) - [E(Y)]^2 = \sum p_i y_i^2 - (\sum p_i y_i)^2 \tag{2.2}$$

to the bootstrap means, i.e. the y_i are the means in column 3, the p_i the probabilities in column 2 of Table 2.2, we easily show that $E(Y) = \bar{x} = (x_1 + x_2 + x_3)/3$, the mean of the original sample of 3. Tedious algebra (Exercise 2.2) establishes that for the bootstrap sample means $Var(Y) = \sum(x_i - \bar{x})^2/9$. The square root of this is the standard error of the bootstrap mean and is consistent with the formula for the bootstrap standard error of the mean quoted in section 2.2. We can establish the general result directly and do so later in this section, avoiding the tedious algebra of Exercise 2.2. Since, for the mean, the bootstrap estimate of standard error is similar to the classical estimate (and indeed equals the maximum likelihood estimator for a normal distribution), bootstrapping shows no advantage here for obtaining the standard error. Any advantage shows for estimators other than the mean, where standard errors are not easy to obtain analytically except under restrictive assumptions about the population distribution.

Because n is small the principal involved in bootstrap estimation of the standard error of the median is easy to illustrate here. The values of the bootstrap medians in column 4 of Table 2.2 take only three values x_1, x_2, x_3, each being associated with more than one of the 10 bootstrap samples. For example, x_1 is the median for each of the samples x_1, x_1, x_1 and x_1, x_1, x_2 and x_1, x_1, x_3 (because we have assumed $x_1 < x_2 < x_3$) and the total associated probability calculated from column 2 is $1/27+3/27+3/27=7/27$. Similarly the median x_2 has probability 13/27 and the median x_3 has probability 7/27, whence, using (2.2) we find (Exercise 2.3) that if Y represents the bootstrap median then

$$E(Y) = (7x_1 + 13x_2 + 7x_3)/27 \text{ and } Var(Y) = (7x_1^2 + 13x_2^2 + 7x_3^2)/27 - (7x_1 + 13x_2 + 7x_3)^2/729.$$

The bootstrap standard error is the square root of this variance.

One can hardly expect to make very meaningful inferences about means or medians of a large population from a sample of 3 by bootstrapping or by any other method. Even with realistic but still relatively small samples (say, $n < 12$ approximately) there is only limited evidence about the nature of the population provided by such scant data and, in particular, when sampling from a large population the sample cdf may not be a good approximation to the population cdf when the sample is small. This calls for caution when using the bootstrap with small samples, but with care useful results may still be attainable with not too small samples. It may, however, be virtually impossible to check the performance of the bootstrap estimate of some relevant quantity such as the standard error of a median against the true value even when we know the source of the data if the sample size is small. For instance, I used the results in Example 2.2 to calculate the bootstrap standard error of an estimate of the median of a normal distribution with true median zero and standard deviation $\sigma = 2$ using samples of 3 from such a distribution. The theoretical standard error of the median is known in this case to be 1.45, derived from the formula for the standard error of the median for a sample of n from a normal distribution, namely se(med) = $1.2533\sigma/\sqrt{n}$, a result given in many statistical text books. I applied the formula for Var(Y), where Y is the bootstrap median for samples of 3 given in Example 2.2, to 10 samples of 3 from this normal distribution and the bootstrap estimated se(med) ranged from 0.024 to 2.663. The mean of all 10 estimates was 1.30, which is tolerably close to the above known value 1.45. In practice taking 10 independent samples of 3 is equivalent to taking one sample of 30, and it would be better to bootstrap this larger sample to make inferences.

The analytic approach used in Example 2.2 soon becomes impractical if n is increased. A general result in combinatorial algebra establishes that the number of distinct bootstrap samples that can be obtained from a sample of n is $^{2n-1}C_n$. Table 2.3 shows how this number increases with n. Also, as is clear from Table 2.2, bootstrap samples no longer have the simple property we met with a permutation test in Example 1.3 that all samples are equally probable. For a few statistics there are formulae to compute the probabilities associated with each bootstrap value that statistic may take. For example, Efron and Tibshirani (1993, section 2.1, problem 2.4) give a formula for the case $n = 7$ (which generalizes to any odd n) for the probabilities associated with all possible bootstrap medians. It is easy to see that if n is odd the only possible values of the bootstrap median are $x_1, x_2, \ldots, x_n$. The situation is more complicated when n is even (Exercise 2.4).

A widely used and important concept when applying the bootstrap is that of **plug-in** estimators. The idea is that if we do not know the population value of a mean, variance, standard error, median or other quantile, correlation coefficient, etc., for some finite or infinite population, we estimate it by the corresponding entity for the sample distribution. If we sample from a distribution with cdf $F(x)$

Table 2.3 Number of distinct bootstrap samples of various sizes

Sample size	Number of bootstrap samples
3	10
4	35
5	126
6	462
10	92378

having mean $E(X) = \mu$ and $Var(X) = \sigma^2$, if we know these values general statistical theory tells us that the mean $\bar{x}$ of a random sample of n is distributed with mean μ and variance σ^2/n. The standard error of the mean is defined as the standard deviation of the distribution of $\bar{x}$, i.e. $se(\bar{x}) = \sigma/\sqrt{n}$. If we do not know μ or σ^2 plug-in estimates of these are the mean and variance of the sampling distribution based on the x_i each having associated probability $p_i = 1/n$, leading to the well known estimators $\mu = \sum p_i.x_i = \sum x_i/n = \bar{x}$ and $\hat{\sigma}^2 = \sum p_i(x_i - \bar{x})^2 = \sum(x_i - \bar{x})^2/n$. The latter is the maximum likelihood estimator for the normal distribution and is related to the conventional estimator s^2 with divisor $n - 1$ by $\hat{\sigma}^2 = (n - 1)s^2/n$. Since $var(\bar{x}) = \sigma^2/n$ the plug in estimator is $\hat{\sigma}^2/n = \sum(x_i - \bar{x})^2/n^2$. The corresponding estimate of standard error is the square root. Because bootstrap samples are all random samples with replacement from a 'population' which is our initial sample the plug-in estimates for means and standard errors for that original sample are our 'population' values for the bootstrap mean and standard error of that mean. In Example 2.2 I showed by considering the bootstrap sampling distribution for the mean, that in a particular case the exact bootstrap values of these quantities based on the complete sampling distribution in Table 2.2 gave these estimators. Disquiet may be felt about bias in some plug-in estimators but I show in section 2.5 that the jackknife, a procedure related to the bootstrap, provides a way to estimate bias.

Estimating a population median by the sample median uses the plug-in principle, but a difficulty arises with the standard error of the sample median because the standard error of the median estimator depends upon the population distribution. However, I show in the next section that for Monte Carlo bootstrap sampling the plug-in principle plays a key role for virtually any type of estimator.

2.4 Using bootstrap samples

In practice bootstrap inferences are nearly always made using Monte Carlo sampling to generate a fixed number, B, say, of bootstrap samples. I denote a typical bootstrap sample by $x_1^*, x_2^*, \ldots x_n^*$ where each x_i^* is equal to some x_j but because sampling is with replacement some x_j may not appear and others may appear more than once among the x_i^*, e.g. if $n = 7$ a typical bootstrap sample might be $x_1^* = x_3$, $x_2^* = x_5$, $x_3^* = x_1$, $x_4^* = x_5$, $x_5^* = x_7$, $x_6^* = x_4$, $x_7^* = x_1$. If B bootstrap samples are generated the bth sample may be written $x_1^{*b}, x_2^{*b}, \ldots x_n^{*b}$. Vector notation provides a convenient shorthand; we write x^* for any sample x_1^*, $x_2^*, \ldots x_n^*$ or x^{*b} for the bth sample $x_1^{*b}, x_2^{*b}, \ldots x_n^{*b}$, $b = 1, 2, \ldots, B$. In some situations, e.g. those concerned with correlation in Chapter 9, each x_i may itself be a vector. For any sample x^* we may obtain a statistic, usually a plug-in estimator, to estimate some parameter or characteristic we are interested in. For generality, I denote the parameter or other population characteristic to be estimated by θ and the statistic used to estimate it from a bootstrap sample x^* by $s(x^*)$ which gives an estimator $\theta^* = s(x^*)$. If the quantity we are estimating is the mean, the statistic $s(x^*)$ will be the sample mean, if it is the median then $s(x^*)$ will be the sample median. Clearly whatever the statistic $s(x^*)$ is, its numerical value will vary from one bootstrap sample to another (because when sampling with replacement each sample may be different) and thus it will have a distribution. As $B \to \infty$ for any statistic $s(x^*)$, its mean will tend to the mean computed for the true bootstrap sampling distribution, although, as I pointed out in section 2.3, this is only known in special cases or can only be worked out for general statistics for small values of n. Practical experience shows that in most situations even for small n both the mean of the B bootstrap samples and the estimated standard error as computed below converge rapidly to the limiting values as $B \to \infty$. Given B bootstrap samples if we denote the mean of the $s(x^{*b})$ by $s(.^*)$, i.e. $s(.^*) = \sum_b [s(x^{*b})]/B$ then the appropriate estimator of the true bootstrap standard error, which I denote by $se_B(s)$, is

$$se_B(s) = \{\sum_b [s(x^{*b}) - s(.^*)]^2\}/(B-1)\}^{\frac{1}{2}}. \tag{2.3}$$

This is the usual estimator of population standard deviation based upon a random sample of B from some population and tends to the true bootstrap standard deviation for the statistic s as $B \to \infty$. In practice the approximation is often good for B as low as 20 providing n is not too small. Even for small n reasonable estimates may be obtained with $B = 100$.

   Simple examples where we have known appropriate parametric procedures for estimation show the strengths of and potential pitfalls in the method.

Example 2.3   In Example 2.2 I established the true bootstrap standard error of the median estimator working with a sample of 3, i.e. x_1, x_2. x_3 to be $se(\mathrm{med}) = [(P_2 - P_1^2/27)/27]^{\frac{1}{2}}$ where

$$P_r = 7x_1^r + 13x_2^r + 7x_3^r, \; r = 1, 2.$$

Table 2.4 Bootstrap estimates of $s^*(.)$ and bootstrap standard error for a sample of 3 taking the values 1, 3, 9

B	First bootstrap run		Second Bootstrap run	
	$s(.*)$	$se_B(s)$	$s(.*)$	$se_B(s)$
20	3.8	2.78	4.4	3.56
50	3.4	2.95	4.5	3.38
100	3.6	2.92	3.9	3.05
200	3.8	2.99	3.7	2.95
300	3.6	2.94	3.7	2.91
500	3.8	2.95	3.7	2.86
1000	3.9	3.00	3.9	2.89
2000			3.9	2.98
∞	4.0	3.05	4.0	3.05

If $x_1 = 1$, $x_2 = 3$, $x_3 = 9$ it is easy to verify (Exercise 2.5) that $P_1 = 109$, $P_2 = 691$, whence se(med) = 3.05. Also the mean of this median estimator E(med) = $P_1/27$ = 4.04. I took two sets of bootstrap samples from the values 1, 3, 9 and in each set I recorded the median (1, 3 or 9) for each sample and also computed the standard error of these estimators using (2.3) for B = 20, 50, 100, 200, 300, 500, 1000 for each set and also for B = 2000 for the second set. The mean values and estimated standard errors are given in Table 2.4, indicating that even with this small sample size the estimates are converging rapidly towards the true bootstrap values. The values for $B = \infty$ in Table 2.4 are the theoretical values obtained above.

Not surprisingly $s(.*)$ is a biased estimator of the sample median which has the value 3. That this should be so is immediately evident from the form of P_1 where x_1 and x_3 have equal coefficients but they themselves are not symmetric with respect to $x_2 = 3$ which is the sample median. However, in all cases the median of our bootstrap estimators was 3, so although the bootstrap estimators of the median show mean bias they do not exhibit median bias. However, even if the bootstrap median is a mean-biased estimator of the sample median, it does not follow that it is a biased estimator of the population median, or if it is biased, that the bias is the same as that for the sample median. This is a limitation of the bootstrap, especially for small samples, because in these circumstances the bias may reflect a characteristic of the sample which itself is not present in the population. This problem becomes less acute as the sample size, n increases. However, many plug-in bootstrap estimators show mean bias and this may be important when using the method to establish a confidence interval, a topic taken up in section 2.6.

In Example 2.2 I showed that the probabilities associated with the possible bootstrap medians x_1, x_2, x_3 were respectively 7/27, 13/27 and 7/27 whence it follows that in, say, 150 Monte Carlo samples the expected number of times we should observe each of these values is $(7 \times 150)/27 = 38.89$, $(13 \times 150)/27 = 72.22$, $(7 \times 150)/27 = 38.89$. In one set of 150 Monte Carlo samples that I used for the data in example 2.3 the observed numbers for each were 33, 80, 37 in reasonable agreement with the expected numbers.

In applications of the bootstrap using only a finite number, B, of bootstrap samples there are two sources of error. The first is the usual sampling error applicable to any sample-based inferences about a population no matter what method of inference is used and this is measured by the standard error; the second is the error specific to bootstrap sampling that is made when we approximate to the true bootstrap standard error using only a finite number, B, of bootstrap samples. The next example sheds light on the way sampling variation is reflected in bootstrap estimates.

Example 2.4 I took 2 random samples of $n = 9$ from a normal distribution with mean $\mu = 0$ and standard deviation $\sigma = 2$. The values, arranged in ascending order within each were:

Sample I	−2.27	−1.74	−1.01	−0.23	0.23	0.40	0.57	0.98	2.20
Sample II	−3.92	−1.01	−0.18	1.14	1.32	1.66	3.70	4.30	4.74

Usual theory indicates that the standard error of the sample median as an estimator of the population median based on a sample of n from a normal distribution is $1.2533\sigma/\sqrt{n}$. In this case se(med) = $1.2533 \times 2/3 = 0.836$. The sample medians are respectively 0.23 and 1.32 and the known population median, zero, lies within two standard errors of the sample median in each case. So using conventional parametric theory, we would not reject the hypothesis of a zero population median. Considering each sample separately I took $B = 20$ bootstrap samples and chose the bootstrap sample medians as the statistic $s(x*)$ and used (2.3) to compute se$[s(x*)]$. For Sample I, my bootstrap approximation gave se$[s(x*)] = 0.65$, and for Sample II, se$[s(x*)] = 1.20$. Again, if we accept these as reasonable approximations to the true bootstrap standard error, both sample estimates of the median, 0.23 and 1.32 lie within two standard error of zero so we should again accept a population median of zero.

However, the first bootstrap standard error is well below, and the second well above, the theoretical standard error. This happens because the bootstrap computations are based entirely upon the sample values and can at best only reflect population characteristics to the extent that these are embedded in the samples.

In this situation, where the samples are from a known population, it is interesting to see how well the samples match population characteristics. Several tests are possible. One may use the classic normal theory tests to determine (i) are the data consistent with $H_0: \mu = 0$, given $\sigma = 2$; (ii) are they consistent with $H_0: \sigma = 2$. One might also carry out a Kolmogorov test, described in section 6.7, to determine if they are consistent with a population $N(0,4)$ distribution (since $\sigma = 2$ implies $\sigma^2 = 4$). None of these tests provide evidence against the null hypothesis, although the exact one-tail p-value in the Kolmogorov test for Sample II is $p = 0.0504$. Thus, although there are no strong grounds for regarding these as unlikely samples, Sample II is approaching an atypical sample. This is reflected in both the second sample mean and the sample standard deviation exceeding their expectation ($\bar{x} = 1.31$, $s = 2.773$). To see that the bootstrap approximation reflects the sample standard deviation let us suppose for a moment that the sample actually came from a normal distribution for which $\sigma = 2.773$. Then the population estimate of the standard error of the median estimator would be $1.2533 \times 2.773/3 = 1.16$, in close agreement with the bootstrap approximate standard error of 1.20. Similarly for sample I we find $s = 1.41$ and assuming our sample came from a normal population where $\sigma = 1.41$ the population estimate of the standard error of the median estimator would be $1.2533 \times 1.41/3 = 0.59$, again in close agreement with the bootstrap estimate 0.65.

When we do not know a population incidental parameter such as σ^2 (often referred to as a nuisance parameter), classic parametric methods of inference are subject to similar influences that reflect how well the sample 'mirrors' population characteristics. In the t-test, for example, the value of s computed for any given sample provides our only information about σ.

In Example 2.4 it was artificial to base inferences on sample medians when the samples were from a known normal distribution. However, in the real world when using the bootstrap one may not know the distributional source of data and often one has no grounds for making strong assumptions about the population.

The next example reflects a more realistic situation where a simple bootstrap might be used.

Example 2.5 The data below are a computer generated random sample of 49 observations from a known distribution but experimenters often meet data like these without such information. I disclose what the distribution is after I have obtained an estimate of the bootstrap standard error of the median. The data are of a type that arises in experimental situations. They could, for example, be response times in hours for some specific reaction for each of 49 mice after administration of a drug (although specification to three decimal places would imply more than usual measurement precision in that situation!), or they might be weights in grams of some impurity in ingots of a metal each weighing 1000 gm.

The data are arranged in ascending order. It is worth doing this – perhaps also drawing a histogram or a box and whisker plot – before an analysis, so as to highlight obvious data features.

0.034 0.167 0.143 0.180 0.346 0.446 0.558 0.593 0.615 0.648 0.650 0.744 0.853 0.913
0.970 1.003 1.009 1.237 1.436 1.537 1.650 1.669 1.778 1.984 1.995 2.054 2.395 2.458
2.579 2.624 2.726 2.741 2.858 2.877 2.998 3.009 3.124 3.451 3.516 3.540 3.657 3.717
4.097 4.171 4.967 6.469 6.902 7.435 8.023

Eye inspection shows the data are skewed. The median is the 25th ordered observation, i.e. 1.995. Half the observations are between 0.034 and 1.995 and half between 1.995 and 8.023. I estimated the bootstrap standard error of the median by taking 25 bootstrap samples and using (2.3). This estimate was 0.4210.

The data were in fact a computer generated random sample from a one-parameter exponential distribution with mean 3. The theoretical population median for that distribution is 2.079. A crude rule of thumb indication used in bootstrapping is that we should accept a hypothesised value of a parameter θ if its plug-in sample estimator lies within 2 standard deviations of that hypothesised value. In this example the difference $2.079 - 1.995 = 0.084$ is appreciable less than the estimated bootstrap standard error 0.4210. There may be unease as to whether 25 bootstrap samples give a reliable estimate of the true bootstrap standard error. There is also an implicit requirement for validity of the above rule of thumb that the true bootstrap standard error is reasonably close to the best estimated standard error if this is known. We saw in section 2.2 that in the case of the mean of a normal distribution this differed by a factor $n/(n-1)$, as does the maximum likelihood estimator in that case.

In the present situation where we have sampled from a known distribution there is an analytic expression for the true standard error of a median estimator from a sample of n, namely

$$se(median) = 1/(4nf^2)^{\frac{1}{2}}$$

where f is the ordinate of the probability density function at the median. For an exponential distribution with mean 3 we have $f = 0.1667$ and when $n = 49$, se (median) $= 0.4285$, close to the estimated bootstrap standard error of 0.4210.

2.5 Bias and the jackknife

The bias, more specifically the **mean bias**, of an estimator t of a parameter θ is defined as

$$bias(t) = E(t) - \theta. \qquad (2.4)$$

Clearly the bias is zero if $E(t) = \theta$.

From Example 2.3 it is clear that the estimated or true bootstrap estimator of a parameter may be mean biased even when the plug-in sample estimator is unbiased. This is particularly true for small samples. In Example 2.3 the sample median is an unbiased estimator of the population median, but the bootstrap estimate of the median had a mean value of 4.1, when the sample median was 3. However, the sample median is only an estimate of the population median and thus is usually not equal to it, so the difference $4.1 - 3 = 1.1$ is only an estimate of bias of the bootstrap estimator. We also saw in section 2.2 that even in a simple situation such as the standard error of a mean that the bootstrap gives a biased estimate of the true standard error. In this latter case it is easy to show that the bias of the square of the standard error is $-\sum(x_i - \overline{x})^2/[n^2(n-1)]$. This follows from (2.4) and the well known result that $\sum(x_i - \overline{x})^2/[n(n-1)]$ is an unbiased estimator of the true variance of the sample mean, i.e., σ^2/n. Biases introduce complications in inferences, in particular when obtaining confidence intervals.

We may estimate biases even if we have no analytic results. If $t(\theta)$ is an unbiased plug-in estimator of a parameter θ and we have B bootstrap samples and the bth plug-in estimator is $t^*(\theta, b)$, $b = 1, 2, \ldots, B$ and $t^*(.) = \sum_b t^*(\theta, b)/B$, then the estimate of bootstrap bias is

$$bias(t^*) = t^*(.) - t(\theta). \qquad (2.5)$$

This bias is often small compared to the standard error of t, except perhaps in small samples. The approach leading to (2.5) is intuitively reasonable but naïve, for unless B is large disquiet may be felt as to how close $t^*(.)$ is to the true bootstrap estimator of θ, and if t has a large standard error unease may also be felt about how well t is estimating θ even though it is an unbiased estimator. In practice B should usually be 200 or more for estimating bias. Better estimators are available. One is given by Efron and Tibshirani (1993, section 10.4).

Bias may also be estimated by a related technique called the jackknife. Historically, the jackknife preceded the bootstrap, and was introduced for bias estimation by Quenouille (1949); the name jackknife was given by Tukey (1958).

The jackknife is applicable to plug-in estimators. Unfortunately it does not work well for estimators like the median where, as will be evident when I define it, the jackknife estimate does not vary smoothly between jackknife samples. Only in the case of the sample mean as an estimator of the population mean is it generally true that the plug-in estimator is unbiased unless further distributional assumptions are made. A well-known example of a plug-in estimator that is biased is $s^2 = \Sigma(x_i - \overline{x})^2/n$ for the population variance σ^2. Since, in this case, $E(s^2) = (n-1)\sigma^2/n = \sigma^2 - \sigma^2/n$ the bias is $-\sigma^2/n$.

I use $\hat{\theta}$ for the plug-in estimator of a parameter θ. To generate the jackknife estimator of θ, the plug-in estimator $\hat{\theta}_{(i)}$ is computed for all samples of $n-1$ observations formed by successively omitting an observation, x_i, $i = 1, 2, \ldots, n$. I denote the mean of the $\hat{\theta}_{(i)}$ by $\hat{\theta}_{(.)}$, i.e. $\hat{\theta}_{(.)} = \Sigma_i \hat{\theta}_{(i)}/n$. The jackknife estimator, $\theta^\dagger$ of θ is

$$\theta^\dagger = n\hat{\theta} - (n-1)\hat{\theta}_{(.)}$$

and the jackknife estimate of bias is

$$\text{bias}(\hat{\theta}) = (n-1)(\hat{\theta}_{(.)} - \hat{\theta}). \tag{2.6}$$

The jackknife estimate of the standard error is

$$\text{se}(\hat{\theta}) = \{[(n-1)\Sigma_i(\hat{\theta}_{(i)} - \hat{\theta}_{(.)})^2]/n\}^{\frac{1}{2}} \tag{2.7}$$

It is immediately evident from (2.6) that if the jackknife estimate of bias estimates the bias exactly, then $\theta^\dagger$ is unbiased, since $\theta^\dagger = \hat{\theta} - \text{bias}(\hat{\theta})$.

The motivation for choosing (2.6) as a measure of bias is that it eliminates bias in the plug-in estimator of variance, and that it also takes the value zero for the unbiased plug-in estimator of the mean. This last is true because $\hat{\theta} = \overline{x}$ and $\hat{\theta}_{(i)} = (n\overline{x} - x_i)/(n-1)$ whence $\hat{\theta}_{(.)} = \Sigma_i \hat{\theta}_{(i)}/n = \overline{x}$. Looked at another way, if the population mean is μ, then $E(\hat{\theta}) = E(\hat{\theta}_{(.)}) = \mu$ whence the result follows. For the plug-in estimator of variance, $\hat{\theta} = \Sigma_i(x_i - \overline{x})^2/n$ we know $E(\hat{\theta}) = (n-1)\sigma^2/n$.

Similarly $E(\hat{\theta}_{(i)}) = E(\hat{\theta}_{(.)}) = (n - 2)\sigma^2/(n - 1)$, whence it is easily established (Exercise 2.7) that the expected bias is $-\sigma^2/n$.

More generally (2.6) does not reduce the bias to zero, nor is $\theta^\dagger$ an unbiased estimator of **any** parameter. However, it will generally reduce bias.

Does the expression (2.7) for the jackknife standard error surprise you? By analogy with the bootstrap or the usual formula for standard errors one might expect a factor n^2 or $n(n-1)$ instead of n in the denominator and no factor $n - 1$ in the numerator. However, for the mean, (2.7) reduces to the usual estimator for standard error. This is easily seen because, as we saw in the discussion of bias, $\hat{\theta}_{(i)} = (n\bar{x} - x_i)/(n - 1)$ and $\hat{\theta}_{(.)} = \sum\hat{\theta}_i/n = \bar{x}$, whence

$$(\hat{\theta}_{(i)} - \hat{\theta}_{(.)}) = (\bar{x} - x_i)/(n - 1)$$

and so

$$\sum_i(\hat{\theta}_{(i)} - \hat{\theta}_{(.)})^2 = [\sum_i(x_i - \bar{x})^2]/(n - 1)^2$$

and in this case (2.7) reduces to the usual expression for standard error, i.e.

$$se(\bar{x}) = \{[\sum_i(x_i - \bar{x})^2]/[n(n - 1)]\}^{1/2}$$

The reason for the factor $(n - 1)/n$ in (2.7) rather than the $1/(n - 1)$ or $1/n$ that one might expect intuitively is explained by Effron and Tibshirani (1993, section 11.2). They point out that an 'inflation factor' like $(n - 1)/n$ is needed because jackknife deviations $(\hat{\theta}_{(i)} - \hat{\theta}_{(.)})^2$ tend to be smaller than their bootstrap counterparts because jackknife samples omit only one data point at a time and are thus less variable than bootstrap samples involving selection with replacement.

An example shows why the Jackknife does not work properly with estimators like the median.

Example 2.6 In Example 2.5 I considered the following sample of 49 values generated from an exponential distribution with mean 3.

0.034 0.167 0.143 0.180 0.346 0.446 0.558 0.593 0.615 0.648 0.650 0.744 0.853 0.913
0.970 1.003 1.009 1.237 1.436 1.537 1.650 1.669 1.778 1.984 1.995 2.054 2.395 2.458
2.579 2.624 2.726 2.741 2.858 2.877 2.998 3.009 3.124 3.451 3.516 3.540 3.657 3.717
4.097 4.171 4.967 6.469 6.902 7.435 8.023

The sample median is $m = 1.995$ and the medians for all Jackknife samples are easily obtained. If the first sample value is omitted the Jackknife median is the median of the remaining 48 values and with the usual convention for the median of an even number of observations it is $m_{(1)} = \frac{1}{2}(1.995 + 2.054) = 2.0245$. A moment's reflection shows that omitting any one observation up to and including 1.984 gives the same value for the jackknife median, i.e.

$$m_{(i)} = 2.0245, 1 \le i \le 24.$$

Similarly,

$$m_{(25)} = \frac{1}{2}(1.984 + 2.054) = 2.019$$

and

$$m_{(i)} = \tfrac{1}{2}(1.984+1.995) = 1.9895, \ 26 \le i \le 49,$$

giving $m_{()} = 2.0072$ and from (2.7) se$(m) = 0.1204$. This grossly underestimates the true se$(m) = 0.4285$ obtained in Example 2.5. The discrepancy arises because the jackknife median takes only 3 nearby values, and two of these are repeated in almost half the samples. So, unlike the bootstrap median which takes a range of values, the lack of variation in the jackknife medians results in the plug-in standard error underestimating the true standard error despite the inflation factor.

This gives a timely warning not to apply data-driven analyses unthinkingly. Using the mean rather than the median avoids this difficulty because the jackknife means will be different for each omitted value.

.There are several modifications of the jackknife. One is the **delete-*d* jack-knife** in which all possible combinations of d observations are omitted. For certain choices of d Efron and Tibshirani (1993, section 11.7) point out that this leads to a consistent estimator of the jackknife standard error for the median.

2.6 Bootstrap confidence intervals

Bootstrapping is widely used to obtain approximate confidence intervals and considerable theory exists not only about their interpretation and properties, but also about refinements of technique for improving properties like coverage. At the basic level Efron and Tibshirani (1993, section 14.2) list 5 different ways to compute confidence intervals using the bootstrap. Each works reasonably well when appropriate, but the difficulty comes in deciding which is appropriate when little is known about the distributional properties of the data, a situation when we are most likely to want to use the bootstrap. An alternative approach to bootstrap confidence intervals which I do not consider is due to Hall (1992) and is described by Manly (1997, Chapter 3).

It is well known that given a sample of n from a normal distribution with variance σ^2 the 95% confidence limits for the mean μ are $\bar{x} \pm 1.96\sigma/(\sqrt{n})$ where $\bar{x}$ is the sample mean, while if σ is unknown the true standard error $\sigma/(\sqrt{n})$ is replaced by an unbiased estimate and the coefficient 1.96 is replaced by the corresponding quantile value of the t-distribution with $n-1$ degrees of freedom. For values of $n > 30$ standard normal distribution quantiles closely approximate the exact t-distribution quantiles. Modifications for other confidence levels are straightforward. In practice this model-driven procedure is widely used even when there is no firm evidence that samples are from a normal distribution, faith in the result depending on hope that the central limit theorem, despite its asymptotic nature, works for most n. We meet examples in later chapters where that assumption is clearly unrealistic.

The above model-driven methods have stimulated two approaches to forming bootstrap confidence intervals. The first is that, for reasonably large samples, approximate 95% confidence limits for a parameter θ may be based on a plug-in estimator θ and are given by $\theta \pm 1.96se^*(\theta)$, where $se^*(\theta)$ is the true bootstrap standard error or an estimate of this based on B bootstrap samples. Clearly for the mean of a sample from a normal population this will give a reasonably good interval because $\theta = \overline{x}$ and we saw in section 2.3 that the bootstrap standard error in this case equals the biased maximum likelihood estimator of the standard error. The only obvious improvements would be to remove the bias. However, as emphasised throughout this chapter, the bootstrap is most useful when we are interested in parameters such as correlation coefficients or means of ratios of variables where analytic theory is either nonexistent or highly distribution dependent. Experience has shown that this simple method does not translate well to such situations. This is hardly surprising, because not only may there be appreciable bias in bootstrap estimators of standard errors, but symmetric confidence intervals are inappropriate if the plug-in estimator of interest has a skew distribution. Examples where skew distributions arise with correlation coefficients are given in sections 9.5.

A second approach to confidence intervals is again stimulated by the role of the t-distribution in normal theory models. It goes one stage further than merely allowing for our not knowing σ^2 and allows for possible nonnormality by forming what is called a bootstrap t-distribution which is unique to that particular data set and problem. The method is more complicated and highly computer intensive, involving further bootstrapping of each individual bootstrap sample to obtain a standard error of the plug-in estimator of the parameter of interest anew for each bootstrap sample. This is called secondary bootstrap sampling and is described in detail by Efron and Tibshirani, (1993, section 12.5). In effect it generates a t-distribution that is unique to a given data set and so must be generated afresh for every data set. It sometimes works well, but its overall performance is disappointing. One reason for this is that, like its normal theory counterpart, it leads to confidence intervals that are symmetric about the point estimator. These may be inappropriate.

In passing it is interesting to note how, for inferences about location even in a one sample problem, the gradual relaxation of assumptions impinges upon both the logical and the computational complexity of hypothesis testing and formation of confidence intervals. When we assume normality and knowledge of σ^2 we may base all inferences on the standard normal distribution irrespective of the data values or sample size. If we assume normality with σ^2 unknown we work with a different t-distribution for each sample size, n, but the test statistic distribution is the same for all samples of a given size. If we drop all distributional assumptions and use a statistic such as the bootstrap t or use permutation test methods the distribution of the relevant statistic is specific to (or putting it another way, conditional upon) the sample data and must be calculated afresh for every sample.

A more fruitful and relatively easy-to-use approach to bootstrap confidence intervals is based upon the distribution of the bootstrap plug-in estimators, $s(x^{*b})$ of each of B bootstrap samples. When B is large, then providing the $s(x^{*b})$ are reasonably free from bias, the distribution of $s(x^{*b})$ should approach the sampling distribution of the plug-in estimator for the parameter under investigation, and an obvious way to estimate a $(1 - 2\alpha)100\%$ confidence interval is to take as the limits the αth and $(1 - \alpha)$th quantiles of the $s(x^{*b})$ distribution. In practice the value of B should be at least 1000. If $B = 1000$ for a 95% confidence interval the limits are the 25th and 975th largest values of $s(x^{*b})$. More generally for a $(1 - 2\alpha)100\%$ interval with B samples the limits are the αBth and the $(1-\alpha)B$th largest sample values of $s(x^{*b})$. For nonintegral values of the quantiles it usually suffices to round to the nearest integer. Confidence intervals formed in this way are called **quantile-based intervals**.

Approximate intervals using these methods have potential defects. Quantile-based intervals may be misleading due to serious bias, and they often undercover the true interval in the sense that a nominal 95% interval obtained this way is in reality only about a 90% interval. This means that if a large number of samples of size n is taken from the original population, and bootstrap 95% confidence intervals are calculated this way for each, only about 90% of such computed intervals will cover the fixed true mean. One of the effects of bias in bootstrap estimates is that the true probabilities associated with each tail are in practice not both equal and which is the greater depends upon the direction of the bias.

These weaknesses of bootstrap quantile-based confidence intervals may be reduced by corrections for bias and undercoverage. Two commonly used methods apply corrections by adjusting the quantiles specifying the limits. These are

- bias corrected and accelerated intervals (abbreviated to Bca);
- approximate bootstrap confidence intervals (ABC).

Both are fully described by Efron and Tibshirani (1993, Chapter 14). The correction for bias is based on a simple bootstrap measure of bias and the acceleration factor makes use of the jackknife. I do not give details, because for most purposes in this book approximate intervals will suffice, but where better intervals are required, or for advanced problems where bootstrapping may be virtually the only way to get intervals, these refinements are important.

The tendency for quantile intervals to undercover arises because bootstrap estimates respond to features of the sample distribution directly but only reflect population characteristics insofar as these are carried over to the sample distribution. Thus bootstrap estimates are a better reflection of the sample situation that they are of the true (and unknown) population situation. Reduced coverage for estimated confidence intervals is one effect of this.

Example 2.7 In Example 2.1 the data set C was from an exponential distribution with mean 1. The data were

0.11 0.12 0.17 0.18 0.22 0.32 0.34 0.37 0.40 0.50 0.51 0.55 1.23 1.84 1.94 2.50 2.86 3.38 3.81 3.86

I formed 1000 bootstrap samples and estimated the median m^* for each. I used MINITAB but most standard statistical packages have the basic facilities for this exercise providing they can produce random samples with replacement and have adequate data sorting facilities to determine the requisite quantiles. For an approximate 95% interval I located the 25th and 975th largest of the 1000 sample medians as 0.320 and 1.890, implying an interval (0.320, 1.890). The interval is markedly asymmetric about the sample median of 0.505. The theoretical standard error of the median for an exponential distribution with mean 1 based on the formula given in Example 2.5 is 0.224. For the 1000 bootstrap samples I used in this exercise $se(m^*) = 0.406$. Is bootstrap estimation breaking down here, or is this an example of the point I made above that the bootstrap reflects properties of the sample directly and those of the population only indirectly, especially when the sample is not very large? Inspection of the ordered sample shows a change in pattern at 0.55; there is a close bunching of sample values below 0.55, then an appreciable gap to the next observation of 1.23. Indeed the interval between 0.55 and 1.23 in which there are no sample values is longer than the interval between the smallest value 0.11 and 0.55 in which there are 10 intermediate values. This pattern is somewhat unusual when sampling from an exponential distribution, but it is the sort of thing that happens from time to time with random sampling. Probabilistic considerations suggest that values greater than 4 should be rare in samples of 20 from a unit mean exponential distribution, whereas values between 0.505 and 1.5 should be fairly common. In this sample there are only 3 values between 0.505 and 1.5, less than expected, and there are more than expected in the interval (1.5, 4) This implies overdispersion of observations in the sample relative to that in the population and this inflates estimates of statistics like the bootstrap standard error of the median, because this overdispersion carries through to bootstrap sampling, as indeed it would to any other method of estimation. It is part of the nature of random sampling that some samples will in this sense show overdispersion and others underdispersion.

To illustrate these effects I took 12 random samples of 20 from an exponential distribution with unit mean and for each calculated an estimated bootstrap standard error of the median based on 1000 bootstrap samples and I also obtained the bootstrap 95% quantile confidence interval for those samples. The first of those samples was the one described above. Table 2.5 gives the confidence intervals and bootstrap standard errors of the median for these samples.

Features in Table 2.5 worth noting are:

Table 2.5 Confidence limits and bootstrap standard errors based on 1000 bootstrap samples for medians of samples of 20 from an exponential distribution with unit mean. The expected value of the median is 0.693.

95% confidence Interval		
Lower limit	Upper limit	se(m*)
0.320	1.890	0.406
0.409	0.870	0.113
0.495	1.152	0.175
0.523	1.544	0.248
0.203	0.776	0.162
0.316	1.244	0.298
0.337	1.540	0.348
0.372	1.087	0.156
0.351	1.543	0.380
0.392	1.059	0.199
0.214	1.228	0.248
0.293	1.340	0.292

- Five of the bootstrap standard errors (final column) are below the theoretical value of $se(m) = 0.224$ and 7 are above.
- As $se(m*)$ increases so, in general, does the width of the confidence interval.
- The lower limits (ranging from 0.203 to 0.523) are less variable than the upper limits (ranging from 0.870 to 1.890).
- All 12 intervals in Table 2.5 cover the population median which is 0.693.

None of these findings should cause surprise. They show how the bootstrap adapts performance to the nature of specific data sets, a feature of most methods of inference. Indeed, as I indicated earlier, the rationale of statistical inference incorporates the notion that there are limitations on how well a sample reflects population characteristics and that is why there must always be uncertainty about inferences. Concepts such as a confidence interval enable us to associate a defined measure with that uncertainty. The characteristic of bootstrap estimates to tailor themselves to sample characteristics is indeed a strength, for it lets us make inferences without specific assumptions about population distributions, inferences based only on a belief that characteristics of a population will be broadly reflected in a sample.

However, I must caution against inappropriate estimates. Avoid intuitive approaches that may be suggested by inappropriate model-driven analogies. For example, do not set 95% confidence limits for the median for the data in Example 2.7 as $m \pm 1.96se(m*)$. This gives the limits $0.505 \pm 1.96 \times 0.4055$ leading to a confidence interval $(-0.290, 1.230)$. This is a nonsense interval, since the median of the exponential distribution must be positive. Using quantile-based confidence intervals derived from bootstrap sampling with the relevant statistic avoids such nonsense intervals.

A further note of caution is needed. Manly (1997, Chapter 3) considers a wide range of bootstrap confidence intervals for the standard deviation for an exponential distribution using a sample of 20 observations from an exponential distribution with unit mean and standard deviation. He compares results for six widely used methods and the performance of all except one of these was far from satisfactory. Manly indicates reasons for the deficiencies, indicating in particular that bootstrap confidence intervals are not always reliable for small samples.

2.7 The two-sample bootstrap

In section 1.6 I considered hypothesis tests primarily for a possible location shift in a two sample problem where, under the null hypothesis, each sample or group came from **the same** population where in some cases that population may consist only of the experimental units. If the hypothesis of no shift were true, then for sample sizes m and n all samples of m from the combined sample of $N = m + n$ observations were equally likely. This enabled us to calculate the exact probability that a random sample of m obtained without replacement yielded the observed value of some statistic, S, and also to calculate the probability of obtaining more extreme values of S that provided stronger evidence against H_0 because they were more likely to be observed under an alternative H_1. From these I computed the relevant exact p-value in the way described in Section 1.6. For an unbiased test when the alternative hypothesis is true we are more likely to get a result in the critical region but what that probability is depends upon the power of the test for that alternative.

The appeal of the Fisher–Pitman test in many experimental situations is that sampling *without* replacement exactly reflects the way the units are allocated randomly to two treatments when there are no further constraints and reflects what in medical parlance is called *placebo* status for the treatments if H_0 holds.

Here is a bootstrap analogue to the permutation test procedure that is discussed by Efron and Tibshirani (1993, section 16.2). It is appropriate when H_0 specifies that the samples are from the same distribution and H_1 specfies a shift in location or indeed one of the other possible alternative hypotheses considered in section 1.6. We proceed as we did for the permutation test by combining the two groups with m, n observations into a single group of $N = m + n$ but now form bootstrap samples by first forming from the group of N observations samples of N with replacement. These samples are then split into samples of m, n where the sample of m consists of the **first** m observations in each bootstrap sample of N and the sample of n consists of the remaining $n = N - m$ observations. The only difference so far is that the Fisher–Pitman test is based on sampling without replacement, but the bootstrap sampling is with replacement. The first m observations in each sample is a bootstrap representation of the first group and the remaining n are a bootstrap representation of the second group. If we form B such samples of N and allocate the observations in them to the first and second group in this way we may use these to get an estimated bootstrap distribution of sample mean differences under H_0.

We may use this estimated bootstrap distribution as the basis for a test of H_0: *identical distribution* against an alternative H_1 that indicates location difference. Here there is a crucial practical difference from the permutation test where we saw that a number of equivalent statistics – first group total or mean, second group total or mean, the difference between group totals or between group means, or the conventional t-statistic were among many statistics, that if correctly used, were equivalent in the sense that they all led to the same p-value.

Denoting the permutation test sums for the first and second group by S_m, S_n this equivalence between the statistics arose because $S_m + S_n$ remained constant over **all** permutations, i.e. for all samples obtained **without** replacement. This is no longer true when sampling with replacement, for then the bootstrap equivalent $S_m^* + S_n^*$ will vary from sample to sample, so many of the equivalent permutation test statistics are clearly no longer appropriate. Under H_0 the difference between population means, θ, say, is zero and an intuitively reasonable statistic for estimation or testing is $\hat{\theta} = \bar{x} - \bar{y}$, the difference between the group means. For bootstrap samples the equivalent statistic is $\theta^* = \bar{x}^* - \bar{y}^*$, but whereas in the permutation test the exact p-value relevant to a one-tail test is given by the proportion of permutations giving the observed or a more extreme value of θ, we may now only estimate the exact p by taking B bootstrap samples and determining the proportion of these for which θ^* is equal to or more extreme than the observed $\hat{\theta}$. Approximation to the permutation test is usually good for moderate to large samples if the data do not blatantly contradict the assumption that any population difference is confined to a location shift. Example 2.8, using data where the t-test should be optimal, shows good agreement between that test, the Fisher–Pitman test, and a bootstrap test based on the statistic θ^* proposed above.

Example 2.8 Consider the two-sample data:

| Group I | 7.6 | 5.3 | 4.8 | 6.0 | 6.6 | 7.1 | 1.8 | 2.8 | 1.6 | 4.2 |
| Group II | 5.1 | 3.4 | 4.6 | 9.4 | 5.2 | 7.4 | 5.9 | 7.4 | | |

The difference between the group means is 1.27 and the one-tail t-test of H_0:*distributions are identical* against H_1:*group II population mean > group I population mean* gives $p = 0.105$.

StatXact gives the exact one-tail $p = 0.107$ for a Fisher–Pitman permutation test. I took 1000 bootstrap samples in the manner described above and computed θ^* as defined above for each and 102 of the computed θ^* were greater than or equal to 1.27, giving an estimate $p = 0.102$.

The close agreement between the three tests in Example 2.8 is reassuring when I disclose that the artificial data used in that example are random samples from normal distributions each with $\sigma = 2$ but with means $\mu_I = 5$ and $\mu_{II} = 6$. Here the t-test is uniformly most powerful if we do not assume the known value of the common population standard deviation; the Fisher–Pitman test may be expected to perform well also since its Pitman efficiency here is unity, although this is an asymptotic property and here we have only a finite sample. The estimated p-value for the bootstrap is comfortingly close. Had the value been very different from that for the other two methods our faith in the bootstrap would have been shaken. In practice in this example most statisticians would happily use a t-test even had they not known how the data were obtained. I pointed out in section 1.5 that a disappointing feature of the Fisher–Pitman test is that it behaves very like the t-test even when the latter is not appropriate, indicating a lack of robustness. The same is true of the bootstrap test developed above. In section 4.1 I show how alternative permutation and bootstrap test and estimation procedures are developed to overcome this weakness of the basic tests used here.

I indicated above that for the bootstrap in the two-sample problem, unlike the permutation test situation, using different statistics leads to different estimates for p, even when we base the estimates on the same set of B bootstrap samples. If, for example, instead of using a statistic $\hat{\theta} = \bar{x} - \bar{y}$, I had chosen the usual t-statistic, t, say, used in the t-test, the procedure would be to compute the plug-in estimator t^* for each of the B bootstrap samples. Unlike the situation in the permutation test the denominator of t^* now changes from sample to sample, depending on the relative frequency with which each data value appears in the bootstrap sample as a consequence of sampling with replacement. In practice in a situation when the samples are from near normal distributions differences between the estimates of p based on the same bootstrap samples using θ or t are seldom greater than the differences one may get by using only one of these estimators, but a different set of B samples. However, when normality is not a reasonable assumption, differences between results with each statistic may be substantial. The quantities t^* are called **Studentized differences**. We meet the concept of studentized variables in other contexts in this book.

In passing, it is interesting to consider the bootstrap in this two sample problem in the contingency table format introduced in Table 1.1, section 1.6. There our permutation test randomization gave rise to all possible tables with fixed row (sample sizes) and column (observed value) totals, the latter being all unity if there were no tied observed values. For each bootstrap sample we may form similar tables but now only the row totals are fixed. The column totals will generally change from bootstrap sample to bootstrap sample as a consequence of sampling with replacement. In effect, this means that bootstrap estimation relaxes the requirement of fixed column totals. I do not pursue this point further here, but over the last 50 years there has been an often-heated controversy about the appropriateness of analysing contingency tables with marginal totals fixed as opposed to allowing either the row or column totals to take different values. In practice there is often little difference in the final result of what are regarded as correct analyses by the exponents of either school. Historically, this controversy predates the use of the bootstrap, but it is interesting to note that in some situations the effective difference in approach parallels that between using a permutation test and using a bootstrap.

A different approach to bootstrapping is needed for the two-sample problem if we start with the null hypothesis that the means are equal and the alternative hypothesis implies mean difference, but drop the assumption of identical distributions under H_0. In these circumstances it is no longer appropriate to pool the samples under H_0 and an obvious bootstrapping procedure is to bootstrap each group separately. The simplest relevant statistics are again $\hat{\theta} = \bar{x} - \bar{y}$ or the corresponding t-statistic, but these are generally only satisfactory with an additional assumption that the variances are approximately equal.

Example 2.9 For the data in Example 2.8 I took 1000 bootstrap samples in each of which the data in Group I and Group II were sampled separately. For each bootstrap sample I computed the statistic $\theta^* = \bar{x}^* - \bar{y}^*$. Unlike the situation in Example 2.8 where the distribution of the θ^* is centered near zero, in this case the distribution is centered about the observed $\theta = 1.27$. For a one-tail test when H_0 specifies a zero location difference the appropriate p-value is estimated by the proportion of θ^* that are negative. I also calculated a 95% quantile confidence interval for θ, determining limits as the 25th and 975th largest θ^*. I repeated this experiment three times, taking $B = 1000$ in each case. Table 2.6 gives the estimated p-value in each of these together with the corresponding values based on Student's t-test which is here optimal because the samples come from normal distributions each with the same variance, assuming here that that variance is not known.

Table 2.6 Bootstrap and t-test p-levels and 95% confidence intervals for Example 2.9

Procedure	One-tail p	95% confidence interval
Bootstrap I	0.085	(−0.51, 3.01)
Bootstrap II	0.064	(−0.28, 3.06)
Bootstrap III	0.069	(−0.35, 2.96)
Student's t	0.105	(−0.79, 3.33)

It is clear from Table 2.6 that the bootstrap results are broadly in agreement with each other but that all give a lower p-value and shorter confidence intervals than Student's distribution. This is consistent with the tendency already mentioned for bootstrap quantile intervals without adjustment to underestimate in the sense that a nominal 95% interval may have a coverage that is little more than 90%. For these data the 90% interval for the t-test is $(-0.43, 2.97)$ supporting this conclusion. Bias corrected and accelerated intervals might reduce the discrepancy.

In section 6.6 I revisit the two-sample bootstrap and consider the situation where the assumption in Example 2.9 that the distributions have the same variance is dropped. In the normal theory parametric situation for testing for location differences this gives rise to the Behrens–Fisher problem which has been a subject of some controversy and various approximate solutions have been proposed. In section 6.6 I consider this type of problem without the restrictive assumption of normality.

2.8 Bootstraps in a broader context

In this Chapter I have, except in section 2.7, considered the bootstrap only for single sample problems. Discussion has also been confined to what is called a nonparametric bootstrap. This may be regarded as a truly data-driven type of bootstrap, even though I applied it to samples from specified distributions for illustrative purposes, using such examples primarily to show the strong and weak points of the procedure, but the method could be applied without such population knowledge.

There is also a parametric version of the bootstrap which may be appropriate if we have partial information about the source of data. If we are confident that our sample comes from some family of distributions, e.g. the one parameter exponential family, we might estimate the parameter for that distribution from our data by maximum likelihood or some other method. Then, instead of bootstrapping our data, we may instead form samples of size n by sampling from an exponential distribution with that estimated parameter value. This is the parametric bootstrap. This method would be unlikely to be used seriously for assessing accuracy of estimates about the mean if we were confident the sample comes from the exponential family because there are classical estimation methods that are then more appropriate. However the method is useful for more complicated estimation problems. For example, inferences about the distribution of ratios x/y of random variables are notoriously difficult as little analytic theory is usually available. If we have a sample of n ratios x_i/y_i where the pairs (x_i, y_i) may reasonably be supposed to be drawn from independent exponential distributions parametric bootstrap samples are obtained by estimating separately the exponential parameter for the X and Y distributions by maximum likelihood or some other appropriate method and then sampling n pairs from the respective distributions and using the ratios from each pair as the parametric bootstrap sample. Details of the parametric bootstrap are given by Effron and Tibsharani, (1993, section 6.5 and Chapter 21).

A field of application where both the bootstrap and permutation methods are widely used is that involving directional data. When birds are released at a fixed point is there a preferred direction of flight? Is there a preferred orientation for fractures in a rock structure? A number of examples are given by Fisher (1993) who devotes his Chapter 8 to a useful introduction to these approaches.

In later chapters I give several examples of the bootstrap including in Chapters 9 and 10 use of the method in problems associated with correlation and with regression where alternative approaches may be hard to implement.

2.9 Cross-validation

I have pointed out that all methods of inference reflect population characteristics only insofar as these are evident in the sample used for making the inferences. I made these comments essentially in relation to parameter estimation. In fields like regression and classification parameter estimation is often only part of a model building process to produce a model that will be used for making predictions when new data become available. What is then of interest is how good these predictions will be when applied to new data. Measures such as residual variance obtained from sample data underestimate errors of prediction. In an ideal world the best way to examine the prediction error for a fitted model would be to use all available data to fit the model and then assess prediction error by taking further samples from the original population and to use these in the fitted model to see how accurately it predicts. However, this may be a costly or even impossible process. The idea of **cross-validation** is to divide the data into two or more portions and to fit a model in turn to all but one of these portions and to calculate the prediction error of the fitted model for prediction in the omitted portion. Each portion is omitted in turn and a combined estimated prediction error obtained. An extreme case especially suitable for small data sets is the so-called **leave-one-out** cross-validation where each observation i, $i = 1, 2, \ldots, n$ is omitted in turn, a model is fitted to the remaining $n - 1$ data and a predicted value obtained for the omitted observation using that model. If, when the ith observation is omitted the predicted value is $y^*_{(i)}$ the cross-validation prediction error is defined as $\sum_i (y_i - y^*_{(i)})/n$. The computational procedure is reminiscent of that for the jackknife, but the objective is different.

Cross-validation is a widely used technique that falls between model-driven and data-driven methods. Its use preceded the bootstrap but its relationship to the latter is discussed by Efron and Tibshirani (1993, Chapter 17) who give numerous references to key papers on cross-validation.

Exercises

In this and later Chapters some exercises are included which assume the reader has access to one or more standard statistical packages such as MINITAB, Genstat, SPSS, etc. These exercises are marked with an *. Such general packages should allow answers or at least partial answers to be obtained easily. Where permutation tests are involved specialist packages such as StatXact or TESTIMATE are ideal, but Monte Carlo approximations are obtainable with most standard packages.

2.1 Construct five-number summaries and box and whisker plots for each of the three data sets in Example 2.1 and consider whether they confirm or illustrate more clearly the features of the histograms in Figure 2.1. [A five number summary consists of the least value, first quartile, median, third quartile and greatest value. Box and whisker plots are a graphical representation of this information described in many modern books on elementary statistics or data reduction including Chatfield (1988)].

2.2 Verify the formula for the variance of the bootstrap estimate of the mean for samples of 3 established in Example 2.2.

2.3 Verify the formulae for the mean and variance of the bootstrap estimate of the median for samples of 3 established in Example 2.2.

2.4 For samples of 4 and for samples of 6 determine all possible values of the bootstrap sample medians.

2.5 For the data in Example 2.3 verify that the bootstrap se(med) = 3.05.

2.6 Use an appropriate test to confirm that the observed numbers of times each median is recorded in 150 bootstrap samples in Example 2.2 is consistent with the expected numbers.

2.7 Confirm that the jackknife estimator of bias (2.6) for the plug-in estimator of the variance of the mean gives an estimate equal to the exact bias.

*2.8 Use an available computer random sample generator to generate samples of 10 from a N(4, 9) and from a N (6, 9) distribution. Use your samples and an appropriate bootstrap procedure to test for a difference between population means. Also generate a sample of 10 from a N(6, 16) distribution and use the same bootstrap procedure to test for a difference in means using this sample and the sample you previously generated for a N(4, 9) distribution. Do the differences in results in the two tests surprise you. Do you think the procedure you used is justified in each case? If not, why not?

*2.9 Chatfield (1988, p. 98) gives the following data for *numbers of worms* $\times$ 10^{-3} in samples of *vaccinated* and *unvaccinated* lambs. Use a *t*-test, a Fisher–Pitman permutation test, a bootstrap analysis and any other method you think appropriate to determine whether it is reasonable to assert that vaccination reduces worm infestation.

Unvaccinated	22	21.5	30	23				
Vaccinated	21.5	0.75	3.8	29	2	27	11	23.5

(Analysis of these data is considered further in section 6.6)

3

Outliers contamination and robustness

3.1 Models and data

No matter how carefully data are collected and recorded they seldom constitute a true random sample from one member of a specific family of distributions such as the normal, Poisson, binomial, exponential, gamma or whatever. Even supposedly random samples from such populations often include 'rogue' observations due to some breakdown in sampling procedure, to limitations of measuring techniques or to errors in measurement or recording, or because the population does not after all conform to membership of one specific family.

The success of model-driven analyses based on assumptions like data forming a random sample from a normal distribution or from a one-parameter exponential distribution, even when there are aberrant observations present, often stems from a robustness in that inferences are little influenced by a small proportion of grossly aberrant observations or even by a large number of small aberrations, such as those induced by round-off or slight imprecision in recording. However, this is not always the case.

Modern computing methods have revolutionized the way statisticians deal with data impurities. I look first at some ways in which faulty data arise and how they might be detected. Two related facets are usually referred to as

- the outlier problem;
- data contamination.

Contamination may lead to outliers, but not all contaminated observations are outliers. However, an outlier is either in some sense contaminated or for other reasons out of line with the bulk of the observations.

3.2 Outliers and contamination

In broad terms for univariate data an **outlier** is an observation so remote from other observations as to cause surprise. Surprise is, by its nature, subjective.

Whether we are surprised by an observation may depend upon what we know about the source of data and what they represent, or if one only has scant information of this sort, upon the position of the observation relative to the rest of the data. I show below that a contaminated observation may or may not be surprising and many contaminated observations are not easy to detect, but again knowing how the data arose and were collected may help in detecting contamination. Example 3.1 illustrates these points. A comprehensive treatise on outliers and contamination and a full discussion of how best to handle them and how they influence statistical inference is given by Barnett and Lewis (1994), the third edition of an authoritative treatise on these matters.

Example 3.1   Here are four data sets

A	0	3	11	9	6	120	8	3	14					
B	1.3	1.9	2.2	1.4	2.7	3.1								
C	140	135	165	135	157	190	155							
D	0.73	1.38	2.12	2.76	2.89	3.20	3.57	3.96	4.41	4.52	4.67	4.70	4.78	4.91	7.73

Without further information than that in the data themselves most people asked to select any observations that might be outliers or in some way contaminated in set A would probably say that 120 was an outlier. They might see nothing surprising about any observation in set B. For set C many would see nothing alarming while others may feel surprised that the observation 190 is appreciably greater than any others, the difference, 25, between it and its nearest neighbour, 165, being almost as big as the difference $165 - 135 = 30$ between 165 and the smallest observation. Having noted that the observations in set D are arranged in ascending order, one might have qualms about the observation 7.73 and perhaps to a lesser degree, about the observations 0.73 and 1.38, as they seem intuitively to be somewhat extreme.

If we are given the additional information that set A represents counts of numbers of aphids on each of 9 rose bushes an entomologist would be unlikely to bat an eyelid, for in many situations involving insect populations, data like these are the norm. Barnett and Lewis (p. 16) quote a more extreme example of data given by Fisher, Corbet and Williams (1943) on numbers of moths of a nocturnal species caught in a light trap at Rothamsted, namely

$$11, 54, 5, 7, 4, 15, 560, 18, 120, 24, 3, 51, 3, 12, 84$$

For set B, while there are clearly no outliers in the sense of extremely large or extremely small values, if we are given the additional information that the data are weights of a growing bird in kilograms measured at fortnightly intervals, a zoologist would have doubts about the validity of the observation 1.4 in the fourth period. A young bird may sometimes suffer a weight loss during the growth process but it is likely to be a smaller loss than this, and if so great a loss were recorded the bird is more likely to be dead a fortnight later than to have returned to a weight suggestive of a normal growth pattern. The more plausible explanation is that this is a contaminated observation where the contamination is probably due to an error in weighing or recording. One possibility would be that a true reading of 2.4 was incorrectly recorded as 1.4. A check on the laboratory procedure may indicate that this could happen quite easily if weights were put in a balance pan and the 1 kg and 2 kg weights were similar in size and design (the difference being mainly in their densities). Here the contaminated observation does not appear as an outlier and the contamination only becomes apparent when we are told that the data are time dependent.

For set C information about the nature of the data may help decide whether there is something aberrant about the observation 190. If we knew them to be heights in cm. of 14 year old girls we would have strong doubts about this observation, although it might not be impossible. On the other hand I would not be alarmed by it if told they were the heights of 7 sunflower plants growing up a garden wall. The sunflower plants in my garden last year varied greatly in height. The observation 157 in this set calls for comment. All the other observations have terminal digit 5 or 0, suggesting that measurements have been made to, or rounded to, the nearest 5 units. This one measurement may have been made with higher precision, but another possibility is that it is wrongly recorded. If it were a misrecording of 155 this is unlikely to have a dramatic effect on any relevant inferences, but if it were another common type of error, a transposition of digits, and the true reading were 175 this would be serious contamination.

If we are told that data set D gives weights in grams of an impurity in 1 kg ingots of zinc selected at random from a large production batch the extreme values here give cause for surprise if past experience has shown, as it very often does in such circumstances, that within any batch the amount of impurity is for all practical purposes normally distributed. In these circumstances there is at least a case for further investigation. Are one or two measurements incorrect? Has some change in the production process led to the impurity distribution being longer tailed than a normal distribution? A common source of data that might be expected to be approximately normally distributed, yet exhibits signs of over-dispersion or slight skewness is that where the greater portion of the data comes from a normal distribution which may be termed the *good* or *true* distribution, but a few come from a distribution that is contaminated by errors. For example, if 11 of the 15 observations in set D were made in one careful laboratory that produced accurate results and these reflected the fact that the impurity was normally distributed with unknown mean μ and unknown variance σ^2, while the other 4 observations were made in a careless laboratory where observations were contaminated by further independent experimental errors with mean zero and variance σ_c^2 the outcome is that such 'contaminated' observations will be distributed with mean still μ, but variance $\sigma^2 + \sigma_c^2$. The presence of even a small proportion of such observations may account for longer tails, even though not all the contaminated observations will lie in the tails. If contaminated observations also shift the mean this will result in some skewness. A shift in the mean may stem from a constant reading error, often called a zero error, in one of several measuring or recording devices, or by a constant level of background impurity in apparatus used for analysing a portion of the specimens. In many situations only one of these contamination factors – often referred to as *slippage in the mean* or *slippage in the variance* operates. If the source of the disturbance is clear the proportion of contaminated observations may be known and if records have been kept of the history of each observation the contaminated observations may be identifiable. In many cases, however, neither the source of individual observations or the nature of the contamination will be known. Set D is, in fact, a computer generated data set consisting of a random sample of 11 *good* observations from a N(4, 1) distribution and 4 *contaminated* observations from a N(4, 9) distribution. This is a situation where not all (even perhaps not any!) contaminated observations may appear as obvious outliers. In set D the contaminated observations were in fact 0.73, 2.12, 2.76 and 7.73. Only the last is clearly an outlier with respect to the N(4, 1) distribution, with the first showing a tendency that way.

Clearly outliers and contaminated observations are a nuisance that must, or should, be taken into account when making inferences. Historically, many early papers about outliers were about rules or criteria for **rejection**. This reflected the earlier dominance of model-driven analyses and an urge to clean-up data to make them consistent with some pre-specified model. If an outlier can be shown to be erroneous rejection or correction (if possible) is appropriate. If, among records of the heights of army recruits one is recorded as 1755 cm this is clearly an error. If it cannot be corrected the observation should be rejected.

When there is no indication that an observation is wrong the appropriate course of action is less clear. Basically it depends upon the population of interest and even more specifically on what questions are being asked about that population. For example, in an experiment to test a drug for reducing blood pressure it may be known that it is ineffective if the recipient drinks alcohol and all recipients might be instructed to not drink alcohol while using the drug. It is likely that a few recipients will ignore that advice. If, in a group of 20 patients given the drug the drops in systolic blood pressure in mm Hg are:

−2, 0, 3, 6, 25, 28, 31, 30, 33, 35, 35, 35, 37, 40, 40, 40, 42, 45, 45, 50

an informed judgement might be that the first four readings were from patients who had almost certainly ignored the alcohol ban. How firmly that view were held is a matter of clinical judgement. If there were strong grounds, based either on past experience or a knowledge of the biochemical reaction, that the drug would almost certainly reduce blood pressure substantially providing alcohol is not consumed, and interest is simply in determining, say, the average reduction when the drug is taken without alcohol it makes sense to reject the first 4 observations. In doing so we might do an injustice because there is no guarantee that all, or any, of the four have consumed alcohol. It would be a matter of clinical judgement whether these participants should be asked if they had taken alcohol. There may be doubts about the truth of answers to that question!

A reason for rejecting these outliers may be that the clinician is interested in the outcome only for those who obey the rules about not drinking. If, on the other hand it were believed that a proportion of the population might not respond as positively as others (perhaps for some indeterminate genetic reason) the interest might be in responses in the population as a whole and the decision might be to include all observations in the analysis. Depending on the inferences one wishes to make, the precise form of the analysis might have to be modified to take account of the heterogeneity induced by, say, a genetic factor.

In the normal distribution situation envisaged in Example 3.1 for data set D where a few observations are contaminated due to laboratory or measurement errors introducing excess variation we sometimes face a dilemma. We may be reasonably confident that the laboratory making these errors is producing results distributed about the correct mean, but with increased variance, i.e. that there is slippage in the variance only. In these circumstances there is still useful information about the mean in the contaminated data, but it is less useful than that from the uncontaminated observations, because the effect of contamination is to increase the estimated standard error of the mean. If we retain the contaminated data this reduces the power of a test like the t-test, or that of a permutation test conditional upon the data such as a Fisher–Pitman test, that behaves like the t-test. If we knew which data were contaminated we could remove them, but this also reduces power compared to what it would be if none of the data were contaminated, because it reduces the sample size. A further complication is that we might not know how many or precisely which data are contaminated. This looks like a no-win situation, but things are not as bleak as they seem, because methods have been evolved that reduce the influence of contaminated observations, especially those that have outlier characteristics. Some of these are described in section 3.5. Some methods are also useful when contamination takes the form of slippage in the mean. The effect of mean slippage, in the case of the normal distribution, is to induce skewness in the sample data. This increases the standard error and introduces bias in the usual estimate of the mean.

So far my remarks about outliers and contaminants have focused on a subjective and intuitive exploratory approach to their existence and detection. Even for normal theory models where we suspect contamination, it is too easy to jump to the conclusion that one or more observations are outliers. Objective criteria are available for deciding whether an extreme observation should be classed as an outlier. These provide our first examples of diagnostic tests.

3.3 Diagnostic tests for outliers

Barnett and Lewis (1994, section 6.3) list 48 tests for outliers in normal distributions alone. Even in this restricted context the situation is complicated because most of the tests proposed are optimal only for fairly specific alternatives to a null hypotheses that all data belong to the same normal distribution where neither, one, or both parameters may be known. Some tests are specific to single suspected outliers, some to potential outliers in one tail only, some to outliers in either tail, some to outliers in both tails, and others to blocks of outliers. A common difficulty is the so-called **masking effect** by which the power of a test for outliers is often reduced by the presence of further outliers or otherwise contaminated observations that are not themselves under test. For illustrative purposes I consider 4 of the 48 tests given by Barnett and Lewis, all but one for the case where μ and σ^2 are unknown. The first two are tests for a single outlier which I take to be the largest value $x_{(n)}$. The argument is easily modified to deal with the smallest value $x_{(1)}$. The statistic used in the first case is the studentized extreme deviation from the mean under H_0, i.e.

$$T = (x_{(n)} - \bar{x})/s \qquad (3.1)$$

where s^2 is the usual sample estimate of σ^2, i.e. $s^2 = \sum (x_i - \bar{x})^2/(n-1)$. Computer programs giving exact p-values for this and the other statistics I derive in this section are, so far as I know, not generally available. However, Barnett and Lewis include tables of minimum critical values at the 5% and 1% significance levels.

Example 3.2 The data in set D in Example 3.1 were

  0.73 1.38 2.12 2.76 2.89 3.20 3.57 3.96 4.41 4.52 4.67 4.70 4.78 4.91 7.73

For these data $\bar{x} = 3.755$ and $s = 1.701$, whence for testing $x_{(n)} = 7.73$ using (3.1) $T = 2.34$. This narrowly fails to indicate significance at the 5% level. The critical value given by Barnett and Lewis (Table XIIIa, p. 485) when $n = 15$ is 2.41.

Because the datum 7.73 is remote from the rest of the data the outcome is disappointing. I have already explained how the data were generated and in terms of the uncontaminated data (with $\sigma^2 = 1$) the value 7.73 is 3.73 standard deviations above the population mean of 4. Failure to achieve significance is partly due to the masking effect of the three other contaminants which inflate the estimate, s, of standard deviation. Further, the statistic (3.1) is tailored to detect outliers in one-tail and is better suited to mean slippage than to variance slippage, e.g. where the contaminated datum is $N(\mu + a, \sigma^2)$ rather than $N(\mu, b\sigma^2)$. We know the latter is the case here. However, with most real data we do not have such prior knowledge and the slight skewness to the right in this data suggests this test is not inappropriate.

An alternative test that works well for an outlier pair, one in each tail, for symmetric contamination of a fairly general form but not too many contaminated observations, uses what is called the Studentized range statistic, i.e.

$$T_R = (x_{(n)} - x_{(1)})/s \qquad (3.2)$$

where s is the same as in (3.1)

Example 3.3 For the data in Example 3.2 we find $T_R = (7.73 - 0.73)/1.701 = 4.115$. Barnett and Lewis (Table XVIIa, p. 494) indicate a 5% critical value of 4.16 when $n = 15$, so the test just fail to firmly establish that the two extreme values are outliers. This may again reflect a loss of power stemming from the presence of two other contaminated observations.

The statistics in (3.1) and (3.2) are both Studentized. A further class of statistics, often called **Dixon statistics**, are based on ratios of the distance of a suspected outlier from a near neighbour to some measure of spread of the data as a whole. When testing for a single outlier some versatility is possible to avoid a masking effect from other potential outliers. If we want to test $x_{(n)}$ as a possible outlier and suspect that $x_{(1)}$ may also be contaminated an appropriate Dixon statistic is

$$T_D = (x_{(n)} - x_{(n-1)})/(x_{(n)} - x_{(2)}) \tag{3.3}$$

Example 3.4 For the data in Example 3.2 $T_D = (7.73 - 4.91)/(7.73 - 1.38) = 0.444$. Barnett and Lewis (Table XIXc, p.498) indicate a 5% critical value of 0.382 when $n = 15$. This test has established significance and the test is reasonable in the light of some suspicion that $x_{(1)}$ is also a possible outlier. If one suspected that $x_{(n-1)}$ was also an outlier this could be replaced by $x_{(n-2)}$ in (3.3) but different tables of critical values are then required.

The situation where the bulk of our sample comes from a normal distribution with known variance, but we suspect contamination leading to a few outliers, sometimes occurs. For data like that in Example 3.2 if we knew that most of the data came from a normal distribution with known variance σ^2 we can base a test for contamination of $x_{(n)}$ on the statistic $T_V = (x_{(n)} - \bar{x})/\sigma$. This standardized variable has not, as its appearance might suggest, got a normal distribution under H_0 because $x_{(n)}$ has not been randomly selected. Critical values for various n are given by Barnett and Lewis (Table XIIIe, p.486). Situations where σ^2 is known are rare, but they may arise when information about σ^2 is available from past experience and one wishes to see if the situation has been changed by possible contamination of part of the data in a new batch.

Example 3.5 I explained that data set D in Example 3.1 was obtained by contaminating a sample from a $N(4,1)$ distribution with 4 data from a $N(4, 9)$ distribution. Thus, assuming that for the good data $\sigma=1$, for testing $x_{(n)}$ we have $T_V = (7.73 - 3.755)/1 = 3.975$. The critical value for significance at the 1% level given by Barnett and Lewis is 3.10 so there is strong evidence that 7.73 is an outlier.

The test may be applied sequentially to the next most extreme value in either tail until eventually a nonsignificant value is obtained. Before proceeding to each test after the first any outliers already detected should be removed and $\bar{x}$ recalculated for the remaining data. The test is powerful and there is virtually no masking effect except insofar as contaminants influence the value of $\bar{x}$. The need to know σ is a practical constraint. A modification is available for the more common situation where we have an independent estimate of σ.

I mentioned that Barnett and Lewis give 48 tests for outliers in samples from normal distributions. These cover many situations both in relation to our know-ledge about the uncontaminated distribution and to the nature of contamination.

Some of the tests furnish an indirect means for detecting outliers in samples from nonnormal distributions providing we can transform the data so that the transformed uncontaminated data is approximately normally distributed. This possibility is discussed by Barnett and Lewis, section 6.3, who also give in their Chapter 6 tests for outliers in a range of univariate distributions, and readers wanting to use these tests will find a wealth of useful information there.

3.4 Accommodating outliers

In the previous section simple examples showed the difficulty in firmly establishing presence of even one outlier, especially when there is possible masking by other contaminated data. Further, we may suspect possible contamination without being sure that it really is present. In these circumstances we would like inference procedures that will in some way over-ride the effects of contamination without losing information of value in the contaminated data. We would also like the safeguard that if there is no contamination the procedures will be almost equivalent to optimum procedure for uncontaminated data. Such procedures are said to be **robust**.

From the 1960s onward rapid development in computing prompted a more critical examination of the widely-made assumption that many model-driven analyses were relatively insensitive to **small departures** from assumptions where small departures covered either minor disturbances in many observations or larger disturbances in only a small proportion of the observations.

Results of these studies were sometimes surprising. Some widely used model-driven analyses were indeed found to be insensitive to such departures; others were shown to be highly sensitive to certain types of departure.

Protection against gross breakdowns in assumptions has exercised statisticians since the pioneering work that led to the Fisher–Pitman test described in section 1.6. But tests proposed by Pitman, however, as we shall see in the next chapter, were often not robust. Some robust permutation analyses were developed by F. Wilcoxon and others from the late 1940s onward.

For location problems there has been controversy among statisticians about the relative merits of the sample mean or median as a measure of location. The arguments have largely centered on **robustness**. It is well-known and intuitively obvious that the sample mean is sensitive to just one extreme observation.

Example 3.6   To estimate the amount of overtime worked in a large industrial plant an accountant checked timesheets of a random sample of 10 employees and recorded the number of hours overtime each worked during the previous 3 months. These were

0   0   6   7   7   9   11   12   14   62

We easily find the sample mean is 12.8 and the median is 8. Had the sample not included the observation 62, but the random sampling mechanism had thrown up in its place an employee who worked only 16 hours overtime the sample mean would have been reduced to 8.2, the median remaining at 8.

The above is a situation not unlike that considered for data set A in Example 3.1 where here the observation 62, despite its remoteness from the other sample values, is not unreasonable in the context. A few employees in some occupations work appreciably more than average overtime. The observation 62 has a large influence on the mean but little effect on the median because if it were replaced by **any** value greater than 8 the median would be unchanged; indeed. the maximum change in the sample median would be reduction from 8 to 7 and this only if the observation 62 were replaced by a value not greater than 7.

This example suggests the median is a more stable estimator of location than the mean. But before concluding that we should prefer the median, there are other considerations. The problem with the observation 62 is that it makes our sample look like one from a distribution skewed to the right (i.e. with a long right tail). Had that observation been replaced by 16 our sample of 10 would look like one that could be supposed to come from a reasonably symmetric distribution, not unlike a normal distribution, the only reservation being perhaps that there might be some bunching of employees working zero overtime. For samples from a normal or near normal distribution one good reason for preferring the mean as a location estimator it that it has a smaller standard error, namely $\sigma/\sqrt{n}$ while that for the median (section 2.3) is $1.2533\sigma/\sqrt{n}$ for a normal distribution.

For the data in Example 3.6 there is not enough evidence to specify with confidence any particular population distribution. In these circumstances a bootstrap estimate of standard errors of both median and mean proves informative.

Example 3.7 The bootstrap standard error of the mean (section 2.2 and 2.3) is $[\sum(x_i - \bar{x})^2/n^2]^{1/2}$, giving 5.367 for the data in Example 3.6. For the median there is no formula for the bootstrap standard error but an estimate based on 25 bootstrap samples was 2.067.

A clear implication is that we do better to use the mean for samples that look as though they may be from a normal distribution but that the opposite appears to be the case with these data, which suggest either that we are sampling from a highly skewed distribution, or that there is a data outlier.

In studies of robustness a key role is played by **influence functions.** I discuss only briefly two simple influence functions. A formal and detailed discussion of influence functions including applications in various fields is given by Barnett and Lewis (1994, section 3.1.3).

Influence functions show, among other things, how an outlier may affect an estimator. The idea was developed by F.R. Hampel and a good account of their role in robust estimation is given in Hampel (1974). The general theory of influence functions involves sophisticated mathematics and is beyond the scope of this book. I consider at this stage only a single outlier in relation to a location estimator (mean or median). I assume we have n uncontaminated observations $x_1, x_2, \ldots, x_n$ that are a random sample from a distribution conforming to some model (e.g. a specific normal distribution) and one additional observation z that is an outlier. I denote the mean of the n 'good' observations by $\overline{x}$ and the mean with additional observation z by $\overline{x}_a$ (a for augmented). The difference $\overline{x}_a - \overline{x}$ may be written

$$\overline{x}_a - \overline{x} = \frac{n\overline{x} + z}{n+1} - \overline{x} = \frac{z - \overline{x}}{n+1} \qquad (3.4)$$

It is clear from (3.4) that the effect of one outlier z upon the sample mean $\overline{x}$ for the uncontaminated observations is a linear function of z. This implies that it is unbounded and tends to infinity if $|z|$ is sufficiently large. It is also clear that the difference $\overline{x}_a - \overline{x}$ is proportional to $1/(n + 1)$ reflecting, as one would anticipate, that the effect for any given z decreases as the sample size increases. The proportion $1/(n+1)$ measures the amount of contamination. Multiplying (3.4) by $n+1$ gives

$$I_n(z) = (n + 1)(\overline{x}_a - \overline{x}) = z - \overline{x} \qquad (3.5)$$

which is the effect per unit of contamination and $I_n(z) = z - \overline{x}$ is called the *finite sample influence function*. As $n \to \infty$, $\overline{x} \to \mu$, where μ is the population mean and the limiting function

$$I(z) = z - \mu \qquad (3.6)$$

is called the *asymptotic influence function*, or often simply the *influence function* as it is the form of greatest interest.

A different analytic approach is needed for the median partly because of the way the sample median changes between an even and an odd number of observations. However, it is easy to see that providing the number of uncontaminated observations, n, is 2 or more the addition of one outlier can only shift the sample median by a bounded amount. That amount, however, will depend upon the population distribution and the sample values. The arguments for odd or even sample sizes differ slightly.

If the number of uncontaminated observations is even, say $n = 2m$ and the observations are arranged in ascending order, we denote, as in section 2.1, the rth **order statistic** by $x_{(r)}$. The sample median of the good observations in this case is $\frac{1}{2}(x_{(m)} + x_{(m+1)})$. If one outlier, z, with **any** value greater than $x_{(2m)}$ is added the effect is to shift the overall median to $x_{(m+1)}$. Similarly if $z < x_{(1)}$ the overall median shifts to $x_{(m)}$. If the number of good observations is odd, say $n = 2m + 1$, the median is $x_{(m+1)}$ and an outlier $z > x_{(m)}$ shifts the median to $\frac{1}{2}(x_{(m+1)} + x_{(m+2)})$, while an outlier $z < x_{(1)}$ shifts the median to $\frac{1}{2}(x_{(m)} + x_{(m+1)})$. Clearly in each case these shifts are bounded no matter how large $|z|$ may be.

Because the uncontaminated sample values reflect the underlying population distribution, the precise values of the order statistics depend upon this distribution so one cannot obtain a simple expression for the asymptotic influence function like that for the mean in (3.6) which was obtained by considering the limiting form of (3.5). However, other approaches are possible. Details are given by Barnett and Lewis (1994, Chapter 4), who obtain an asymptotic influence function for the median. It is

$$I(z) = \frac{\text{sgn}(z - m)}{2f(m)} \tag{3.7}$$

where m is the median for the good observations that are distributed with probability density function $f(x)$ and $\text{sgn}(z - m) = +1, 0, -1$ depending upon whether $z >, =, < m$. Clearly $I(z)$ is bounded for a continuous distribution.

The supremum of $|I(z)|$ given by (3.7) is $1/2f(m)$, whereas for the mean the supremum is unbounded. This supremum is called the **gross error sensitivity**.

Example 3.8   For a sample from a $N(\mu, \sigma^2)$ distribution the effect of one contaminant upon $\bar{x}$ is unbounded, while the effect upon the median cannot exceed $1/2f(\mu) = \sigma\sqrt{(2\pi)}/2 = 1.253\sigma$, since the median of the normal distribution of the good values is μ.

It is clear from the form of the influence function that the mean may break down as an estimator of location with just one contaminated observation, for its possible effect is unbounded. For the median the effect of one such observation is bounded. It is not difficult to see that if $n = 2m + 1$ is odd that if two contaminated observations are added, whatever their magnitudes, the greatest effect on the median is to shift it from $x_{(m+1)}$ to $x_{(m)}$ or to $x_{(m+2)}$. The argument extends to 4 contaminated observations with bounds at most $x_{(m-1)}$ to $x_{(m+3)}$ and proceeding in this way, when $2m$ contaminated observations are added the bounds are at most $x_{(1)}$ to $x_{(2m+1)}$. Only when there are more than $2m$ contaminants do the possible values of the median include some of the contaminated observations and therefore become unbounded if possible contaminated values are unbounded. In this situation there are $2m+1$ good and at least $2m+1$ rogue observations, so that at least 50 per cent of the observations must be contaminated before the median becomes unbounded. This is known as the **breakdown point**. It is sometimes expressed as a proportion, i.e. $\frac{1}{2}$, rather than as a percentage. For the mean the breakdown proportion is $1/(n+1)$. The concept of breakdown point is due to

Hodges (1967) and is further explored by Hampel (1971). It is an important measure of robustness because in many practical situations there is a strong suspicion – or even strong evidence – that appreciable proportions of observations may be contaminated. A situation where this may be so is that where some observations are made by one experimenter or laboratory and the rest by another, a possible situation giving data like set D in Example 3.1, where one observer (or laboratory) makes less precise measurements than the other. For variance slippage the over-all sample mean is an unbiased esitmator of μ, but as pointed out in section 3.3 the slippage increases the standard error. If we reject contaminated observations (which we can only do if we known which they are) the reduced sample size results in a loss in power.

Often contaminated data are presented for analysis without our knowing which observations come from each source, or only the proportions from each source may be known. In a situation like this the idea of an estimator that gives greater weight to the good observations is appealing if the weighing can reduce the influences of contamination. This aim inspired some of the estimators proposed in the next section.

3.5 Some robust estimators

Some robust estimators are data-driven while others are adapted from models to allow for potential data peculiarities. Sometimes the two approaches lead to the same estimator.

A typical example in the first category is the **trimmed mean**. This is a device used for location estimates when there are outliers, or it is thought that the sample comes from a long tailed-distribution. The observations are arranged in ascending order and then, in its simplest form, we reject the top t % and the bottom t % of the observations, where t is a prechosen number. Sometimes we elect to reject different proportions, or else fixed numbers rather than proportions, in each tail. Using percentages, common choices for not too small samples are $t = 10$, i.e. removal of the top and bottom deciles, or for somewhat larger samples, $t = 25$ when the trimming points are the first and third quartiles. The extreme case is $t = 50$, corresponding to using the median as the estimator.

For a true random sample from a $N(\mu,\sigma^2)$ distribution trimming tends to increase the estimated standard error because it reduces the number of obser-vations, but this may be off-set by the sample effectively now being one from a truncated normal distribution giving a reduced sum of squares of deviations from the mean, but this leads to underestimation of σ^2. However, with contaminated observations the effect of trimming is likely to be beneficial both in reducing possible bias and in lowering the standard error of the estimate. Bootstrapping is often used to explore the benefits of trimming. The key effect of trimming is to gives zero weight to extreme observations and so if an appropriate choice of t, the percentage trim, is made, it reduces the influence of outliers to zero.

An alternative to trimming is **Windsorization,** where extreme observations are shrunk to the value of the remaining observation of greatest magnitude in each tail, thus reducing, but not eliminating, their influence. The rationale for this procedure is that very often an outlier contains some useful information about location but it may be unduly influential without adjustment.

A more subtle approach is one where the data are allowed to influence the degree of trimming or Windsorization. This concept, applicable to other estimators as well, gives rise to what are called **adaptive procedures**. I do not describe this approach here, but it is important and useful and is discussed by Barnett and Lewis (1994, section 3.2.4 and elsewhere).

Example 3.9 The following data may be a sample from a distribution with longer tails than a normal distribution because there is bunching of values near the mean and then a fairly wide spread to a couple of values in each tail. I explore the relative merits of the sample mean, sample median, 25% trimmed mean and 25% Windorized means as location estimators.

$$-3 \quad 11 \quad 21 \quad 24 \quad 25 \quad 25 \quad 27 \quad 29 \quad 31 \quad 32 \quad 44 \quad 57$$

There is evidence of longer tails than those for a normal distribution but little indication of asymmetry and unless there were theoretical grounds outwith the data itself (e.g. past experience with similar experimental material or knowledge about how the data were generated that suggested they might come from, say, a double exponential or a Cauchy distribution) it is hard to nominate any specific population distribution. Since the standard errors of the trimmed or Windsorized means are both distribution dependent, the bootstrap comes into its own as a method of comparing estimators. I estimated the bootstrap standard errors by calculating each statistic using the same 40 bootstrap samples. For the mean the theoretical bootstrap standard error can be obtained using the formula $\sum(x_i - \bar{x})^2/n^2$ from section 2.2, giving the value 3.94. My estimate from 40 bootstrap samples was 3.95. Table 3.1 gives the value obtained for the above data using each estimator and the estimated bootstrap standard error.

The point estimates are broadly comparable as one would expect for reasonably symmetric data, but the standard error of the mean is roughly twice that for the more robust estimators.

Table 3.1 Four location estimates and their estimated standard errors based on 40 bootstrap samples for the data in Example 3.4

Estimator	Mean	Median	25% trimmed mean	25% windsorized mean
Estimate	26.92	26.00	26.83	27.17
Bootstrtap s.e.	3.95	2.00	1.82	2.18

For these data the standard errors of the median, trimmed and Windsorized means differ only slightly and on this basis there is little to choose between them. Since the point estimates agree reasonably one should consider the implications of the mean having a higher standard error. Clearly this s.e. is high because of the influence of the tail observations (remember what the influence function tells us about any one such observation). This further implies that if we took other samples of 12 from the underlying population we may get extreme values (perhaps more in one tail than in the other) that would have a greater influence on our mean estimator than upon other estimators, which all automatically downweigh tail values.

The mean, median, trimmed means and Windsorized means are expressible as linear functions of the order statistics $x_{(i)}$ and are sometimes referred to as **L-estimators** (L for linear).

The general form of an L-estimator is

$$\mu^* = \sum_i w_i x_{(i)}.$$

If $w_i = 1/n$, $i = 1, 2, \ldots, n$ then μ^* is the sample mean. If n is odd, say $n = 2m+1$, and we put $w_{m+1} = 1$, $w_i = 0$ if $i \neq m + 1$ then μ^* is the sample median. If n is even and $n = 2m$ the weight for the median estimator are $w_m = w_{m+1} = \frac{1}{2}$ and $w_i = 0$ otherwise. For a trimmed mean, if k obervations are trimmed from each tail then $w_i = 0$ if $1 \leq i \leq k$ or if $n - k + 1 \leq i \leq n$ and $w_i = 1/(n - 2k)$ otherwise. For Windsorization if k values in each tail are shrunk to $x_{(k+1)}$ and $x_{(n-k)}$ respectively then $w_i = 0$ if $1 \leq i \leq k$ or if $n - k + 1 \leq i \leq n$, and $w_{k+1} = w_{n-k} = (k + 1)/n$, otherwise $w_i = 1/n$. Clearly, since they include the mean, L-estimators as a class are not in themselves necessarily robust. They tend to gain robustness when extreme order statistics are down-weighted.

In section 3.3 I briefly discussed several of many distribution-based tests for outliers, but because of masking many are, for example, notoriously bad at detecting more than one outlier in the same tail; others tend to miss a pair of outliers in opposite tails. A simple but reasonably robust distribution-free test is to classify any observation x_0 as an outlier if

$$\frac{|x_0 - \mathrm{med}(x_i)|}{\mathrm{med}[|x_i - \mathrm{med}(x_i)|]} > 5 \tag{3.8}$$

where $\mathrm{med}(x_i)$ is the median of all observations in the sample, including the suspected outlier. The denominator in (3.8) is a measure of spread called the **median absolute deviation** (MAD). Choice of 5 as a critical value is somewhat arbitrary but is motivated by the fact that if the remaining observations have approximately a normal distribution it picks up a suspect observation more than 3 standard deviations from the mean.

Example 3.10   Consider the data in Example 3.9, i.e.

$$-3 \quad 11 \quad 21 \quad 24 \quad 25 \quad 25 \quad 27 \quad 29 \quad 31 \quad 32 \quad 44 \quad 57$$

We first test the observation furthest from the median using (3.8). If this is not an outlier we proceed no further; if it is, we test the next most extreme observation in either tail, proceeding until we find an observation that is not an outlier.

The median is 26 and the absolute deviation from this of the datum -3 is $|-3-26|=29$ and similarly the other absolute deviations are 15, 5, 2, 1, 1, 1, 3, 5, 6, 18, 31. Ordering these we easily find that their median, the MAD is 5.

The most extreme deviation has magnitude 31 and since $31/5 = 6.2$ we class this as an outlier. The next most extreme observation has absolute deviation 29 and since $29/5 = 5.8$ we also class this an outlier. The test detects no further outliers. The data look somewhat nonnormal, but the above test is not one for normality. I give a test for normality in section 6.7, and in fact that test does not reject a hypothesis of normality for these data. But one must bear in mind that acceptance of a null hypothesis does not mean it is correct – we do make errors of the second kind! The Cauchy distribution has very different properties from those of the normal distribution, yet samples from it, especially small ones, often look like and are consistent with, tests for their coming from a normal distribution.

There are many reasons why observations may be genuine but atypical. Huber (1977) suggests that it is not uncommon for up to 10% of observations to be from a different and often longer tailed distribution than the bulk of the observations, yet both may have the same mean. In section 3.3 I pointed out this may occur with measurements by different observers or in different laboratories. Diverse sources of raw materials, or taking observations at different times when, for example, climatic differences may increase variability, are other factors that may introduce often unnoticed, or hard to detect, changes in variability.

Useful as computers are, they are sometimes sources of gross data errors that often go undetected, e.g. those caused by hitting two keys simultaneously when entering data, recording perhaps a correct 3.6 as 34.6. Visual checks or printing out summary statistics (e.g. means, ranges, maximum and minimum data values) may detect such errors, but in large data banks with thousands or perhaps millions of entries some such mistakes usually slip through. Decimal points in the wrong place are common errors, especially with certain data-formatting conventions.

Grouping and rounding effects may introduce nonnormality to otherwise near normal data; the effect is usually less severe than gross data errors, but in some contexts as shown in Sprent (1993, Example 2.8) such effects cannot be ignored.

Even with basically reasonable models we may find that real data include one or two gross departures or a large number of small perturbations. This has led to a search for procedures that are almost optimal if there are no such upsets and are little affected by a few disturbances.

One such class of estimators are the Huber–Hampel *M*-estimators, the 'M' because they are like optimal maximum likelihood estimators when these are appropriate and are little disturbed by gross departures of a few, or small perturbations in many, observations.

Using M–estimators in non-trivial problems requires adequate computer programs. I demonstrate the basic idea for estimating the mean of a symmetric distribution. In this case the maximum likelihood estimator is the sample mean. In particular, for a normal distribution given n observations $x_1, x_2, . . . , x_n$ this is equivalent to least squares estimation where we choose our estimator $\hat{\mu}$ as the value of μ that minimizes

$$U(x, \mu) = \sum_i (x_i - \mu)^2.$$

In words, $\hat{\mu}$ is the value of μ that minimizes the sum of squares of deviations of the observed x_i from μ. Mathematically $(x - \mu)^2$ is known as a **distance measure** because it provides a measure (in the form of the square) of the distance of a data point x from μ. Other distance measures are possible, e.g. the absolute distance $|x - \mu|$. Functions like U, here the sum of such measures, are called **distance functions**. While $U(x, \mu)$ is minimized when $\hat{\mu} = \overline{x}$ the sample mean, the function $V(x, \mu) = \sum_i |x_i - \mu|$ is minimized by setting $\mu^* = \text{med}(x_i)$ the sample median. As indicated in this and the previous chapter for samples from the normal and many other distributions we prefer $\hat{\mu}$ to μ^* because it has smaller variance for fixed n.

However, as shown in section 3.4 even one outlier has more effect on $\hat{\mu}$ than it has upon μ^*. The influence function for any one observation given by (3.6) shows that $\hat{\mu}$ is linearly dependent upon the distance of an outlier z from μ, i.e. an outlier lying 6 standard deviations from the mean has twice the influence of one lying only 3 standard deviations from the mean.

Huber (1972) and others developed M-estimators to cope with situations where a few outliers were possible but one wanted to make inferences valid for the remaining data for which a specific maximum likelihood estimator was available. These estimators have a built-in mechanism for reducing the effect of outliers, if indeed there are outliers, yet are almost identical with the maximum likelihood estimator when there are no outliers. In this way they differ from some robust L-estimators such as the median or some trimmed means because, while these work well when there are outliers, they are often appreciably less efficient in the sense of having greater standard errors when there are no outliers.

To describe the idea behind Huber's M-estimator for location I generalize the idea of a distance function. Some continuity and differentiability properties are needed for a straightforward approach; some distance functions e.g. $V(x, \mu)$ lack these. I do not discuss such niceties, which can often be surmounted in practice.

It is well known that to find any minimum of $U(x,\mu)$ regarded as a function of μ, we differentiate U with respect to μ and equate the derivative to zero, leading to the *normal* or *estimating* equation

$$\sum_i[-2(x_i - \mu)] = 0 \qquad (3.9)$$

with solution $\hat{\mu} = (\sum_i x_i)/n = \bar{x}$, the sample mean.

A distance function is defined as a function $d(t)$ such that for any t

(i) $d(t) \geq 0$
(ii) $d(t) = d(-t)$
(iii) the derivative $\psi(t) = d'(t)$ is a non-decreasing function of t for all t.

Condition (iii) is relaxed for some M-estimators.

For estimating the mean, μ, of a symmetric distribution Huber (1972) proposed a function $d(t)$, where for some fixed $k > 0$,

$$\begin{aligned} d(t) &= \tfrac{1}{2}t^2 & \text{if } |t| \leq k \\ &= k|t| - \tfrac{1}{2}k^2 & \text{if } |t| > k. \end{aligned}$$

The derivatives $\psi(t) = d'(t)$ are

$$\begin{aligned} \psi(t) &= -k & \text{if } t < k \\ &= t & \text{if } |t| \leq k \\ &= k & \text{if } t > k \end{aligned}$$

We estimate μ as μ^* that satisfies the normal equation

$$\sum_i \psi(x_i - \mu) = 0. \qquad (3.10)$$

If $k \to \infty$, then for all x_i, $\psi(x_i - \mu) = x_i - \mu$ and (3.10) gives the sample mean as the appropriate estimator, equivalent to maximum likelihood in the normal case. For finite k the form of $d(t)$ shows that for $|x_i - \mu| \leq k$ we minimize a function equivalent to that for least squares, while if $|x_i - \mu| > k$ we minimize a linear function of absolute differences. We have to choose k. In this simple problem practical experience suggests that a useful choice is one such that the interval with end-points given by $\text{med}(x_i) \pm k$ contains between 70 and 90 per cent of the observations. I discuss the choice of k and its implications in more detail after Example 3.11. In general (3.10) requires an iterative solution and when k is chosen one may proceed by writing (3.10) as

$$\sum_i \frac{\psi(x_i - \mu)}{(x_i - \mu)}(x_i - \mu) = 0 \qquad (3.11)$$

Writing $w_i = \psi(x_i - \mu)/(x_i - \mu)$, (3.11) becomes $\sum_i w_i(x_i - \mu) = 0$ with solution

$$\mu^* = \sum_i (w_i x_i)/(\sum_i w_i) \qquad (3.12)$$

However, the weights are functions of μ^* so we must start with an estimate μ_0 of μ^* and with this calculate initial weights w_{i0} and use these in (3.12) to calculate a new estimate μ_1. This in turn is used to form new weights w_{i1}, say, and the cycle continues until convergence which is usually achieved after a few iterations. For moderate or large samples, calculation is tedious without a suitable computer program, but the method is easily illustrated for small data sets; by suitable choice of these sets we may also see how it compares with other possible estimators, robust or otherwise. The method works well with only one outlier that effectively destroys what is otherwise a reasonable sample from a symmetric distribution.

Example 3.11 Obtain and compare Huber *M*-estimators of location for the three data sets

A 0 2 2 3 5 6 8   B 0 2 2 3 5 6 20   C 0 2 2 3 5 6 121

The median of each set is 3; the means are 3.71, 5.43 and 19.86. If the observations other than the greatest are in each case regarded as satisfactory, the median is a reasonable estimator, but the mean is barely acceptable for *B* and unacceptable for *C*. The trimmed mean with the first and last observation trimmed is in each case 3.6. Discarding the outlier (last observation) in each case gives a mean of 3 (all data sets then being identical). I detail the iterative steps for *M*-estimation for set *B*. If we take $k = 3$ and the median, 3, as our first estimate μ_0 all but the last observation are in the interval with end points 3 ± 3. Since we set $\psi(x_i - \mu) = x_i - \mu$, then if $|x_i - \mu| \leq 3$ it follows that the data points 0, 2, 2, 3, 5, 6 all have weights $w_i = 1$ in (3.12) while for $x_{(7)} = 20$ we have $\psi(20 - 3) = k = 3$ whence $w_{(7)} = 3/(20 - 3) = 3/17$, and so (3.12) gives

$$\mu_1 = [(0 + 2 + 2 + 3 + 5 + 6)\times1 + 20\times(3/17)]/[6 + 3/17] = 3.49.$$

Clearly for μ_1, since $k = 3$, values of x between 0.49 and 6.49 all have weights $w = 1$, while for $x_{(1)} = 0$, $w_{(1)} = 3/3.49 = 0.86$ and similarly $w_{(7)} = 3/16.51 = 0.18$, whence we find that $\mu_2 = [18 + 20\times0.18]/6.04 = 3.58$. The next iteration, starting with $\mu_2 = 3.58$ gives $\mu_3 = 3.59$, and a further iteration brings no change. Using a similar procedure for sets *A* and *C* gives respectively $\mu_A{}^* = 3.60$ and $\mu_C{}^* = 3.70$ (Exercise 3.1).

Different amounts of trimming or Windsorization give different estimates, so too do different choices of *k*. However in practice if the estimators are truly robust there should be little difference between rational choices, where by rational I mean choices that will downweigh rogue observations appreciably, but give strong support to good observations. Clearly the *M*-estimator used here is less sensitive to outlying observations than is the sample mean and influence function studies confirm this, for the estimator reduces the influence of the outlier, giving it a role more like that in estimating a median. Generally speaking *M*-estimators have Pitman efficiency close to 1 when the maximum likelihood estimator on which they are based is appropriate for the bulk of the data. For samples from a symmetric distribution with a few outliers they generally have higher Pitman efficiency than the median estimator.

The first aim in choosing *k* is to downweigh outliers. However, if *k* is too small only a few observations near μ_0 get full weight and the estimator may be strongly influenced by any rounding or grouping effects in the fully-weighted observations.   A further unsatisfactory feature of arbitrary choice of k for estimators like the Huber estimator is associated with a change of scale. If we replace all x_i by $y_i = a + bx_i$ where a and b are constants then the least squares estimate of the mean of the Y population is $a + b\hat{\mu}$. A similar relationship holds for sample medians, but not for the M-estimator use in Example 3.11. This desirable property may be recovered by replacing $\psi(x - \mu)$ by $\psi[(x - \mu)/s]$ where s is some measure of spread. Experience has shown one appropriate choice of s is the median absolute deviation (MAD) divided by 0.6745. This divisor makes the corresponding s a consistent estimator of σ if the underlying distribution is normal.   The weights in (3.12) are modified by using $(x-\mu)/s$ in the denominator also. This modification is a standardization of the data and if it is used then setting $k = 1.5$ gives an estimator with desirable properties for downweighting outliers without overemphasizing the influence of moderate grouping or rounding. Since at stage, j, say, of the iterations k is now relevant to the values of $(x_i - \mu_j)/s$ where μ_j is the jth iterative estimator of μ, we must recalculate $(x_i - \mu)/s$ for each sample value at each iteration. There is no change in s or k between iterations.

   In Example 3.11 the M-estimator used gave acceptable point estimators but clearly some measure of accuracy such as a standard error is required to compare performance with other estimators, or indeed other choices of k. The bootstrap is one way of obtaining an estimated standard error.   An alternative is the jackknife.   Another possibility is to use an M-estimator of σ^2, and to base the standard error of the mean on this estimator.

   Hampel, in Andrews et $al.$ (1972) and others, have suggested alternative distance functions that ignore (i.e. give zero weight to) very extreme observations. Effectively, Huber's distance function limits the influence of an extreme value to what it would be if it were at a distance k from μ^*. Hampel's proposed function gives even less influence to moderately extreme values and zero weight to very extreme ones. Andrews (1974) proposed M-estimators using yet another distance function that seemed particularly appropriate for regression. As I show in section 11.3 this is a situation where outliers are particularly hard to detect visually.

   L-estimators make use of the ordered data, while M-estimators deal with the data as given, but clearly there is a common underlying motivation in that each type of estimator gains robustness by downweighting extreme observations and indeed a Windorized mean may be looked upon as an M-estimator for the mean with the weights of extreme observations all set equal to the unaltered observations of greatest magnitude in the relevant tail.

   Another type of robust estimator called an R-estimator, is based on ranked data, a subject I discuss in Chapter 4, and an important example is given in section 4.4.

Exercises

3.1 For the data sets in Example 3.11 confirm the estimated means given by the M-estimator considered in that example.

3.2 Perform what you consider appropriate tests for outliers in each of the following data sets

Set A	3.2	6.1	4.3	7.9	5.1	22.2	19.4	11.3	8.4	6.4	2.1
Set B	−1279	433	−229	8883	754	348	−534	11	555	0	74	−102

[If some of the tests suggested in section 3.3 are used you will need to refer to tables such as those given in Barnett and Lewis (1994)].

*3.3 For data set B in Exercise 3.2 explore possible robust methods such as trimmed means, Windorized means, the median for estimation of location, using bootstrap methods where you consider these appropriate.

*3.4 Use an available computer random sample generator to generate a sample of 12 from a N(4, 9) distribution and a sample of 4 from a N(5, 16) distribution. Combine the two samples and regard the N(4, 9) data as good sample values and the N(5, 16) data as contaminated. Use the various appropriate techniques described in this chapter (and any others you can find and consider suitable) to compare their performance for estimating the mean of the distribution of the good values.

*3.5 Dale et al (1987) [reproduced in Hand et al (1994), Set 299] give data for plasma beta concentrations in pmol/l of 11 runner who collapsed after a Tyneside Great North run. Perform any tests for outliers you consider appropriate and using bootstrap or other methods compare and comment critically upon the mean, median and various trimmed or Windsorized means as estimates of population location

66   72   79   84   102   110   123   144   162   169   414

4

Location tests for two independent samples

4.1 The Fisher–Pitman test robustness and ranks

I consider two-independent-sample location problems before the conceptually simpler one-sample case because most investigations are comparative and the former provide more realistic examples in this context. Permutation and bootstrap methods are both well-developed and in this chapter I cover the relationship between these and consider other robust approaches.

Neither the t-test nor the Fisher–Pitman test on raw data are necessarily robust against patterns of extreme observations like those arising when sampling from long-tailed or highly skewed distributions, from a mixture of distributions, or with some forms of contamination. Some tests with lower Pitman efficiency than the t-test when assumptions for the latter hold have higher efficiencies than either that or the Fisher–Pitman test when samples are from clearly nonnormal distributions.

Example 4.1 In Example 1.1 I gave the following data for numbers of references in a random sample of papers published in the journal *Biometrics* in the years 1956–57 and for a further sample of papers published in 1990, explaining there why the data were of interest.

1956–7	2 3 6 6 6 9 9 10 14 15 16 72	
1990	6 8 9 10 16 18 20 22 22 23 23 26 28 59	

I give formal tests for normality in section 6.7, but inspection suggests these data are decidedly nonnormal, the values 59, 72 imply in each case a long upper, or right, tail. I indicated one reason for this nonnormality in Example 1.1 A distribution-free test may be appropriate to test for a right-shift (increase) in the 1990 population median relative to that for 1956–7, i.e. to test whether the median difference

$$\theta = (\text{median in 1990}) - (\text{median in 1956–7})$$

is positive. An appropriate test is a one-tail test for $H_0: \theta \le 0$ against $H_1: \theta > 0$.

The data are not continuous (being necessarily integer valued) and there are several ties, but this is a situation where, in practice, a t-test is often applied, somewhat unthinkingly, not only because of nonnormality or discontinuity, but because for skew distributions a test for median shift is usually preferred to one for a difference between means. The mean, if it exists, and the median coincide in symmetric distributions, but not more generally. For the t-test, available in any general statistical program, $t = 1.069$ (24 degrees of freedom). In a one-tail test this corresponds to $p = 0.1478$, or a 'significance' level of 14.78%, well above the conventional 5%, so we would not reject H_0. For the Fisher–Pitman test the exact $p = 0.1729$ (obtainable from, e.g. StatXact or TESTIMATE) and the asymptotic approximation based on (1.6) is $p = 0.1431$, close to the t-test probability.

Ties tend to increase the probability associated with the observed first sample sum, S_1, in a permutation test. Here that probability, given in output by StatXact or TESTIMATE, is 0.0063. If the alternative convention (section 1.7) of sharing this probability between the critical and the acceptance regions is used this reduces the exact value only slightly to $p = 0.1729 - \frac{1}{2}(0.0063) = 0.1697$. Clearly we do not reject H_0 at the levels suggested by these tests, for all levels are of the order of 15% and in near agreement with each other. A 95% confidence interval for the difference, *1990 mean – 1956-7 mean* based on the t-distribution is $(-6.8, 20.2)$. The exact Fisher–Pitman permutation test interval which I found by linear interpolation starting with the t-interval limit for the location parameter is slightly shorter, being $(-5.6, 17.2)$. In other situations the difference may be less marked and the Fisher–Pitman test will not necessarily lead to a shorter interval.

Examine the data. The numbers of references per paper look greater in 1990. The sample median is 9 in 1956–7 and 21 in 1990. The means are 14 and 20.7. The skewness is hardly surprising for reasons given in Example 1.1. The blending of papers on new developments with review papers result in a mixture of distributions with one dominating. This accounts for the high sample values 72 and 59. Such outliers (here valid observations, not errors in counting) relative to the bulk of the data influence the outcome of both the t-test and the Fisher–Pitman test, effectively reducing their power. In the former they inflate the denominator in the t-statistic and in the latter they create large jumps in the statistic S_1 as they enter or leave the calculated value for different permutations.

A more appropriate test and estimation procedure starts from the idea that if the samples are not from identical populations (the assumption under H_0 in the Fisher–Pitman test), then in view of the skewness, one should use a test based on difference between **medians** rather than the difference between means which is in effect what the t-test or Fisher–Pitman test does. It does not seem to be widely realised that one may base a permutation test on the median difference or on trimmed or Windsorized means although this is specifically pointed out by Efron and Tibshirani (1993, Chapter 15). The reference set of permutations is the same as that for the conventional Fisher–Pitman test but care is needed in choosing a test statistic. The difference between medians is an appropriate statistic. In practice, the bootstrap appears to be used more widely than a permutation test in these circumstances, perhaps because computing facilities for *exact* median-based permutation tests are not so widely available. However, writing macros for Monte Carlo approximations in most standard statistical packages is no more difficult than writing these for the corresponding bootstrap analysis.

Example 4.2. A median-based permutation test may be applied to the data in Example 4.1. An appropriate test statistic is the difference, $T = median\ (1990) - median(1956\text{-}7)$. To obtain the exact p-value for a one or two-sided test against the hypothesis of identical distributions this statistic is computed for all permutations and, as for the mean, the p-value for a one-tail test of the alternative that the median difference is positive is given by the proportion of permutations giving a median difference equal to or greater than that observed. Unless a program is available to perform this exact test a Monte Carlo approximation must be used. For these data I used a simple MINITAB macro to form 1000 Monte Carlo samples from the permutation distribution and computed the value of T for each and then found the proportion, p, of these T with median difference greater than or equal to the observed median difference $T = 21 - 9 = 14$. This process gave an estimated one-tail $p = 0.011$, i.e. 11 of the 1000 samples gave $T \geq 14$. Three more samples of 1000 gave estimates $p = 0.004$, 0.010 and 0.002.

These results indicate robustness of the median estimator, the Monte Carlo p suggesting a true p-value close to, or perhaps less than, 0.01, in contrast to a value near $p=0.15$ suggested by the mean-based tests in Example 4.1

Unlike the Fisher–Pitman test in section 1.6 where the first sample sum, the first sample mean, the second sample sum, the second sample mean, the difference between sample means and even the conventional t-test statistic all give the same p-value, this is no longer true for medians. A moment's reflection shows, for example, that knowing the first group median for any permutation does not automatically tell us what the second group median will be. To determine that we need to know which observations are in at least one of the samples. The statistic T defined above is appropriate as it uses median information from both samples.

The permutation test for medians is not satisfactory for small samples because then many permutations give the same value of T and discontinuities in possible p-values are greater than they are for the mean statistic. This problem is less acute if we use less drastically trimmed means than it is for the median (remember that the median is a 50% trimmed mean).

Example 4.3 For the data in Example 4.1 an aim of trimming might be to exclude the outlying 72 and 59 from computation of the trimmed means. A moment's reflection makes it clear that with only two such outliers if we trim two observations from each tail in each sample, then in any permutation these outliers will be excluded from the computed trimmed mean. With this trimming, in similar manner to Example 4.2 I chose as the test statistic $T = trimmed\ mean\ (1990) - trimmed\ mean\ (1956\text{-}7)$ and computed an approximate one-tail p using a Monte Carlo sample of 1000 permutations, which gave a one-tail $p = 0.006$. A further sample of 1000 gave the same p-value. The p here represents the proportion of samples in 1000 where the difference between trimmed means is equal to or greater than the difference between trimmed means for the data which, with 2 observations trimmed from each tail, is $T = 9.525$.

Because the outliers are in the upper tail a nonsymmetric trimming is appealing. Thus we might trim only the two greatest values from each sample formed by permutation. Windsorization is another possibility worth exploring. This is equivalent to using a mean estimator where the extreme observations are not ignored but adjusted to have values more in line with the bulk of the data.

In section 2.7 I outlined the relationship between permutation tests and bootstrapping, pointing out an essential technical difference that while permutation tests use sampling without replacement the bootstrap uses sampling with replacement.

Example 4.4 Bootstrapping using the median parallels the procedure in Example 4.2 except that sampling is with replacement. The bootstrap samples are formed in the way described in section 2.7 and applied in Example 2.8. For the data in Example 4.1 I formed 1000 bootstrap samples and for each I computed $T^* = \theta_1^* - \theta_2^*$ where θ_i^*, $i = 1,2$ are the medians for the first and second bootstrap samples obtained by resampling with replacement from the original combined sample data and allocating the first 12 resulting values to sample 1 and the remaining 14 to sample 2. In 1000 bootstrap repetitions 17 gave $T^* \geq 12$, the observed median difference, giving $p^* = 0.017$ as a bootstrap estimate of p. A second sample of 1000 gave $p^* = 0.010$, in broad agreement with the estimated p for the corresponding permutation test.

I also carried out a bootstrap sampling experiment for the trimmed mean used in Example 4.3 where 2 observations were trimmed from each tail of each bootstrap sample. For 3 sets of 1000 samples I found $p^* = 0.028, 0.028, 0.033$ where these are computed from the proportion of samples in 1000 where the trimmed mean difference equals or exceeds the observed sample difference of 9.525. These p^* all exceed the corresponding permutation estimates. This is because when sampling with replacement we may get more than two of the upper tail outliers 72 and 59 in one, or even in both, of a pair of bootstrap samples and these will lead to a greater proportion of fairly extreme values of the test statistic computed for bootstrap samples because only two of these extreme values are trimmed out. In this situation slightly more severe trimming should reduce the relevant p^*.

In section 2.7 I considered an alternative approach to bootstrapping in the two-sample problem where, instead of combining the samples, resampling with replacement was carried out separately for each of the two data groups. This leads to sets of bootstrap estimates of a statistic that are centered about the observed value for the group differences in location. I illustrated the method there for mean differences, but it extends easily to trimmed mean differences or median differences, and also to Windsorized mean differences. For a one-tail test of significance where the alternative hypothesis specifies a positive location difference the exact p-value is approximated by the proportion of B bootstrap samples that take zero or negative values. The procedure may also be used to obtain confidence intervals in the way indicated in Example 2.9 for the mean difference, that approach generalizing in an obvious way to median, trimmed mean, or Windsorized mean differences. For validity, the null hypothesis need only specify location equality rather than the stronger requirement of identical distributions needed for the bootstrap analogue of the combined sample permutation test.

Example 4.5   For the data in Example 4.1 I applied the procedure used in Example 2.9 first to the median difference using the statistic $T = \theta_1 - \theta_2$, where θ_i, $i = 1,2$ are the medians for the first and second data sets. T^* denotes the corresponding bootstrap sample statistic. In 1000 bootstrap samples 12 gave a zero or negative value of T^* giving a one-tail $p^* = 0.012$. Another sample of 1000 gave $p^* = 0.014$. The 95% two-sided confidence limits are given by the 25th and 975th ordered bootstrap values of T^*. For the two samples of 1000 I used, these gave confidence intervals (2.0, 17) and (1.0, 17), confirming the indication of a positive median difference.

I repeated this exercise for a trimmed mean where the two observations in each tail for each bootstrap sample were trimmed (the same trimming as that in Example 4.4). Here for two Monte Carlo samplings with $B = 1000$ I found the one-tail $p^* = 0.060$ and 0.042 respectively giving only a slight suggestion of significance in a one-tail test.   The two-sided 95% confidence intervals were (−1.925, 16.6) and (−0.1, 16.925).

The close parallels between a permutation-test and the bootstrap suggest that bootstrapping using the difference in sample means should give similar results to the Fisher–Pitman test which we saw in Example 4.1 in turn gave similar results to the t-test. Again, for two sets of bootstrap samplings with $B = 1000$ in each case this was confirmed with the one-tail $p^* = 0.153$ and 0.122 and two-sided 95% percentile-based confidence intervals (−8.30, 17.01) and (−5.86, 17.20).

4.2 Ranks

The examples in section 4.1 showed for one data set how choice of an appropriate L-estimator can give robust estimates for location differences. An historically earlier and highly successful way to achieve robustness by reducing outlier influence is to transform data to ranks. The rank analogue of the Fisher–Pitman test is the **Wilcoxon rank-sum test**, now almost as familiar as Student's t-test as a basis for inference in the two-sample location problem.   To use it each observation is replaced by its rank in the combined samples data arranged in ascending order from least to greatest.   When there are ties the intuitively reasonable convention is to score these by their mid-ranks.   Thus, if the six smallest observations are ranked from 1 to 6 but we then have two tied observations we give each the mean of the ranks 7 and 8, i.e. 7.5.   Similarly, if there were three such ties at this position we would give each the mean of the ranks 7, 8 and 9, i.e. each is ranked 8.   With this convention the relevant permutation test is equivalent to the Fisher–Pitman test with ranks or mid-ranks replacing the original data. Thus any computer program for the Fisher–Pitman test may be used for Wilcoxon's test by replacing the raw data by ranks.

Example 4.6 For the data in Example 4.1 the Wilcoxon rank-sum test uses the ranked data and this is easily verified to be:

1956–7	1	2	4.5	4.5	4.5	9	9	11.5	13	14	15.5	26
1990	4.5	7	9	11.5	15.5	17	18	19.5	19.5	21.5	21.5	23	24	25

StatXact has three programs that may be used for the Wilcoxon rank-sum test – reflecting the way tests are inter-related. The recommended one is specifically for the Wilcoxon rank-sum test and only requires the original data, transformation to ranks being performed automatically by the program. Alternatively transformation to ranks may be done manually as above, and the Fisher–Pitman test applied to these ranks. Thirdly, we may use a linear-by-linear association test. To do that the data are entered as a $2 \times c$ contingency table with $c = 17$ (because there are 17 different rank or mid-rank scores in the two samples). Column scores are these ranks and the row scores either 0, 1 or 1, 0 depending on which sample ranks are used in the summation. The format is like that described for the Fisher–Pitman test in section 1.6 except that some of the cell entries are now greater than unity because of data ties. For the one-tail test all these programs give $p = 0.0065$, indicating a significance level of 0.65%. The first two tests are also available in TESTIMATE and give the same result. Sometimes different software or even valid alternative programs in one package lead to slightly different results due to rounding, or because ties may be treated differently.

The value $p = 0.0065$, contrasts sharply with $p = 0.1478$ for the t-test and $p = 0.1729$ for the Fisher-Pitman test using the original data, but is close to the Monte Carlo sampling results I got for the permutation test or bootstrap tests based on median differences. Like the situation there, the Wilcoxon test is robust against extreme values. Although the Fisher–Pitman test performed poorly here it sometimes does better than the t-test if there are just one or two outliers or extreme observations providing these are consistent with the alternative hypothesis specified in H_1 (in this case that the second sample has the greater median or is shifted to the right). In the above example the high value 72 in the 1956–57 sample is contrary to that trend.

Both the Fisher–Pitman and the Wilcoxon rank-sum tests are valid not only as tests for location shift in otherwise identical populations, but are valid and have good efficiency in a wider context to test hypotheses about **dominance** of one distribution over the other. For example, we may have samples from two distributions with cumulative distribution functions $F(u)$, $G(v)$ that are identical under H_0, but under H_1 for all x, $G(x) \le F(x)$ (or $G(x) \ge F(x)$) with strict inequality for at least some x. If $G(x) \le F(x)$ this is described as the G distribution being stochastically larger than F because it implies that the G distribution is shifted to the right. For this situation the corresponding probability density functions may look like those in Figure 4.1. Apart from the dominance requirement under H_1 the two distributions may be quite dissimilar, e.g. one may be symmetric, the other skew. Under H_0, however, the distributions are identical. Location difference is an important special case of dominance in which, apart from the location shift, the distributions are otherwise identical under H_1. In section 1.6 I pointed out that two common models were the fixed effects model and the random effects model. In the random effects model the treatment difference τ varies between units and has a distribution. If this distribution has mean $\mu > 0$ and the variance is not too large this model typically leads to a dominance situation. Indeed the dominance situation arises with no stronger an assumption than that the treatment difference on any unit in the group showing dominance is always non-negative and in some

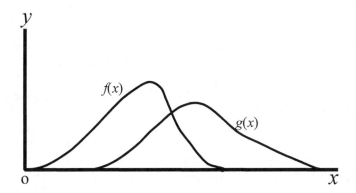

Figure 4.1 Probability density functions $f(x)$, $g(x)$ for which $G(x) \leq F(x)$ for all x with strict inequality for at least some x.

cases at least, positive, without any further specification of that effect. Under either the location difference or the dominance specification of H_1, higher observed values or higher ranks should dominate in one sample as opposed to a fairly even spread of ranks under H_0. The hypothesis test in a dominance case is sometimes expressed in the form H_0:Pr$(Y > X) = \frac{1}{2}$ against H_1:Pr$(Y > X) \geq \frac{1}{2}$, with strict inequality for at least some X. In section 6.5 I discuss models where location shift and dominance that involve scale changes are inseparable.

The Fisher–Pitman test is a conditional test (section 1.7) because the value of the statistic S_1 implying a given *p-value* is **conditional** upon the data. The Wilcoxon test, being based on ranks is, for the no-ties case unconditional, because it depends only on the order of the sample values and **not** directly on the values themselves except insofar as the latter determine the order. With ties, the possible values of the statistic are conditional upon the position of the ties in the rank ordering, but the effect of this conditioning is often small in practice. If the numbers of observations in each sample are m, n respectively one may use either S_m or S_n, the respective sample rank sums, as a test statistic. These statistics may be modified as we did in the Fisher–Pitman statistic by subtracting from S_m (or S_n) the mean for samples of m (or n) ranks randomly selected from a set of $m + n$ ranks. However, more common alternative statistics used in this case are other linear function of S_m or S_n, namely $U_m = S_m - \frac{1}{2}m(m + 1)$ or $U_n = S_n - \frac{1}{2}n(n + 1)$. Tables in Neave (1981), Lindley and Scott (1995) and elsewhere give values of these latter statistics required for significance at nominal 5% or 1% levels for various values of m, n assuming no ties. A few tables also state corresponding exact levels. Programs like StatXact and TESTIMATE make such tables virtually obsolete and I do not recommend their use (especially if they give only nominal 5% or 1% levels) unless appropriate software is unavailable.

Often basically continuous data are recorded only to a certain accuracy, which may be to the nearest integer, and except in very small samples or in some almost pathological cases such rounding does not have a great influence on the results, but care is required. An example of a problem that may arise from rounding (small perturbations of many observation) in a different context is given by Sprent (1993, Example 2.8). Rounding is irrelevant to allocation of ranks except when it induces ties that modify the analysis, but the effect of this is usually small, unless there are many rounding-induced ties. In a context of contamination rounding is a typical example of small perturbations to many observations.

4.3   Validity of inferences

In making inferences from tests like those in Examples 4.1 and 4.6 it is often assumed that the samples are random samples from populations subject only to constraints specified in H_0 or in H_1 and that each is one of an infinity of random samples that could be drawn from such populations. The logical difficulty here is that the data in Example 4.1 were not drawn from populations giving a potentially infinite number of possible samples, but from small finite populations, namely the specific papers that appeared in *Biometrics* in 1956–7 and in 1990. In practice most people obtain and use data like these to make wider inferences. For example, somebody may express the view that there was a tendency for papers on biometry written in the 1950s to contain fewer references than those in the 1990s and might want to use these data to back up this belief. To do this calls for the notion of *superpopulations* of papers on biometry written in the 1950s and of papers on biometry written in the 1990s. To test a hypothesis about increasing numbers of references in the later period, then implicit in our decision to choose papers in issues of *Biometrics* in specific years as a data source is an assumption that these finite populations are behaving effectively like random samples from all papers in biometry published in the relevant decades. This may be reasonable, but without further validation, believing it is little more than an act of faith. If we accept that the act is justified, inferences are relevant to the superpopulation of all biometry papers written (or in the latter period, if you are reading this before December 31, 1999, perhaps still to be written) in those decades. My comments on the range of validity of inference depending upon how data are obtained apply to all methods of inference and I expand upon them after Example 4.10.

4.4   An alternative form of the rank sum test

Mann and Whitney (1947) showed how to calculate U_m or U_n defined in section 4.2 directly from the data without assigning ranks. These statistics are easier than a rank sum to calculate manually and have practical advantages. Whichever statistic is used the name **Wilcoxon–Mann–Whitney test** is often used to cover both the Mann–Whitney and the Wilcoxon rank-sum forms. I use the abbreviation **WMW** for either test unless I need to distinguish between formulations.

Clearly $S_m + S_n = \frac{1}{2}(m + n)(m + n + 1)$, the sum of ranks from 1 to $N = m + n$, so that once one of S_m, S_n has been calculated it is easy to determine the other. It also follows from their definitions that U_m, U_n both have minimum possible value zero and that

$$U_m = mn - U_n. \tag{4.1}$$

Computer programs conventionally quote the value of S or U corresponding either to the sample with fewer observations or to the sample which yields the lower value. Users should check which is calculated for any given problem or in any given program, although all give the same p-value.

Example 4.7 shows how to calculate the Mann–Whitney statistic.

Example 4.7 Sprent (1993) gave data for the number of pages from a random sample of 10 books on biology and for a random sample of 12 books on management science in the Dundee University Library. The question of interest was whether books on one topic tended to be longer or shorter than those on the other. The numbers of pages in the sampled books were:

Biology (Sample I)   143 173 226 233 250 287 291 303 634 637
Management (Sample II) 50 164 198 221 225 302 328 335 426 534 586 618

Ordering values in each sample from least to greatest is not essential, but it helps the counting process. To obtain U_n we count the number of times each observation in Sample I is exceeded by an observation in Sample II. The sum of these counts gives U_n directly.

The first observation, 143, in Sample I is exceeded by 11 observations in Sample II (i.e. all except the first). Next, 173 is exceeded by 10 observations in Sample II and proceeding this way, we finally note that the observations 634, 637 in Sample I are not exceeded by any observation in Sample II. In summary, $U_n = 11 + 10 + 7 + 7 + 7 + 7 + 7 + 6 + 0 + 0 = 62$. From (2.1) we deduce that $U_m = 120 - 62 = 58$.

In Exercise 4.1 I ask you to rank these data to find the value of the Wilcoxon rank sum statistic S_m and to verify for these data the relationships between S_m, S_n, U_m, U_n given above.

StatXact confirms that $U_m = 58$ and shows that under H_0, $\Pr(U_m \le 58) = 0.4614$ so there is no question of rejecting H_0 since there is an almost equal probability of getting a lower or higher value if H_0 holds. StatXact and TESTIMATE both also give the alternative but equivalent statistic $S_n = 140$.

The Mann-Whitney procedure is modified for ties. An observation in one sample that ties with an observation in the other is scored as one-half, i.e. when computing U_n a second-sample observation the same as one in the first scores ½.

An alternative scoring system is to score 1 (as above) for a greater observation, but zero for a tie (instead of ½) and -1 (instead of 0) for a lesser observation. While U_n takes values (either integers or sometimes integer $+½$ with ties) in the range 0 to mn, U'_n takes integer values between $-mn$ and mn. In particular, if the sample sizes are m, n and we use the test statistic U_n, then if there are in total r values in the second sample that exceed first sample values, and t ties, we easily establish that there are $s = mn - r - t$ values in the second sample that are less than first sample values. Our scoring system with 1 for an excess and ½ for a tie and zero for a deficiency gives $U_n = r + ½t$, and the corresponding statistic with scores 1, 0, -1 gives $U'_n = r - s$. It is easily verified that $U'_n = 2U_n - mn$, i.e. U'_n is a linear function of U_n, and is here an equivalent statistic. The Mann–Whitney scoring system is motivated by a dominance concept stated as $H_1: Pr(X < Y) \geq ½$ *with strict inequality for at least some X.* Clearly under this alternative second sample values will tend to be greater than first sample values leading to high values of S_n or U_n.

The scoring system -1, 0, 1 is used in some computer software to avoid fractions which may induce rounding errors. I use it in an application in Example 4.15 where I find it less error prone than counting only excesses and ties. Also, in Chapter 12, I show that the WMW procedure is a special case of more general procedures, some of which conventionally use the 1, 0, -1 scoring system and others the 0, ½, 1 system.

An important practical case involving ties arises with grouped data in a table of counts in ordered categories.

Example 4.8 Consider again the data in Example 4.1 on numbers of references in published papers. Suppose we are given only the numbers of papers containing 0—9, 10—19, 20—29, $\geq$30 references, i.e.

No. of references	0—9	10—19	20—29	$\geq$30
1956—7	7	4	0	1
1990	3	3	7	1

We may use the Mann–Whitney procedure to compute U_n, scoring ties as ½. StatXact or TESTIMATE deal with such data using the WMW program (or alternatively that for the Fisher–Pitman permutation test, or in the case of StatXact, also the linear-by-linear association test program). Care is needed to assign appropriate scores using other than a WMW program which assigns scores automatically. If calculating U_n manually note that in the column headed *0—9* for *each* of the 7 data for *1956—7* there are 3 ties in *1990* contributing a score of 1.5 and also $3 + 7 + 1$ greater values giving a total contribution of $1.5 + 3 + 7 + 1 = 12.5$. Since there are 7 data in the cell $(1,1)$ the total contribution from all 7 is 7×12.5. Proceeding this way for all columns we find that $U_n = 7 \times 12.5 + 4 \times 9.5 + 0.5 = 126$ and hence that $U_m = 12 \times 14 - 126 = 42$. When there are numerous ties it is unwise to use tables based on the no-tie case, even for a test at a nominal level. For this example StatXact indicates significance at an exact 1.0% level in a one-tail test (compared with 0.65% for the corresponding test with the complete data). There is generally some loss of power with grouping because we use less information, but sometimes only grouped data are available.

Unlike the normal theory *t*-test, the WMW test requires no assumption of continuity in the original data, although the central limit theorem may reasonably justify relaxing the continuity assumption for the *t*-test. I used the Mann–Whitney formulation in Example 4.8, but the test may also be done using Wilcoxon rank sums where the observations in each column classification are given an appropriate mid-rank, e.g. the 10 counts in the group 0–9 are each given the mid-rank 5.5, and so on. These scores could also be used in a linear-by-linear association test as column scores, the row scores being 0 and 1.

Confidence limits for a location shift in continuous data may be obtained at appropriate WMW exact levels. The method in section 1.8 for the Fisher–Pitman test is not directly applicable, because this would treat the ranks as data and give intervals relevant to ranks. In principle, a confidence interval for the original data could be obtained by a trial and error method, adding or subtracting various constants d, say, to all values in one of the samples and noting the rank sums for the amended data until a value of d is attained which leads to a rank sum that just indicates significance at the relevant level. This tedious procedure is avoided by the Mann–Whitney formulation and the method used is an example of an **R-estimator**, a type of robust estimator mentioned in passing in section 3.5.

For samples of m, n with typical sample values x_i, y_j computation of U_n effectively uses the sign of each $d_{ij} = y_j - x_i$. If this difference is positive it scores 1, if negative it scores zero, if zero (a tie) it scores ½. If we subtract a sequence of constants c from each observation in the second sample, then clearly when c passes through a value such that $y_j - c - x_i = 0$ there is a sign change in the corresponding d_{ij} and so the value of U_n changes, i.e. it changes when $c = y_j - x_i$ for any i, j. If we know the value of U_n, say t, required for significance in a two-tail test at a prespecified $100\alpha\%$ significance level, then to obtain a $100(1-\alpha)\%$ confidence interval we set the lower limit at the $(t+1)$th smallest d_{ij} and the upper limit at the $(t+1)$th largest d_{ij}.

StatXact 3 includes a program for confidence intervals. An alternative approach for approximate 95% or 99% confidence intervals is to use tables to find

Table 4.1. Paired differences for numbers of pages in biology and management books.

	143	173	226	233	250	287	291	303	634	637
50	93	123	176	183	200	237	241	253	584	587
164	−21	9	62	69	86	123	127	139	470	473
198	−55	−25	28	35	52	89	93	105	436	439
221	−78	−48	5	12	29	66	70	82	413	416
225	−82	−52	1	8	25	62	66	78	409	412
302	−159	−129	−76	−69	−52	−15	−11	1	332	335
328	−185	−155	−102	−95	−78	−41	−37	−25	306	309
335	−192	−162	−109	−102	−85	−48	−44	−32	299	302
426	−283	−253	−200	−193	−176	−139	−135	−123	208	211
534	−391	−361	−308	−301	−284	−247	−243	−231	100	103
586	−443	−413	−360	−353	−336	−299	−295	−283	48	51
618	−475	−445	−392	−385	−368	−331	−327	−315	16	19

relevant values of U_m or U_n for significance at the required nominal level and then to find the requisite number of largest and smallest d_{ij}.

Example 4.9 shows a manual procedure for generating the d_{ij} and indicates how it may be abridged to give only those we need. Some general statistical packages (e.g. MINITAB) will generate the values and some recent software releases give nominal (though sometimes only asymptotic) confidence limits.

Example 4.9 Table 4.1 is formed from the data in Example 4.7. The first row are the ordered data for Sample I (biology books) and the first column are the ordered data values for Sample II (management books). The entries in the body of the table are the differences between the datum value at the head of that column and the datum value in the corresponding row. The table has a pattern; the differences of largest magnitude are at the top right (positive) and bottom left (negative).

The first entry in the body of the table is computed as 143 − 50 = 93. If calculating the entries manually, and especially if only the entries to determine the confidence limits are required, it is best to start at the top right or bottom left of the table, where the entries of greatest magnitude occur, since only a limited number of these are required. As a computing aid and a check, all differences between corresponding entries in any two given rows or columns are the same, e.g. entries in the first and second columns all differ by 30, and those in the last two rows all differ by 32, those in the first and fourth columns differ by 90. Entries increase as we move across columns from left to right and decrease as we move down rows from top to bottom.

It may be verified from StatXact or TESTIMATE for the above sample sizes that $U_n = 29$ for significance in a two-tail test at the 4.26% level. Thus, for exact (100 − 4.26) = 95.74% confidence limits we eliminate the 29 greatest and least differences in Table 4.1. These are easily obtained by inspection and the remaining largest and smallest entries, 103 and −185, indicate a 95.74% confidence interval (−185, 103). Confidence intervals at other levels can be obtained. For example, $U_n = 19$ corresponds to an exact two-tail test at the 0.56% level. The corresponding 99.44% interval is easily seen from Table 4.1 to be (−284, 211).

Approximate confidence intervals can be obtained using a two-sample bootstrap as in Example 4.5, treating the difference in sample medians as a robust estimator. For 2000 bootstrap samples using this approach I got the quantile based 95% confidence interval (−231, 147.5), somewhat longer than the nominal 95% interval obtained in Example 4.9. However the very wide 99.44% interval obtained in that example suggests a long-tailed distribution of the location difference. Nevertheless, the bootstrap result here is disappointing, especially with the known tendency for unadjusted quantile based intervals to undercover the true interval.

For large samples asymptotic results are useful for testing and estimation. For testing for a zero location difference we use the results (1.4) to (1.6). Applying them to ranks gives algebraic simplification particularly in the no-ties situation. If we replace z_i by ranks s_j, $j = 1, 2, 3, \ldots, n + m$ and use the sum of the first-sample ranks, S_m, tedious but straightforward algebra reduces (1.6) to

$$Z= \frac{S_m - \frac{1}{2}m(m + n + 1)}{\sqrt{[mn(m + n + 1)/12]}} \tag{4.2}$$

To obtain the corresponding result for S_n interchange m and n throughout. This and many other asymptotic results are improved by a continuity correction to allow for approximating to a discrete function of S_m by a continuous variable Z. The appropriate correction is to add ½ to S_m if Z is negative and to subtract ½ from S_m if Z is positive. Ties do not alter the numerator in (4.2), but the denominator is changed. This is best computed from (1.5) with the z_i equal to the assigned rank and mid-rank values. With ties the values of S_m are no longer confined to integer values, and some integer values of S_m may not be realizable, so using a continuity correction and what that correction should be is questionable. However, in practice for moderate sample sizes with only a few ties using a correction of ½ as above often gives a better approximation. Modern programs for exact permutation tests or Monte Carlo approximations avoid the need for asymptotic tests except when $N = m + n$ is large and neither m nor n is very small, so that when asymptotic results are really needed the effect of a continuity correction is often small. Using the relationship between S_m and U_m it is easy to verify for the no-tie case that (4.2) reduces to

$$Z= \frac{U_m - \frac{1}{2}mn}{\sqrt{[mn(m + n + 1)/12]}} \tag{4.3}$$

Approximate $100(1 - \alpha)$% two-tail confidence limits for a location difference are obtained for given sample sizes m, n by setting Z in (4.3) at the value required for significance at the 100α% level and computing the relevant U_m. Rounding is usually needed to obtain an integer value. The appropriate values of the d_{ij} must then be calculated and since the asymptotic result is only reliable for large samples an appreciable number of d_{ij} may be needed, making a program such as that in MINITAB virtually essential if one does not have a full program for confidence intervals like the one in StatXact.

A commonly-used point estimate of the location difference θ is the median of all the d_{ij} given by a table like Table 4.1. This estimator is the **Hodges-Lehmann** estimator. Its properties were discussed by Hodges and Lehmann (1963) and it is a classic example of an R-estimator and is easily seen to give the value -18 in Example 4.9.

Example 4.10   To extend the investigation into whether more references occur in recent papers in biometry (Example 4.1) I wanted to know if the same trend occurred in general statistical papers. I counted the numbers of references in **all** papers concerned with statistical theory, practice or methodology appearing in Series A of the *Journal of the Royal Statistical Society* in each of the years 1959 and 1993. That Journal also publishes other material such as book reviews and annual reports of the Society's affairs, but I excluded these from the counts. Relevant numbers were:

1959	1 2 7 10 12 14 17 17 19 23 34
1993	2 3 4 8 17 21 22 22 23 26 28 34 38 40 41 46 56 74 140

These data were analysed in a similar way to that in Examples 4.1 and 4.6. Despite the apparent outlier, 140, in 1993 the t-test here indicated significance at the 2.75% level in a one-tail test. The exact Fisher–Pitman test did even better giving significance at the 0.86% level, but the asymptotic test was very close to the t-test result. This broadly satisfactory result, despite the outlier, is not as illogical as it may first seem and if one explores the permutation distribution in detail one sees that the several high values in the 1993 sample strongly support a dominance hypothesis which is reflected in the exact Fisher–Pitman statistic value. However, the strong non-normality implied by the outlier detracts from the asymptotic result by inflating the denominator in (1.6), effectively slowing the rate of convergence to normality that is implicit in the central limit theorem. StatXact also allows us to compute a confidence interval for the median difference based on the WMW test and this turns out to be (3,27) at an approximate 95% level (see also Exercise 4.4). The Hodges-Lehmann point estimator of the difference is 14. The exact one-tail $p = 0.0083$ in the WMW test, being little different from that for the Fisher–Pitman test.   Comparing this with the situation in Examples 4.1 and 4.6 we see that it is not always obvious whether a robust method is essential! When there is doubt it is safer to use a robust method.

For these data I also obtained a 95% bootstrap quantile-based confidence interval for differences between the medians taking $B = 2000$ samples and computing the median difference for each sample.   This gave an interval (4,28) in close agreement with the above WMW/Hodges-Lehmann interval. The one-sided bootstrap estimate of p was $p^* = 0.006$ again in close agreement with the values for the WMW test and the Fisher–Pitman test.

In the above example the data were for all papers published in the named journal in the relevant years. In this sense they are observational data for complete populations whereas the data in Example 4.1 were random samples from finite populations. Each sample there consisted of approximately one fifth of all papers published in the stated years. The complete population data in Example 4.10 means that any inferences to a wider population depend for validity upon an act of faith whereby we assume these published papers are in effect equivalent to random samples from populations that we might describe as 'all papers like those that appear in the *Journal of the Royal Statistical Society* in the given years'. If we accept some such assumption our permutation tests are relevant exact tests and any confidence interval is appropriate to the median difference between populations. Such an interval has no meaning so far as the actual papers appearing in the Journal in the given years are concerned. That median difference is a fixed constant, the difference $26 - 14 = 12$ between medians for the finite populations of published papers; these populations only become 'samples' when we consider them to come from hypothetical superpopulations of similar kinds of papers.

We could calculate a confidence interval for the median difference in Example 4.1 using the methods in this section (Exercise 4.3), but care is needed in its interpretation. In that example we took a random sample of approximately one fifth of all published papers in the relevant issues of *Biometrics*, but an interval computed in the way described above would be relevant to an assumed superpopulation for which we regarded the subpopulations of all papers published in *Biometrics* in the relevant years as samples. Sampling theory covers the somewhat different procedures for inferences from samples to finite populations in parametric analyses, introducing what is usually known as a finite population correction involving a sampling fraction; see, e.g., Cochran (1953, section 2.5). Again, I emphasize that restrictions upon the scope of validity of inferences when sampling from finite populations, or when dealing with observational data, apply both to distribution-free and to parametric inference. I pointed out in section 1.4 that there is a large statistical literature on the implications and pitfalls of analysing and making inferences from data that are not from either designed experiments with proper random allocation to treatments, or that do not consist of random samples (perhaps stratified or modified in other ways to increase precision) from large or potentially infinite populations. Much of this work is based on commonsense principles. One should, for instance, appreciate in Example 4.10 that the journal in question publishes mainly papers on social statistics or papers of general statistical interest read at meetings of the Society. Patterns of numbers of references might be quite different even in the same years in a journal publishing papers on the mathematical theory of statistics. Even in the chosen issues of the journals, patterns may vary dramatically from year to year, especially if there has been a change in editorial policy between the years being considered. Some statisticians think that hypothesis testing or calculating confidence intervals is inappropriate for observational data. My view is that it can make sense providing the limitations are born in mind. Those opposed to it have yet to find satisfactory alternative approaches, while most admit that some sort of analysis of such data is of practical importance. An interesting alternative and nonprobabilistic approach to the study of observational data is given by Draper *et al* (1993), although it is not clear that this overcomes sometimes justifiable reservations about how wide is the validity of any conclusions.

4.5 Transformation of ranks

In Example 4.6 using ranks reduced the influence of outliers where these, by their nature, tended to hide a location difference in a *t*-test. Ranks have an added advantage, relative to the Fisher-Pitman test, of giving an unconditional test. When assumptions to validate the *t*-test hold, a WMW test has a slightly lower Pitman efficiency of $3/\pi$, or in percentage terms approximately 95.5%. If the initial data are simply the ranks, the WMW test is, by nature, appropriate. The above and many other efficiency comparisons for nonparametric tests relative to the *t*-test were given by Hodges and Lehmann (1956). Despite the slightly lower efficiency with normality, for samples from a range of nonnormal distributions the WMW test has appreciably higher Pitman efficiency than the *t*-test. In particular, for samples from the exponential distribution the WMW test is 3 times as efficient as the *t*-test. However, there are other tests that are more efficient than WMW for samples from some distributions. When there is doubt about what the population distribution really is, the WMW test often provides a good safety net for tests of a location shift or dominance if these are characteristics of interest.

Are there monotonic transformations of ranks, other than back-transformation to the original data, that give higher efficiencies or other benefits? A transformation that gives under H_0 an approximation to samples from a normal distribution has intuitive appeal. Since many asymptotic results involve an approximation to normality it is reasonable to hope that convergence of exact to corresponding asymptotic results might be more rapid for data so transformed.

Ranks themselves are like samples from a uniform distribution. More formally, for a continuous uniform distribution over $(0, N)$ the rank r in the combined samples of $N = m + n$ observations corresponds to the $r/(N + 1)$th quantile of that distribution. This suggests a transformation of ranks to corresponding quantiles of the standard normal distribution might be appropriate. We easily make this transformation using tables of the standard normal cumulative distribution function (cdf). Thus the observations 3.5, 5.9, 2.3, 22.7 133.5 are ranked 2, 3, 1, 4, 5 and the corresponding quantile scores are the 2/6th, 3/6th, 1/6th, 4/6th and

5/6th quantiles of the standard normal distribution. The score for the 1/6th = 0.1667th quantile is the x value such that the standard normal cdf, $\Phi(x) = 0.1667$. Tables or software give $x = -0.97$. Similarly the 2/6th quantile is $x = -0.43$ and clearly the 3/6th quantile (median) is 0.00. By symmetry the 4/6th and 5/6th quantiles are $x = 0.43$ and $x = 0.97$. Quantile normal scores were proposed by van der Waerden (1952;1953) and are often called **van der Waerden** scores. They may be used as data in any program for the Fisher–Pitman test, but StatXact and TESTIMATE have programs that perform the transformation automatically. Often permutation tests based on van der Waerden scores give similar results to rank exact tests. However asymptotic results using (1.6) with these scores may be closer to the permutation test results for smaller samples than would be the case for ranks or for the original data if these were clearly nonnormal. With ties at, say, rank s and rank $s +1$ one could assign a quantile corresponding to the mid-rank, but a more logical score is the mean of the van der Waerden score for rank s and that for rank $s + 1$, although in practice the difference between this mean and the mid-rank quantile is usually small.

Instead of quantiles of the standard normal distribution we may use **expected normal scores,** where the ith rank is replaced by the expectation of the ith order statistic for a standard normal distribution; the ith order statistic being the ith smallest value in a sample of N observations, denoted by $x_{(i)}$, (section 2.1). Fisher & Yates (1957, Table XX) gave expected normal scores corresponding to ranks for $N \le 50$. In practice there is usually little difference between results using van der Waerden scores or those using expected normal scores. An example of the use of the former with the data in Example 4.7 is given in Sprent (1993, p. 121). Expected normal scores or van der Waerden scores may be used in a linear-by-linear association program in place of the raw data.

Transformations appropriate for analysing survival data are considered in section 4.7–4.9.

An interesting family of transformations that effectively ignore some data and yet often combine high Pitman efficiency with simplicity were proposed by Gastwirth (1965) in a paper that dealt not only with the problem of location shifts but also with scale differences. I discuss these in section 6.6.

4.6 Using minimum data – the median test

The **median test** uses minimal information. We only need to know the numbers of observations in each sample and how many observations in one sample are below the combined sample median. It is sensitive to location shifts in otherwise identical population distributions, tending to lose power when the shapes of the population distributions differ fairly markedly, especially if one sample is small. Simplicity is paid for in a loss of both efficiency and power relative to tests — parametric or nonparametric — that may be more appropriate when full data are available. It is one of a wider class of tests based on quantiles that are discussed more fully by Gibbons and Chakraborti (1992, section 7.4).

The basic idea is that if two populations have the same median, then in each sample the number of observations above (or below) that common median has a binomial distribution with parameter $p = \frac{1}{2}$. We do not know the value of that common median. However, under a null hypothesis of identical distributions a sensible estimate is the median, M, of the combined sample of $m+n$ data. The test with minor modifications was proposed by Westenberg (1948) and by Brown and Mood (1948) and is sometimes referred to as Mood's test. If M is the combined sample median the relevant probabilities under H_0 of numbers above and below M in each sample are now conditional upon the numbers below M in each sample totalling to the numbers below M in the combined sample. If $m + n$ is even and if no sample value equals M, in the combined sample the number of values above and below M will each be $\frac{1}{2}(m + n)$. If $m + n$ is odd at least one sample value equals M. In any case where one or more sample values equal M, a common but by no means universal convention is to omit such values and proceed with a reduced sample. An argument for doing this is that such values are uninformative about any departure from the hypothesis of no difference in medians. If there are only a few ties I prefer this to somewhat *ad hoc* alternatives like allocating values equal to M either randomly or by some arbitrary rule, to the *above* or *below* category. If there are many such tied values, and especially if these dominate in one sample there is a stronger case for random allocation of these ties because otherwise there may be an appreciable loss of power due to an effective reduction in sample size. Conditioning on the total numbers above and below M in the combined sample determines the appropriate permutation test.

Example 4.11   Consider the data on numbers of references in samples of papers from two different volumes of *Biometrics* given in Example 4.1. Inspection shows that the median of all 26 sample values is $M = 14.5$ and that in Sample I there are 9 values less than M and 3 values greater than M, while in Sample II there are 4 values below M and 10 values above M. This information is summarized in Table 4.2.

Under H_0, we expect approximately the same number above and below the median in each sample.

Table 4.2 is a conventional 2×2 contingency table with fixed marginal totals, the row totals being determined by the sample sizes and the column total by the rule that M is the combined sample median. The test is based on all possible categorical tables arising from permutations of the original data that give rise to these marginal totals. It is easy to see, and indeed is well-known that given the entry in any one cell of such a 2×2 table the others are uniquely determined when marginal totals are fixed. Clearly possible entries in the first row and column

Table 4.2 Numbers of observations below and above the combined sample median M in Example 4.1

	Below M	Above M	Total
Sample I	9	3	12
Sample II	4	10	14
Total	13	13	26

must be between 0 and 12 inclusive. If the population medians are identical we expect values near the middle of this range. A low value in cell (1,1) suggest Sample I comes from a population with a higher median than that for Sample II and a high value (as here) suggests the opposite.

To carry out a formal test we need to know how many of the possible permutations of the original data give rise to each possible value in cell (1,1). Let us assume temporarily that if H_0 is true then M is the value of the true median of both populations. This may not be so, but it is the best estimate of a common median under H_0 available from these data. I show below that this assumption is not critical to the argument. In a sample of 12 independent observations (Sample I) each is equally likely to be above or below the median and the number below has a binomial distribution with $n = 12$, $p = \frac{1}{2}$ so that the probability of observing r below the median is $p_1 = {}^{12}C_r (\frac{1}{2})^{12}$ and similarly that of getting $13 - r$ below the median in a sample of 14 (Sample II) is $p_2 = {}^{14}C_{13-r} (\frac{1}{2})^{14}$. We need the probability of obtaining these outcomes conditional upon observing 13 below the median in the combined sample of 26 and that probability is $p_c = {}^{26}C_{13} (\frac{1}{2})^{26}$. Thus, by the usual rules of conditional probability the required probability is $P = p_1 p_2 / p_c = {}^{12}C_r {}^{14}C_{13-r} / {}^{26}C_{13}$. The probability $p = \frac{1}{2}$ cancels out in the expression for P. This will happen if the binomial analogues of p_1, p_2, p_c are written down for any probability p, i.e. P is independent of p. Thus if H_0 is true we do not need to assume that M is the population median. That P holds irrespective of p has far-reaching implications in nonparametric inference and in Chapter 12 we use this result and extensions in many situations. I generalize this result before completing this example.

If we replace the sample sizes 12, 14 by m, n then clearly

$$P = {}^{m}C_r \times {}^{m}C_{\frac{1}{2}(m+n)-r} / {}^{m+n}C_{\frac{1}{2}(m+n)}. \tag{4.4}$$

Modification is needed if some observations equal the combined sample median and if these are discarded or otherwise allocated to groups, but the final test procedure is similar. It is not hard to show that for the particular marginal totals involved in this problem that (4.4) is a special case of a more general result that under an assumption of no association between row and column categories in a 2×2 table, the probability of observing the following outcome with the indicated fixed marginal (i.e. row and column) totals, namely

Row	Column	1	2	Row total
1		a	b	$a+b$
2		c	d	$c+d$
Column total		$a+c$	$b+d$	$a+b+c+d$

is

$$P = \frac{(a+b)!(c+d)!(a+c)!(b+d)!}{(a+b+c+d)!\, a!\, b!\, c!\, d!} \tag{4.5}$$

The factorials in the numerator of (4.5) are those for marginal totals, and those in the denominator correspond to the grand total and the individual cell totals. These probabilities define a **hypergeometric distribution**. Calculating P for moderate counts is feasible with a pocket calculator and most general statistical software packages compute them for all tables with the given marginal totals and perform what is usually called Fisher's exact test for independence in a 2×2 contingency table (section 12.2).

Example 4.11 (continued) In this example to test H_0:*medians are equal* against H_1: *Sample II is from a population with median greater than that for the Sample I population,* I use as the test statistic the cell count in the first row and column (i.e. the number in sample I below the median) and the critical region for rejecting H_0 will correspond to high values of this statistic. We observed 9 from 12 below the combined samples median, so we require the probability of observing 9, 10, 11 or 12 in this cell to determine the size of the critical region.

Any program for Fisher's exact test for a 2×2 table indicates that the probability of observing 9 or more in cell $(1,1)$ is 0.0236. Some programs provide individual probabilities of observing each possible number in this cell, found by substitutions in (4.5). Denoting this relevant number by Y, in particular we find :

$$\Pr(Y = 9) \ = 0.0212$$
$$\Pr(Y = 10) = 0.0023$$
$$\Pr(Y = 11) = 0.0001$$
$$\Pr(Y = 12) = 0.0000$$

The distribution of Y is symmetric, so if we require a two tail test the significance level corresponding to the observed value $Y = 9$ is $p = 2 \times 0.0236 = 0.0472$, our result thus indicating significance at the 2.36% level in a one-tail test and at the 4.72% level in a two-tail test.

Since this test uses less information than the WMW test, it is not surprising that the significance is not so clear cut as in Example 4.6, or even as in Example 4.8. Interestingly, we see that **after** selecting the over-all sample median as the criterion for allocation to cells in the 2×2 table, we may look upon the resulting problem as a heavily tied example of the WMW test. All values below the common median in each sample are regarded as ties at a common 'low' value and all above the median as ties at a common 'high' value. We can then compute the WMW statistic as we did in Example 4.8, or use the WMW programs in StatXact or TESTIMATE and obtain identical exact results to those from the Fisher test. However, the commonly used asymtotic results differ in the two cases because the conventional asymptotic result for WMW is a normal approximation which is not equivalent to the asymptotic approximation used in the Fisher exact test.

Indeed, standard results for the mean and variance of the hypergeometric distribution give, when there are no observations equal to the common median, and Y is the number of Sample I values below the combined median

$$E(Y) = \tfrac{1}{2}m$$

$$\mathrm{Var}(Y) = \tfrac{1}{4}mn/(N-1)$$

and asymptotically

$$Z = \frac{Y - E(Y)}{\sqrt{\mathrm{Var}(Y)}} \tag{4.6}$$

is a standard normal variable. A continuity correction of 0.5 is desirable, this being added to Y if $Y < E(Y)$ and subtracted if $Y > E(Y)$.

For Example 4.11 it is easy to verify that this gives $Z = 1.93$, indicating a one-tail significance level of 2.6%, slightly greater than that given by the exact test, but the agreement is reasonable with these small samples. The modifications needed to (4.6) if some observations equal the combined median are given by Gibbons and Chakraborti (1992, section 7.4)

One should not neglect relevant information except for pressing reasons. It is worth noting, however, that if the conditions for the WMW test do not hold, the median test is sometimes preferable. For example, it is easily verified that if, in the data used in Example 4.1 we replace the first four observations in Sample II by 0,0,1,1 the analysis in Example 4.11 is unchanged. However, a WMW test using the amended data gives a one-tail $p = 0.0984$. The increase is attributable to the amended values in Sample II all being less than *any* Sample I values, suggesting that the concept of dominance for the alternative hypothesis (and location shift in otherwise identical populations is simply a special case of this) is no longer appropriate. This inappropriateness results in a loss of power for the WMW test. The median test is less sensitive to this not-too-marked breakdown of the concept of populations identical in shape or of one dominating the other.

Unless both samples are fairly large the median test statistic may take only a small number of possible values; indeed the maximum possible number of values is one more than the smaller sample size. This leads to marked discontinuities in possible significance levels if one sample is small even when the other sample is large. Further, in that situation there may be a dramatic loss of power compared to that for the WMW test. Situations with large differences between sample sizes are not uncommon in clinical trials where there is often a large control group of patients treated with a standard drug but only a small number of patients treated with a new drug in short supply.

   The following example indicates that the median test, because of discontinuities in *p*-values, may fail to pick up a difference that the WMW or other appropriate test may find highly significant.

Example 4.12 A response (e.g., lowering of blood-pressure, weight gain, reduction in pain) is recorded for 4 patients treated with a new drug and for 60 receiving a standard treatment. Patients are allotted to treatments at random. A strong response is favourable. In the event, the measure for each of the 4 who receive the new drug exceeds that for any patient given the standard drug, so any sensible test should indicate superiority of the new drug. In the WMW test those receiving the standard drug are ranked 1 to 60 and those receiving the new drug are ranked 61, 62, 63, 64. Either Statxact or TESTIMATE indicates significance in a one-tail test at a level less than 0.001%. In fact, the precise level is 0.00016% (Exercise 4.7). If the same data are analysed using the median test, clearly for the sample receiving the new drug all values are above the combined sample median and STATXACT gives a one-tail significance level of only 5.66%; thus the result would not attain significance at the 5% level.

   Discontinuity in available exact levels here means that the median test cannot detect differences as significant at a level below 5.66% unless one uses the convention of splitting the probability of the statistic attaining exactly its observed value equally between the critical region and the acceptance region. Even then the best possible significance level is 2.83%.

   The loss of power with the median test with such unbalanced samples arises because the information that the new drug is **markedly** superior is thrown away. Had the rankings of the new drug among the 64 patients been 33, 34, 35, 36 we would have the same 'data' for the median test and it would again indicate significance at the 5.66% level. A full investigation of the power loss of the median test relative to the WMW test in circumstances like those considered here was made by Gastwirth and Wang (1987). With new drug rankings 33, 34, 35, 36 the WMW would no longer be appropriate because the data would clearly not support an alternative hypothesis of dominance.

   Confidence limits for a population median difference based on the median test are easily obtained if we have the full measurement data. If the true difference as $\delta = \theta_2 - \theta_1$, where θ_1, θ_2 are the population medians, an exact $100(1 - \alpha)\%$ confidence interval for δ is (d_1, d_2) where the end points d_1, d_2 are chosen so that if we change each observation in sample II (or sample I) by more than these amounts we would reject at the $100\alpha\%$ level a hypothesis H_0: *the medians of population I and amended population II are equal.* For given sample sizes it is easy to work out with modern computer software the numbers below the median in the first sample that will just give significance.

Example 4.13 Consider the data in Example 4.7 for books on biology and management. The data were

Biology	143	173	226	233	250	287	291	303	634	637		
Management	50	164	198	221	225	302	328	335	426	534	586	618

It is easily verified (Fisher exact test) that given samples of 10 and 12, significance in a two-tail test at the 3% level follows in a median tests if the number in the first sample below the median is 2 or less (or by symmetry 8 or more). The next step is to determine what constants d_1, d_2 must be subtracted or added to the second sample values to bring about a situation of 2 or 8 below the median in the first sample. Clearly we will not get a significant result at the 3% level if the combined samples median after adjustment to the second sample values lies between 226 and 303, but when this combined median lies outside these values the significance requirements are met. Since, to set the combined median at or above 226 there must be a total of 11 observations below the combined sample median this implies we must subtract a quantity from each second sample value so that the ninth largest value, here 426, is reduced to 226, i.e. we must subtract 200. Alternatively to set the combined median at or above 303 we need 9 observations in the adjusted second sample to be above 303 so we must add $303 - 221 = 82$ to all second sample values to achieve this. This implies that the exact 97% confidence interval for the difference $\theta_2 - \theta_1$ based on the median test is $(-82, 200)$. In Example 4.9 using the WMW procedure we found (changing signs) a 95.74% confidence interval $(-103, 185)$. Make sure you understand why I changed the signs!

There is a slight shift between the intervals in Examples 4.9 and 4.13, as well as a small difference between exact confidence levels, but the lengths of the intervals are almost the same. The data suggest that the samples are from distributions skewed to the right, but that one may not quite dominate the other. Thus the WMW analysis may not be wholly appropriate and this may partly explains why each procedure gives confidence intervals of similar length, for the test using less information may be less sensitive to a breakdown in assumptions.

An alternative median test suggested by Mathisen (1943) is the **control median test**. Sample I may represent a standard procedure and is designated the **control sample**. The test statistic is V, the number of Sample II values less than the Sample I median, M_1. If the sample sizes are m, n respectively and m is even and if no Sample I value is equal to M_1, then, if we write $r = 2m$ and there are j Sample II values less than M_1 it follows that in the combined sample there will be $r + j$ values less than M_1, and $r + n - j$ greater than M_1 (assuming no Sample II value equals M_1). Under the null hypothesis that both samples are from the same population, arguments similar to those used to obtain (4.4) establish that

$$\Pr(V = j) = {}^{r+j}C_j \times {}^{r+n-j}C_{n-j} / {}^{m+n}C_n \qquad (4.7)$$

Small values of V suggest that the median of Population II is greater than that of Population I and large values of V suggest the opposite. In a one-tail test in the first situation the critical region consists of the observed or smaller j and the corresponding p-value may be computed using (4.7) and summing over the relevant values of j.   In the second case the critical region consists of the observed and larger values of j.   In a two-tail test the critical region is comprised of both large and small values of j.   Only slight modifications to the above arguments are needed if m is odd or if some sample values equal the Sample I median.   The modifications needed if the median of Sample II is taken as the control median are virtually self evident (Exercise 4.5)

Example 4.14.   For the data on numbers of references considered in Example 4.1 it is easily verified that the 1956–7 median is 9. Two observations in that sample and one in the 1990 data have the value 9 and I follow the convention suggested for the previous median test of dropping such values. For the data without these values we find $m = 10$, $n = 13$, $r = 5$, $j = 2$, whence (4.7) gives $\Pr(V = 2) = 0.080$. The less likely outcomes are $j = 1$ or $j = 0$ and the associated probabilities are $\Pr(V = 1) = 0.032$ and $\Pr(V = 0) = 0.007$ whence $\Pr(V \leq 2) = 0.119$ so we would not reject H_0 in a one-tail test. It is sometimes argued that, in a situation like this, one may take either sample as the control sample. By taking the 1990 sample as the control sample we avoid the occurrence of ties since the median for that sample is 21. Further there is only one 1956–7 sample value greater than 21 and by a symmetry argument we might perform a test using the number of observations in that sample above the 1990 median. Doing this gives a p-value (i.e. $\Pr(V \leq 1)$ of $p = 0.0316$ (Exercise 4.6)

Inequality of p-values depending on the choice of control sample is an apparent weakness of the control median approach, despite the fact that it has Pitman efficiency 1 relative to the median test based on the combined sample median. Asymptotic normal approximations are available (Gibbons and Chakraborti, 1992, p.208) for the control median test, but these also lack symmetry.

An interesting application of median tests is given by Gastwirth (1968). In many life-testing or survival experiments observations are collected sequentially and are ordered as they consist of times to failure of each item in preselected samples of m and n. The observations are ordered not only in each sample but in the combined samples. If we only want to know whether median time to failure differs between populations corresponding to the two samples, it is possible to terminate the experiment before all items have failed, still getting exactly the same p-value or decision if one is using a median test or a controlled median test as one would get if the experiment were completed. This may lead to a considerable saving in experimental resources. A median test can always be so arranged that the critical value of the test statistic is the number in one of the samples that are below the hypothesized median under H_0, and without loss of generality we may take this to be Sample I. Further, for either the median or the controlled median test, given m and n one may work out *a priori* for testing at any preassigned significance level the critical value for the number of Sample I values that must be below the combined sample median. If the median test is used the test may be terminated either when this number is exceeded in Sample I, for we then accept H_0, or if it has not been exceeded by the time we have obtained $\frac{1}{2}(m+n)+1$ observations we may then terminate the test for at that stage we know that any further observations in Sample I will be above the overall median estimated for the combined sample.

A broadly similar argument applies for the controlled median. If we take Sample II to be that for which the median is controlled, the stopping rule is now that we stop either if the critical number in sample I is exceeded, or if it has not already been exceeded, when we have obtained $\frac{1}{2}n+1$ Sample II values, for then no more Sample I values will be below the control median.

For either the median or the control median situation we accept H_0 if the relevant critical value is exceeded first, otherwise we accept H_1. Clearly the decision would not be changed if we completed the experiment by waiting for all items to fail, since our test only requires numbers below the combined median or below the control median to be known. Gastwirth shows that asymptotically the average number of failures that have to be observed for a given m, n before a decision is reached is less for the controlled median test than for the median test. He also proposes a slight modification to the control median test that further reduces the expected number of observations needed before a decision can be reached.

If the population distributions differ in shape the situation for the median or controlled median test is somewhat analogous to that in the classic Behrens–Fisher problem in location tests for samples from two normal populations with different variances and is therefore often referred to as the generalized Behrens–Fisher problem. A modified median test has been given by Fligner and Rust (1982) for that situation. The generalized Behrens–Fisher problem has also been considered by several writers including Pothoff (1963) for different but symmetric population distributions leading to modfication of the WMW test. I consider a bootstrap approach to this problem in section 6.6

4.7 Analysis of survival data

Nonparametric and distribution-free methods are widely used in analyses of survival or failure-time data. Two characteristics of such data have stimulated development of special tests.

- Distributions of survival times in clinical trials, or of times to failure of structures or machines in engineering or industrial contexts, are typically long-tailed or skewed to the right.
- Some units are often lost to the study before a response of interest (e.g. death of a patient or failure of a machine) occurs.

The first characteristic reduces the role of the normal distribution compared to the part it plays in many applications and parametric tests for survival data often assume data are samples from exponential, Weibull or other long-tail distributions. The case for distribution-free methods is strengthened because, for these distributions, the distribution-based parametric tests often lose power rapidly when assumptions are violated.

Withdrawal of subjects before the response of interest occurs is called 'right-censoring' to indicate that we are losing an observation towards the right-tail of a distribution. This is the only type of censoring I discuss so I simply refer to **censoring**. Censoring may occur at a fixed time because it is decided to terminate an experiment before all subjects have shown the response of interest (often death or failure) or it may occur when a predetermined number, or proportion, of the units in the study or among those subjected to a particular treatment, have responded. In section 4.6 in discussing the median test, another reason for right censoring was proposed, namely when criteria for a decision that cannot be changed by taking further observations are met. It then makes economic sense to discontinue the experiment unless data are required for other purposes.

A further common type of censoring occurs if subjects withdraw (often at random) before the response of interest is observed. In clinical trials where survival time following a treatment is the response such censoring occurs if a patient leaves the district, fails to attend follow-up clinics, dies from a cause other than the condition being treated (e.g. by accident or due to another illness). There are added analytical complications if this type of censoring is treatment dependent and at a rate that differs between treatments. e.g. if for one treatment patients withdraw because of serious side-effects. That complication can be allowed for in an analysis, but I do not cover this more difficult problem. It was discussed briefly by Peto and Peto (1972) and has been by many others since in specific situations. I assume here that censoring is either due to a decision to terminate the trial at a given time after commencement or else that it occurs randomly across all treatments. Random censoring also occurs in industrial contexts; for example, when a machine fails for a cause other than that under investigation or is withdrawn from a study for safety reasons, or because testing equipment fails.

Altman (1991, Chapter 13) gives an excellent account of the practical background to survival studies in medical research highlighting some potential pitfalls when analysing survival data, and deals with some aspects I do not cover including the modelling of survival data using regression type models emphasizing in particular proportional hazards regression analysis that allows one to take account of the effect of factors like age on survival prospects.

For simplicity I refer to the response of interest as **failure** because it is commonly mechanical breakdown or death of an individual, although the response need not necessarily be an adverse event (it could be the disappearance of some clinical symptom, or complete recovery). The measured variable will usually be time that elapses between a given origin (typically time of diagnosis or commencement of treatment or installation of a new machine component) and the event *failure*. Some of the procedures might sensibly be applied also to measurements from other long-tailed distributions that do not involve time (e.g. the numbers of references in published papers considered in Example 4.10).

If censoring is independent of treatment one might ignore censored observations and use the WMW or other appropriate test on the remaining data. This is valid, but usually results in a loss of power and efficiency and is clearly rejecting potentially useful, even if 'incomplete' information. Several modifications of the WMW procedure have been proposed to deal specifically with censoring without making specific reference to samples being from long-tailed distributions. I discuss two of these before passing to an application relevant directly to samples from distributions with long right tails when there may or may not be censoring.

4.8 Modified WMW tests

Two proposed modifications to the WMW test procedure differ in the way they deal with censored observations where the censoring is not treatment dependent. Each may be used to explore shifts in location or dominance of one population distribution over the other. In clinical trials the alternative hypothesis often specifies dominance rather than location shift, for a promising treatment will often prolong life or time of remission sometimes for lengthy periods if successful, while hardly altering the prospects for other patients, thus often changing the shape of the survival distribution. On the other hand in engineering applications, introduction of a new component may simply increase a machine life by a fairly constant amount for all units, and here a location shift hypothesis may be appropriate.

An early approach to this problem was one by Gehan (1965*a*, *b*). A comprehensive, but in many ways intuitive, account of some theory behind analysis of censored data is given by Peto and Peto (1972). Various writers have suggested refinements and modifications to these and other basic models, extending them to deal with ties and treatment-dependent censoring.

For the WMW and other tests I have shown that numerically different scoring systems (e.g. rank sums or Mann–Whitney counts) lead to alternative but equivalent statistics, and the choice between several equivalent scoring systems for censored data by various writers has been even more diverse and complicated. I use a modified Mann–Whitney rank sum type of scoring for one test and a seemingly different system for the other, but I then show that the latter leads to what is effectively a modified Wilcoxon rank-sum statistic. I do this because each is the more easily understood in its own context.

A censored observation clearly has a longer survival time than a unit that fails at or before that censoring. This is the partial information in a censored observation. How does one compare the survival times for a censored unit and one that fails later; or survival times for units censored at either the same or at different times? The Gehan and the Peto approach differ in the way they tackle these problems. Gehan's modification of the WMW test is often called the **Gehan–Wilcoxon** test. In essence Gehan regarded two censored observations as giving no definite information on survival times relative to each other. Nor did he regard a censored observation as providing information on whether the subject does or does not have a greater survival time than that of a later failure. Peto and Peto on the other hand make a specific assumption about the average survival time for any censored observation. Some people regard Gehan scoring as ultra-conservative and that of Peto and Peto as speculative.

Example 4.15 The following data are given by Dinse (1982). An asterisk here and throughout this section indicates a censored observation. The data are survival times in weeks for 10 patients with symptomatic and for 28 patients with asymptomatic lymhhocytic non-Hodgkins lymphoma. Do survival time patterns differ between the two groups? Whether the difference is regarded as a median shift or one of dominance, a modified WMW test allowing for censoring seems reasonable.

Symptomatic	49	58	75	110	112	132	151	276	281	362*
Asymptomatic	50	58	96	139	152	159	189	225	239	242
	257	262	292	294	300*	301	306*	329*	342*	346*
	349*	354*	359	360*	365*	378*	381*	388*		

We test

    H_0: *the survival time distributions are identical*

against

    H_1:*$G(x) \leq F(x)$ with strict inequality for some x where $F(u)$, $G(v)$ are the survival time cdfs for symptomatic and asymptomatic cases.*

The Gehan–Wilcoxon test is easily explained and carried out using an analogue of the alternative WMW statistic U_n' given on p.86. We calculate for each observation in the first sample the difference between the number in the second sample that definitely have a longer time to failure and the number that definitely have a shorter time to failure. For an observed first sample failure, all later failures or items censored at the same or a later time in the second sample definitely have a longer time to failure and each is scored as +1; items in the second sample that fail earlier definitely have a shorter time to failure and each is scored as −1, but an earlier censored item may or may not fail earlier and is counted as a tie and so scored as zero. For any censored item in the first sample all censored items in the second sample and any later failures in that second sample are regarded as ties and scored as zero while earlier failures in the second sample are scored as −1. The sum of these scores is the statistic U_n'. TESTIMATE provides a program for an exact permutation test in this case.

For the above data clearly the score associated with the entry 49 in Sample I is 28 since all 28 Sample II observations represent later failures. For the second entry 58 the score is 25 since 26 second sample failure times are greater than 58 and one is less (the remaining one being a tie which is scored as zero). The final entry of 362 receives a score −16 since all second sample failures have a lower value and the censored values are regarded as ties. The complete score U_n' is

$$U_n' = 28 + 25 + 24 + 22 + 22 + 22 + 20 + 4 + 4 - 16 = 155.$$

The percentage significance level is 0.40%, i.e. p=0.0040. The value of U_n' is confirmed by TESTIMATE, which uses a different but equivalent scoring system described in the manual, and also gives, as well as the exact permutation test level above, an asymptotic estimate of the significance level of 0.42%.

With these data the apparently higher level of censoring with asymptomatic cases looks to imply that censoring is condition-related. This is not necessarily so because survival times are generally longer for asymptotic cases and there may well have been clinical reasons to discontinue the follow-up among survivors at some stage slightly in excess of 300 weeks no matter whether these cases were symptomatic or asymptomatic. These data have been used here only to illustrate the procedure, and a statistician faced with the analysis of this type of clinical data should enquire if it is reasonable to assume censoring is not treatment or condition dependent before making that assumption. The Gehan–Wilcoxon test does not require strong assumptions about the form of the survival functions but when censoring is severe the mild assumptions about the life expectation of censored units may result in an appreciable loss of power relative to tests that give more weight to censored data. The reason for this is demonstrated by Prentice and Marek (1979) using heavily censored data. The source of the difficulty is that censored data contribute a large number of zeros to the calculation of U_n' and since large values of $|U_n'|$ correspond to small p the effect of censoring is often to dilute the value of U_n' that might be obtained if the censoring had not taken place.

If the distribution function for survival times, T, is $F(u)$ this implies that for an individual the probability of failure at or before time t is $\Pr(T \le t) = F(t)$. The opposite event is survival beyond time t, and the relevant probability is given by the survival function $H(t) = \Pr(T > t) = 1 - F(t)$. This is the proportion of the population that survives beyond time t. In analysing survival data $H(t)$ is often of more direct interest than $F(t)$. In survival studies $F(0-) = 0$, $F(t)$ is a monotone increasing function of t, and $F(\infty) = 1$, so it follows that for $t \geq 0$, $H(t)$ decreases monotonically from 1 to zero. Corresponding to any population with distribution function $F(t)$ there is a sample, or empirical distribution function, $S(u)$, which is a step function, this being the function that plays such a prominent role in bootstrapping. For a sample of n distinct ordered failure times $t_1, t_2, t_3, \ldots, t_n$ the sample distribution function has a step of height $1/n$ at each sample value and remains constant between these values. More formally,

$$S(t) = 0 \text{ if } t < t_1,$$
$$S(t) = 1/n \text{ if } t_1 \leq t < t_2,$$
$$S(t) = 2/n \text{ if } t_2 \leq t < t_3,$$
$$\ldots \ldots \ldots$$
$$S(t) = (n-1)/n \text{ if } t_{n-1} \leq t < t_n,$$
$$S(t) = 1 \text{ if } t \geq t_n.$$

The corresponding sample survival function is

$$R(t) = 1 - S(t)$$

and is clearly a step function that decreases from 1 to 0 with a step of height $1/n$ at each sample value, remaining constant between each such step. It represents the proportion of the sample surviving beyond time t. Because of the discontinuity at each observed t_r, it is sometimes convenient to represent $R(t)$ by its mean value at that step; more formally this implies that at t_r we set

$$R(t_r) = \tfrac{1}{2}\{[1 - (r-1)/n] + [1 - r/n]\} = (n - r + \tfrac{1}{2})/n. \qquad (4.8)$$

Allocating this score $R(t_r)$ to an observed t_r is a sensible scoring system in that it assigns to each observation an estimate of the probability of survival up to or beyond that time. It may look complicated, but any linear transformation of scores leads to a statistic equivalent to the sum of these scores. Thus, if we replace the score (4.8) by $W(r) = n + \tfrac{1}{2} - n.R(t_r)$ we find $W(r) = r$, the Wilcoxon rank-sum score. Thus, for samples of m, n with no tied and no censored observations, to test the hypothesis that they have identical survival distributions against the alternative of a location shift or dominance of survival times in one population a Wilcoxon rank sum test is appropriate. This is taking a sledgehammer to crack a nut. Because of the relationship between sample survival function and the sample cumulative distribution function (one is a linear function of the other) the original arguments used to justify validity of the WMW test for a location shift or dominance translates in a straightforward way to survival functions. If there are ties in the data the use of mid-ranks is appropriate, because the survival function has a step of height s/n if there are s tied values for any given t.

The above contorted exercise suggests a way to develop a modified test for censored data. When there are censored data, steps in $R(t)$ are confined to values of t at failure times, but when there is censoring between two consecutive failures the steps are no longer equal. In the Peto modification of the WMW test we also allocate scores to censored data, but in a different way to that for known failure times.

I describe the procedure algebraically, adding simple numerical illustrations at several stages. With censored data we estimate the proportion surviving, and hence the appropriate $R(t)$ at each failure by considering the proportion of the sample that **certainly** survives beyond that time. Several estimators have been suggested for this, but one commonly used is the Kaplan–Meier estimator, proposed by Kaplan and Meier (1958) in the context of life tables. I denote by $R(t-)$ and $R(t+)$ the values of $R(t)$ at times immediately before and after t. Since $R(t)$ remains constant between failures $R(t-)$ and $R(t+)$ are identical except at any time, t, corresponding to one or more failures. If there are exactly $s \geq 1$ failures at time t_r then, given $R(t_r-)$ the Kaplan–Meier rule for calculating $R(t_r+)$ is

$$R(t_r+) = R(t_r-)[1 - s/r(t_r)] \tag{4.9}$$

where $r(t_r)$ is the number of failures at or after time t_r **plus** the number of observations censored **at or after** that time. In other words it is the number of units at risk immediately prior to any failure(s) at time t_r.

Example 4.16 For the data

$$2 \quad 5 \quad 6^* \quad 7^* \quad 11 \quad 11 \quad 16 \quad 68^*$$

determine the values of $R(t)$ before and after each failure. Remember, an asterisk denotes a censored observation.

There are steps at 2, 5, 11, 16. At 2, $R(2-) = 1$, whence by (4.9), $R(2+) = 1(1 - 1/8) = 7/8$. This implies the usual single step of 1/8. Now $R(2+) = R(5-)$ whence $R(5+) = (7/8)(1 - 1/7) = 6/8$, again indicating a step of 1/8. Since steps only occur at exact failure times, $R(5+) = R(11-)$, whence $R(11+) = (6/8)(1-2/4) = 3/8$, since there are two failures at 11 and only 3 failures plus one censored observation then or later. The step at 11 is thus $6/8 - 3/8 = 3/8$. Similar at 16 we have $R(16+) = (3/8)(1 - 1/2) = 3/16$. All we can say about the behaviour of $R(t)$ for values of $t > 16$ is that at some unknown $t \geq 68$ when all censored observations have finally failed, then $R(t)$ falls to zero.

Remembering how we formed our scores in the no-tie, no-censoring, situation an obvious thing to do at the steps is again to take the mean of the relevant $R(t+)$ and $R(t-)$. All we know about a censored observation is that it will fail between when it is censored and infinity, when $R(t)$ is zero. Peto and Peto suggested that a reasonable score to assign to a censored observation was the mean of $R(t)$ at censoring and its final value zero. Thus for the data in Example 4.16 the value at the time of censoring of the observations 6*, 7* was $R(5+) = 6/8$, so the score for each is $\frac{1}{2}(6/8 + 0) = 3/8$.

Example 4.17 Using the scoring system described it is easy to check the relevant scores, i.e.

Observation	2	5	6*	7*	11	11	16	68*
Score	15/16	13/16	6/16	6/16	9/16	9/16	9/32	3/32

Denoting any score by s, the linear transformation $r = (17 - 16s)/2$ gives scores 1, 2, 5.5, 5.5, 4, 4, 6.25, 7.75. The sum of these scores is 36, the same as the rank sum for a combined sample of 8 in a WMW test, had there been no censoring.

Example 4.18. Applying the Peto–Wilcoxon test to the data and the hypotheses specified in Example 4.15, the relevant scores for the combined sample may be worked out in the way shown above. This is tedious and TESTIMATE calculates relevant scores directly, although it uses a linear transformation of those given by the method described here. It performs an exact test for significance directly, giving also the asymptotic result. The one-tail permutation test indicates significance at the 0.42% level and the asymptotic test at the 0.46% level, in close agreement with the results in Example 4.15. In many practical cases the two tests give similar results; as already mentioned, differences are most likely when the censoring is severe, when the Peto–Wilcoxon test tends to be more powerful. That situation does arise in practice. In the example on survival times given by Prentrice and Marek (1979) there were 281 dogs in the combined samples, 83 in one and 198 in the other and all but 29 observations were censored. Such situations are not uncommon in real investigations!

StatXact gives what is called a generalized Gehan–Wilcoxon test. This is similar to the Peto–Wilcoxon procedure but uses a slightly different scoring system that is essentially one suggested by Kalbfleish and Prentice (1980, p.147) but this is modified to take account of ties by averaging the scores that would be given if the ties had been arbitrarily split, a commonly used procedure in other contexts for ties, see e.g. the discussion of the van der Waerden test scores in section 4.5. Details are given in the StatXact manual. The StatXact test usually gives results close to those for the Peto–Wilcoxon test.

While the above modifications of the WMW test are valid for exact tests if censoring is not treatment dependent, an implicit assumption is that under H_0 the survival time distribution for censored observations is the same as that for uncensored ones. This may not be the case if censored observations have some common characteristic unrelated to treatment, but which is not shared with uncensored observations. For example, if many, or all, who drop out randomly have a history of drug or alcohol addiction their life expectancy may be different to that of those not so addicted. However, if the allocation to treatments of such patients is random and if under H_0 failure time distributions for such patients are the same for each treatment (though not necessarily the same as that for patients not in this sub-category) the analysis is still valid.

The following example looks to contradict the statement just made. Suppose that we could obtain exact failure times for censored units and these turn out to be rather different from those assumed in, say the Peto–Wilcoxon model, then the effect on significance level may be dramatic.

Example 4.19 Consider the following survival time data

Sample I	11	20	27	35	41			
Sample II	13	15*	17*	18*	32	47	56	61

In the appropriate one-tail Peto–Wilcoxon test for dominance or upward shift of mean for the population corresponding to the second sample, TESTIMATE indicates significance at the 4.97% level. If the censored units belong to a subgroup with poorer life expectancy than the other subjects suppose that further enquiries show their true failure times are 25, 26 and 29. Given this information the WMW test now gives a one-tail significance level of 21.76%!

Because all censored observations fall in one sample this need not imply that censoring is not independent of treatment, for with only three random drop-outs the probability is 0.25 that these will all be in the same sample if sample sizes are equal, and may even be greater otherwise. If the censored observations have a failure time distribution under H_0 which differs from that for uncensored observations, the Gehan-Wilcoxon or the Peto-Wilcoxon methods in fact protect us against making incorrect inferences when all observations in this category fall, as a result of randomization, in one sample and are then censored.

Example 4.19 indicates that without censoring we might lose a potentially significant result because of a particular randomization. When certain sub-groups have a potentially different distributions for a measured variate precision can sometimes best be increased by considering appropriate covariates, a matter I discuss in section 7.7. If censored observations form a sub-groups that might (with hindsight) have been measured by an appropriate covariate (e.g. alcohol consumption), the censoring may act as a surrogate for such a covariate.

*4.9 Improved tests for survival data

The family of exponential distributions is relevant to some survival data, but more sophisticated distributions such as the Weibull distribution also play a prominent role. Some writers suggest that the exponential distribution has an importance in this field akin to that of the normal distribution in classic measurement studies.

*This section gives a slightly more advanced treatment of a specialist topic and may be omitted at first reading.

A parametric test, based on the F-distribution [see e.g. Cox (1964)] exists for testing a hypothesis of equality of means of two populations exponentially distributed, but this is sensitive to departures from the exponential distribution. Earlier, Savage (1956) proposed a transformation of ranks analogous to the normal scores transformation, but designed to give a sample something like one from a one-parameter exponential distribution with unit mean. The motivation was that comparisons on this basis may be more rational for survival data than transformations to attain approximate normality. Cox (1964) points out that this transformation also may help to overcome sensitivity to departures from the exponential distribution which could seriously reduce power of relevant parametric tests when applied to the raw data.

Peto and Peto (1972) showed that a test based on a generalization of this transformation to allow for ties and censored observations in a certain way was a locally most powerful test of hypotheses about location shift or dominance for a particular class of survival functions. A corresponding but differently motivated approach by Mantel (1966) and others leads to equivalent tests based on what is now widely referred to as a **log-rank transformation** although the original scores are often referred to as Savage, or exponential, scores. Although now in almost universal use, the term log-rank transformation is misleading because the transformation is not to the logarithm of ranks, but rather to a scoring system based on an approximation to the logarithm of a certain data-based transformation involving a function of ranks.

If there is no censoring and there are no ties, expected values of the order statistics for a unit exponential distribution in the combined sample of all observations are used as scores (the Savage scores). For a combined sample of $N=m+n$ observations Savage scores take the form:

$$s_r' = 1/N + 1/(N-1) + \ldots + 1/(N-r+1) \text{ where } r = 1, 2, \ldots, N.$$

It is easily verified that the sum of all s_r' is N and an equivalent score is

$$s_r = 1 - s_r' \tag{4.10}$$

and these latter scores sum to zero. Thus, if we use the sum of these scores for either sample as our statistic then under H_0 the resulting statistics, S_m or S_n, each have expectation zero.

With censoring or tied data there are several basically equivalent ways of calculating an appropriate statistic but it is not always easy to see why they are equivalent. To give some indication of why two commonly used practical approaches are indeed equivalent I look more closely at the structure of Savage scores in the no-tie and no-censoring situations. To do so requires more algebra than I have used previously, but I illustrate the implications and use of various results by simple numerical examples with small data sets.

The scores in the form (4.10) provide a useful starting point to look at the structure of the statistic S_m, the sum of these scores for Sample I which consists of m observations in the two-independent sample case. If Sample II has n observations, the combined sample has $N = m + n$ observations. Because

$$s_i = 1 - \sum_{j=1}^{i}[1/(m + n - j + 1)],$$

then if S_m is the sum of all s_i corresponding to failures in Sample I it is easy to see that S_m may be written

$$S_m = m - \sum_{j=1}^{m+n}[m_j/(m + n - j + 1)] \tag{4.11}$$

where m_j is the number of Sample I failure times having rank j or higher, i.e. the number of survivors immediately prior to the failure at time $t_{(j)}$. In particular, $m_1 = m$ and $m_j = 0$ when j exceeds the highest ranked observation in Sample I. The denominator in the summation term in (4.11) is the number of survivors in **both** samples immediately prior to failure time $t_{(j)}$ so that if n_j is defined for Sample II analogously to m_j for Sample I it follows that the denominator term may be written $m_j + n_j$, whence (4.11) becomes

$$S_m = m - \sum_{j=1}^{m+n}[m_j/(m_j + n_j)] \tag{4.12}$$

A study of (4.12) shows that the terms in S_m may be arranged in two groups; the first consists of m terms each corresponding to a failure in Sample I and each taking the form

$$s_{1j} = 1 - m_j/(m_j + n_j) \tag{4.13}$$

and the second of n terms each corresponding to a failure in sample II and each taking the form

$$s_{2k} = - m_k/(m_k + n_k) \tag{4.14}$$

Clearly (4.12) is the sum of all s_{1j}, s_{2k}.

Example 4.20 For two samples

Sample I	27	33	39	
Sample II	34	36	41	53

the ranks are

Sample I	1	2	5	
Sample II	3	4	6	7

the Savage scores using the form (4.10) for sample I are

$$s_1 = 1 - 1/7, \quad s_2 = 1 - 1/7 - 1/6 \text{ and } s_5 = 1 - 1/7 - 1/6 - 1/5 - 1/4 - 1/3$$

and adding these gives $S_m = s_1 + s_2 + s_5 = 3 - 469/420 = 1.8833$.

To use (4.11) directly we do not need the ranks because we obtain from the original data at each failure time the number of survivors in each sample immediately before that failure. Thus immediately prior to the failure at $t = 36$ there is still 1 survivor in Sample I and 3 in sample II, giving a contribution $1/(1 + 3) = 1/4$ to the sum. Since $m = 3$, (4.11) may be written

$$S_m = 3 - 3/7 - 2/6 - 1/5 - 1/4 - 1/3 - 0 - 0 = 1.8833 \qquad (4.15)$$

as above. Similarly the terms in (4.13) and (4.14) are easily seen to be

$$s_{11} = 1 - 3/7, \quad s_{12} = 1 - 2/6, \quad s_{15} = 1 - 1/3, \quad s_{23} = -1/5, \quad s_{24} = -1/4, \quad s_{26} = s_{27} = 0.$$

Adding these gives 1.8833, as in (4.15).

The log-rank test is widely used because it is a locally most powerful test for certain hypotheses about what are called **proportional hazard** models relevant to survival functions that are Lehmann alternatives defined below. These include exponential and Weibull distributions. Here is a brief description.

Given a continuous distribution of survival times $F(u)$, the **hazard rate** $h(t)$ at time t is the limiting probability of failure in the small interval $(t, t + \delta t)$ conditional upon survival at least to time t. If $f(u)$ is the probability density function corresponding to the cdf $F(u)$ this conditional probability is

$$h(t) = f(t)/[1 - F(t)]$$

The denominator is the survival function $H(t)$. Two families with survival cdfs $F(u)$, $G(v)$ are **Lehmann alternatives** if their survival functions satisfy the relationship

$$1 - F(t) = [1 - G(t)]^\alpha \qquad (4.16)$$

for some $\alpha > 0$. Differentiating with respect to t gives

$$f(t) = \alpha[1 - G(t)]^{\alpha - 1} g(t),$$

and writing $j(t)$ for the hazard rate corresponding to $G(t)$ this implies

$$h(t) = \alpha j(t) \qquad (4.17)$$

showing that the ratio of the hazard rates is constant; hence the name *proportional hazards*. Clearly one-parameter exponential survival time distributions satisfy (4.16) and (4.17). If two such distributions have means λ_1 and λ_2, then $F(t) = 1 - \exp(-t/\lambda_1)$, $G(t) = 1 - \exp(-t/\lambda_2)$, whence $1 - F(t) = \exp(-t/\lambda_1)$ and also $1 - G(t) = \exp(-t/\lambda_2)$ so (4.16) holds with $\alpha = \lambda_2/\lambda_1$.

For a proportional hazards model we may want to test $H_0: \alpha = 1$ (implying identical survival distributions or equal hazard rates) against a one- or two-tail alternative specifying some other value of α. Testing leads naturally to the log-rank test for the no-tie and no censoring situation and we may extend it to deal with censoring and ties. Peto and Peto (1972) show this to be a locally most powerful test for proportional hazard alternatives.

For no censoring and no ties, if the samples from the two populations consist of m, n observations respectively then under $H_0: \alpha = 1$ the hazard rate for the identical populations at any given time are the same, so immediately before the failure at time $t_{(j)}$ (i.e. the jth failure) if the numbers at risk in each sample are m_j, n_j then the probability that the jth failure is in Sample I is $m_j/(m_j + n_j)$ and that it is in Sample II is $n_j/(m_j + n_j)$. I now associate a counter d_j with the jth failure and set $d_j = 1$ if that failure is in Sample I and $d_j = 0$ if it is in Sample II. The counter d_j is often called an indicator function. Clearly

$$\Pr(d_j = 1 | m_j, n_j) = m_j/(m_j + n_j)$$

and

$$\Pr(d_j = 0 | m_j, n_j) = n_j/(m_j + n_j)$$

and it follows that $e_j = E(d_j | m_j, n_j) = m_j/(m_j + n_j) = \Pr(d_j = 1 | m_j, n_j)$.

Now d_j (either 0 or 1) are the observed numbers of failures **in Sample I** at each of the observed failure times for the combined samples and e_j are the corresponding expected numbers. An intuitively reasonable test statistic is the sum of the differences between these observed and expected numbers, i.e.

$$S_m = \Sigma d_j - \Sigma e_j = m - \Sigma m_j/(m_j + n_j)$$

where the summation is over all $m + n$ failures. This is exactly of the form (4.12) and the terms for which $d_j = 1$ contribute components of the form (4.13) and those for which $d_j = 0$ contribute components of the form (4.14).

Modifications for censoring or ties are relatively straightforward. We calculate conditional expectations relevant to each observed failure, replacing m_j, n_j by m_j', n_j' where these are now the numbers still certainly at risk immediately before the jth failure, i.e. after excluding all earlier failures and any observations that have previously been censored. If there are c_j tied failures at the time of the jth failure the corresponding d_j is the number of those failures that are in Sample I and the corresponding $e_j = c_j m_j'/(m_j' + n_j')$. Because we only compute the components of S_m at actual failure times there are now fewer than $m + n$ components.

Example 4.21 Determine the log-rank statistic S_m for testing $H_0: \alpha = 1$ against alternative hypotheses specifying some other value of α in a proportional hazards model given

Sample I	2	4	8*	11		
Sample II	3	5	9	11	15*	18

Table 4.3 Computation of the log-rank statistic with censoring and ties

Failure time	m_j'	n_j'	c_j	d_j	e_j
2	4	6	1	1	0.4000
3	3	6	1	0	0.3333
4	3	5	1	1	0.3750
5	2	5	1	0	0.2857
11	1	3	2	1	0.5000
18	0	1	1	0	0.0000
	Totals			3	1.8940

Table 4.3 is a convenient way to set out the computations. Complete the first column from the data, then complete the remaining columns one row at a time. Check the computations carefully to be sure you follow the procedure. The censored observations do not appear directly in the table, but their presence is reflected by changes in m_j', n_j' and although there are two failures at $t = 11$ this only requires one table entry as the factor c_j takes account of the tie and enters into computation of the corresponding e_j. Clearly $S_m = 3 - 1.8940 = 1.106$.

TESTIMATE computes this statistic with the opposite sign; this sign difference could arise either by summing differences in the form *expected–observed* or by using a form corresponding to the one used here for calculating S_n because the statistics satisfy the identity $S_m + S_n = 0$. In an exact one-tail test $p = 0.1762$ implying 'significance' only at a 17.62% level!

The above method gives a useful algorithm for computing the log-rank sum statistic with censored and/or tied observations but it does not provide a score for each observation, irrespective of censoring. Peto and Peto (1972) derived log-rank scores for individuals using an estimator of the survival function proposed by Altshuler (1970). Such scores may be used in a program for a Fisher–Pitman test or a linear-by-linear association test. In my notation their method leads to the following scores:

- If the jth ordered observation is a failure, or if there are c_j tied failures at that time, each is given the score

$$s_j = 1 - \Sigma c_i/(m_i{+}n_i)$$

where the summation is over all failures at or before time $t_{(j)}$. When $c_i = 1$ there is a single failure. Earlier censored items are not included in this count.
- For all censored items between a failure at time $t_{(j)}$ and the next failure (if any) each is given the score

$$s_k = - \Sigma c_i/(m_i + n_i)$$

where the summation is over all failure times up to and including that at time $t_{(j)}$. Censored items prior to the first failure are scored as zero.

Table 4.4 Log-rank scores for the censored data in Example 4.22

Sample	Failure or censoring time	Score
1	2	$1 - 1/10$
2	3	$1 - 1/10 - 1/9$
1	4	$1 - 1/10 - 1/9 - 1/8$
2	5	$1 - 1/10 - 1/9 - 1/8 - 1/7$
1	8*	$-1/10 - 1/9 - 1/8 - 1/7$
2	9*	$-1/10 - 1/9 - 1/8 - 1/7$
1	11	$1 - 1/10 - 1/9 - 1/8 - 1/7 - 1/2$
2	11	$1 - 1/10 - 1/9 - 1/8 - 1/7 - 1/2$
2	15*	$-1/10 - 1/9 - 1/8 - 1/7 - 1/2$
2	18	$1 - 1/10 - 1/9 - 1/8 - 1/7 - 1/2 - 1$

Example 4.22 We obtain the scores s_j, s_k for the data in Example 4.21. These are set out in Table 4.4. From Table 4.4 we extract the sum of the scores for the Sample I units, giving

$$S_m = 1 - 1/10 + 1 - 1/10 - 1/9 - 1/8 - 1/10 - 1/9 - 1/8 -$$
$$1/7 + 1 - 1/10 - 1/9 - 1/8 - 1/7 - 1/2 = 3 - 4773/2520 = 1.106,$$

the same total as that in Example 4.21.

The scores in Example 4.22 may be used in (1.6) to give an asymptotic test statistic. In this case the mean is zero, since the sum of all scores is zero.

If StatXact is used for the log-rank test, slightly different scores are obtained because the program deals differently with ties, using a procedure that I recommended for allocating scores such as expected normal or van der Waerden scores, i.e. initially assuming the ties are separated by a small amount and then allocating to each tied value the mean of these separate scores. This has a carry-over effect to other scores but the test results are generally very close to those obtained using Altshuler scores or my earlier argument using the difference between observed and expected scores at each failure time.

Sometimes, especially with small data sets, it is difficult to decide which of two or more test procedures, e.g., WMW, van der Waerden scores or a log-rank test is optimal. Using one test when another is appropriate may result in a substantial loss of Pitman efficiency. Gastwirth (1985) proposed tests applicable in a wide range of circumstances where we may be reasonably confident that one of a finite set of models is appropriate without clear evidence about which is optimal. The tests he proposed are robust over the family of possible models in the sense that no other test has asymptotically a higher minimum efficiency relative to the optimum test for each model in the family. He calls such tests maximin efficiency robust tests (MERT). StatXact includes a program for calculating relevant scores for such tests in two independent sample problems. These scores may then be used in a general permutation test program, e.g. as data in a Fisher–Pitman test program. The theory invokes advanced ideas and the reader should consult Gastwirth's paper for a wide-ranging theoretical discussion and the StatXact 3 manual for technical details. In essence the tests are based on linear functions of the test statistics used in the competing optimal tests for the different models and the appropriate linear function takes account of the correlations between the different statistics or the scores used in them. The method provides protection against a poor choice of test and is particularly useful with censored data. MERT test procedures generally have a relatively high efficiency when compared to the optimum test that one would use if the true model were known.

In Example 4.6 I analysed the data on numbers of references in papers in Biometrics using the WMW test which would be appropriate for a shift in location with otherwise identical population distributions. However, because of the long tails there may be temptation to use a log-rank test. This turns out to be somewhat inappropriate because the optimal model for that test, the proportional hazard model, involves both location and scale changes, and the latter are not clearly indicated by the data. Indeed for that data the one-tail log-rank test gives $p = 0.0572$ compared to $p = 0.0065$ for the WMW test. A MERT combination for these two tests may be performed using StatXact and it gives $p = 0.0177$, closer to the WMW figure than to the log-rank result.

Software like StatXact or TESTIMATE let us carry out tests quickly with a wide range of scores so the safety-net feature of the MERT tests is perhaps less useful than it would be without those facilities. Tests applicable to alternative model situations are also discussed by Andersen *et al* (1982).

Exercises

4.1. For the data in Example 4.7 calculate the Wilcoxon rank sum for the first and second samples and verify the relationships given in the text between the statistics S_m, S_n, U_m, U_n. Check that $U_m=58$ directly in the way I computed U_n in the example.

4.2 For the data in Example 4.8 show that the value of the Wilcoxon rank sum for the first sample is 110

4.3 Obtain a confidence interval with confidence level at least 95% for the data in Example 4.1 making the best use of any available computer software and basing your work on the WMW test theory. How completely and satisfactorily you can do this will depend upon the software available to you. Obtain also an appropriate confidence interval based on the median test in section 4.6.. (It is feasible to obtain this latter interval without computer software for an exact test).

4.4 In Example 4.10 I quoted approximate 95% confidence intervals based on the WMW procedure for numbers of references for different years in the Journal of The Royal

Statistical Society, series A. The exact confidence level may not be quite the one aimed for, because at the end-points of the interval (the confidence limits) there is a sign change in one of the paired differences. Consider how you might check the exact levels by carrying out with appropriate software such as StatXact or TESTIMATE a test of zero median differences after adjusting values in one sample to align the medians in the two samples

4.5   Explore the modifications needed to the control median test [i.e. to (4.7)] if the median of Sample II rather than that of Sample I is used as the control.   Consider also the modifications to the statistic needed if one or more values in either sample is equal to that of the control median.

4.6 Check the computation of the value $p = 0.0316$ quoted in Example 4.14

4.7 In Example 4.12 a moment's reflection shows that if the patients treated with the new drug receive the four highest ranks, there is only one permutation (that observed) that contributes to p in the relevant one-tail test. Thus confirm the exact p-value quoted in Example 4.12.

*4.8   Analyse the data in Example 4.15 using the logrank test.

*4.9   In Example 4.10 I pointed out that the Fisher–Pitman test performed quite well despite the skewness of the data in the 1993 sample. Explore the behaviour of that test if the value 74 in that sample is transferred to the first sample. Compare its performance in this situation to that of the WMW test.

*4.10 Carter and Hubert (1985) gave data for percentage variation in blood sugar over one hour periods in each rabbit in two groups of nine given two different dose levels of a drug. Is there evidence of a median response difference between levels?   Obtain the Hodges–Lehmann estimate of the response difference and a confidence interval for this difference at an exact level as near as possible to 95%.

Dose level I	0.21	−16.20	−10.10	−8.67	−11.13	1.96	−10.99	−15.87	−12.81
Dose level II	1.59	2.66	−6.27	−2.32	−10.87	7.23	−3.76	3.02	15.01

*4.11 I give below the numbers of words with various numbers of letters in 200 word sample passages from the presidential addresses to the Royal Statistical Society by W. F. Bodmer (1985) and J. Durbin (1987). Is there acceptable evidence of a difference between the average lengths of words used by the two presidents?

Numbers of letters	1–3	4–6	7–9	10 or more
Bodmer	91	61	24	24
Durbin	87	49	32	32

*4.12 After Example 4.3 I suggested that a permutation test based on trimmed means in which only the top 2 observations in each sample were trimmed might be expected to perform quite well for the data used in that example. Use available statistical software to carry out Monte Carlo studies of the performance of a test using this basis for trimming. Carry out a similar study using Windsorization in which the two most extreme observations in each tail of each sample are put equal to the remaining relevant extreme observation.

*4.13 Peto *et al* (1977) give the following data for survival times in days of two groups of cancer patients under two different treatment regimes. An * represents a censored observation. Investigate the evidence for a difference in survival times. Do you consider bootstrapping may be a useful alternative to the methods developed in section 4.8 and 4.9?

Treatment A   8 852*   52 229 63   8 1976* 1296* 1460* 63 1328* 365*
Treatment B 180 632 2240* 195 76 70   13 1990*   18 700 210 1296 23

Do you think any of the data are outliers? If so, what action, if any, do you think should be taken prior to analysis?

*4.14 Burns (1984) gave the following data in a study of motion sickness incidence. Subjects were placed in a simulator, each of two groups being submitted to a different type of motion. The data are times in minutes at which each subject first vomited. The experiment was stopped after 120 minutes and subjects who had not vomited at that time are censored. Some subjects who requested that the experiment be stopped before they vomited are also censored and the time of censoring is indicated for these. Is there evidence that the survival curves differ? These data are analysed more fully by Altman (1991, Chapter 13) and are also given with a more detailed explanation of the experiment in Hand *et al* (1994, Set 18).

Motion A   30 50   50* 51 66* 82   92 and 14 subjects censored at 120
Motion B   5   6* 11   11 13   24 63   65 69 69 79 82 82   102 115 and 13
          subjects censored at 120

*4.15 Pocock (1983, Table 14.6) and Hand *et al* (1994, Set 423) give data for survival times in months for patients suffering from a form of hepatitis for a control group and for patients treated with prednisolone. Patients marked * were still alive at the stated time and no further information is available about them. Is there evidence of a difference in the survival patterns?

Control   2 3 4 7 10 22 28 29 32 37 40 41 54 61 63 71 127*   140* 146* 158*
          167* 182*
Treated   2 6 12 54 56* 68 89 96 96 125* 128* 131* 140* 141* 143 145*
          146 148* 162* 168 173* 181*

*4.16 While the log-rank test is optimal for the proportional hazards model it is generally not optimal for detecting other kinds of differences between survival curves. Stablein and Koutrouvelis (1985) give the following data for a gastrointestinal tumor study in which some patients received chemotherapy only and others received both chemotherapy and radiotherapy. For a data set of this size an asymptotic test should give similar results to an exact test. An inspection of the data suggests that rather than a proportional hazard model a more appropriate model might allow for a cross-over of the survival curves for it appears that survival times are shorter for the combined treatments among survivors who die relatively early, but that survival times are longer among those who are relatively long-term survivors. Stablein and Koutrouvelis propose a modified test that is sensitive to such crossing hazards. After trying the asymptotic log-rank tests you may want to study the proposals in their paper and carry out their proposed test.

Chemotherapy 1 63 105 129 182 216 250 262 301 301 342 354 356 358 380 383
          383 388 394 408 460 489 499 523 524 535 562 569 675 676 748 778 786
          797 955 968 1000 1245 1271 1420 1551 1694 2363 2754* 2950*
Chemotherapy + radiotherapy 17 42 44 48 60 72 74 95 103 108 122 144 167 170
          183 185 193 195 197 208 234 235 254 307 315 401 445 464 484 528 542
          567 577 580 795 855 1366 1577 2060 2412* 2486* 2796* 2802* 2934* 2988*

5

Location tests for single and paired samples

5.1 The single sample location problem

Statisticians seldom meet genuine one-sample problems other than as useful textbook devices to illustrate basic ideas about inference. For real single-sample data, tests or estimation concerning location are usually straightforward and only exist in a meaningful way if the data are a random sample from a larger population or when this is not so, if one accepts the existence of such a population and it is reasonable to regard the observations as having the properties of such a sample. The comparative experiment situation described in section 1.6 where units in a strictly finite population (e.g. all the units currently available) are allocated randomly to distinct treatments no longer has any relevance. However, I show in section 5.6 that if a different but appropriate randomization procedure is used to allocate two treatments to units in a comparative experiment, then the investigation of certain location hypotheses about differences between these treatments reduces to a single-sample problem and this is perhaps the most important reason for studying the latter.

5.2 The single sample permutation test

For a sample of m observations from a normal population with unknown mean and variance the one-sample t-test is a uniformly most powerful test for hypotheses about the population mean, μ, and confidence limits for an unknown μ may be based upon this test. Pitman (1937a) proposed a permutation test that provides exact p-values when the sample comes from any **symmetric** continuous distribution. Like the Fisher–Pitman two-sample test introduced in section 1.6 it has Pitman efficiency 1 when the sample is from a normal population, but also, like the t-test, it is not always robust against departures from symmetry. If θ_0 is the hypothesized population median (for a symmetric distribution this is equal to the mean if the distribution has a mean but a few symmetric distributions, e.g. the Cauchy distribution, have no mean), subtracting θ_0 from each sample value reduces the problem to making inferences about a parameter $\theta = 0$. The rationale of the test (which I call the Fisher–Pitman one-sample test) is easily demonstrated for testing the hypothesis $H_0:\theta = 0$ against one- or two-sided alternatives.

For samples from a symmetric distribution with median (or mean) zero, any sample value x_i, with magnitude $|x_i|$ is equally likely to be at a distance x_i **above** or **below** zero, i.e. to have the corresponding positive or negative value. Thus if H_0 is true, one expects a fairly even mixture of positive and negative values in a sample of size m.

A preponderance of positive values among the x_i suggests a population median greater than zero, and a preponderance of negative values a median less than zero. The reference set for the permutation test is all 2^m allocations of signs $+$ and $-$ to the m observations. When testing H_0 against a one-sided alternative $H_1:\theta > 0$ a preponderance of positive signed values favours H_1; for an alternative $H_1:\theta < 0$ a preponderance of negative signed values favours H_1; for $H_1:\theta \neq 0$ a preponderance of either positive or negative signed values favours H_1. Using arguments like those in section 1.6 for choosing a statistic in the two independent sample Fisher–Pitman test, it is easily seen that appropriate (and equivalent statistics) for testing are S, the sum of all sample values, or S_+ the sum of all sample values with a positive sign, or S_- the sum of the samples values with a negative sign. In Exercise 5.1 I ask you to verify that the usual one-sample t-statistic may also be used in the permutation test, but then the standard t-tables are not valid for testing. For a sample of m, on permuting the signs, 2^m algebraically distinct sums are obtained, although some may be numerically equal. I illustrate the procedure for a trivial numerical example.

Example 5.1. A sample of 4 from a symmetric distribution takes the values

$$-2.1, \quad 0.3, \quad 0.5, \quad 0.8.$$

Determine the exact p-value for testing the hypothesis $H_0:\theta = 0$ against $H_1:\theta > 0$. Table 5.1 gives relevant calculations in a table. The first row gives the values of each $|x_i|$. Each remaining row gives one of the $2^4 = 16$ sign combinations assignable to the x_i and the remaining columns give the values of the statistics S, S_+ and S_-. It is easily verified that $S = S_+ - S_-$.

Using arguments similar to those in section 1.6 for the two independent sample Fisher–Pitman permutation test, clearly in this example the minimum possible p-value for a one-tail test is 1/16 and for a two-tail test is 2/16. More generally, for samples of m these values are $1/2^m$ and $1/2^{m-1}$. When $n = 4$, conditional upon the given $|x_i|$ we would only reject H_0 at the $(1/16) \times 100 = 6.25\%$ significance level in the relevant one-tail test if all the data had positive signs. The appropriate exact one-tail p-value associated with our observed data is $p = 9/16$.

Two-sided (or one-sided) confidence intervals corresponding to permissible values of p are obtainable in a manner reminiscent of that used in section 1.8 for

Table 5.1 Permutation distribution for a given sample of 4

0.3	0.5	0.8	2.1	S	S_+	S_-
+	+	+	+	3.7	3.7	0.0
−	+	+	+	3.1	3.4	0.3
+	−	+	+	2.7	3.2	0.5
+	+	−	+	2.1	2.9	0.8
−	−	+	+	2.1	2.9	0.8
−	+	−	+	1.5	2.6	1.1
+	−	−	+	1.1	2.4	1.3
−	−	−	+	0.5	2.1	1.6
* +	+	+	−	−0.5	1.6	2.1
−	+	+	−	−1.1	1.3	2.4
+	−	+	−	−1.5	1.1	2.6
−	−	+	−	−2.1	0.8	2.9
+	+	−	−	−2.1	0.8	2.9
−	+	−	−	−2.7	0.5	3.2
+	−	−	−	−3.1	0.3	3.4
−	−	−	−	−3.7	0.0	3.7

*This sign configuration corresponds to that of the data.

the two-sample independent Fisher–Pitman test. Given a sample of m observations $x_1, x_2, \ldots, x_m$ we consider the set of derived variables $y_i = x_i - d$ as d runs from $-\infty$ to ∞. As in section 1.8 we are interested in the values of d at which the statistic S (or S_+ or S_-) changes its position among the 2^m permutations. Clearly when $d = -\infty$ or indeed lies between $-\infty$ and the least x_i, all y_i will be positive and S for that set of y_i will take its maximum value among all permutations . However as d increases through the least x_i a negative value enters the sum and the order of the sum among all possible S is decreased. Pursuing this argument it is easy to see that the ordering among permutations changes whenever d passes through the mean of **any** subset of r of the x values where $1 \leq r \leq m$. In Exercise 5.2 I ask you to verify that this results in a total of 2^m-1 changes in S, the number required to change S from the greatest value to the least value among those for all permutations.

Example 5.2. Obtain a two-sided confidence interval with confidence level at least 90% for the median of the these data:

$$1.3, \quad 1.8, \quad 2.0, \quad 2.7, \quad 3.4, \quad 3.9$$

There are $2^6 = 64$ permutations in the relevant reference set and if we include the three largest and the three smallest sums in the critical region when testing a hypothesis about a given θ we would reject H_0 in a two tail test at the $(6/64) \times 100 = 9.375\%$ significance level. By arguments similar to those used in Example 1.6 we find a relevant 90.625% confidence interval is that between the third smallest and the third largest change points among those for all sub-sets of the given data   Clearly the three largest change points occur when d equals

$$3.9 \text{ and } (3.4 + 3.9)/2 = 3.65 \text{ and } 3.4$$

and the three smallest when d equals

$$1.3 \text{ and } (1.3 + 1.8)/2 = 1.55 \text{ and } (1.3 + 2.0)/2 = 1.65$$

whence the required interval is $(1.65, 3.4)$.

The number of permutations in the reference set grows rapidly as m increases and in real situations good software is needed for tests or to form confidence intervals. StatXact and TESTIMATE both provide programs for one- or two-tail hypothesis tests; exact tests are available for small or moderate sample sizes. StatXact also provides a Monte Carlo program that gives good approximations to the exact p-value (together with a confidence interval for the true p) and both packages give an asymptotic approximation. Currently I know of no published program for determining a confidence interval in this case although one may use trial and error methods in repeated hypothesis tests subtracting various values of d from given data (see Example 5.3 below). Especially for small data sets there may be some doubt about the 'status' of the end-points of a confidence interval as these values correspond to values of S that may be tied. One check is to test the hypothesis that the median takes these precise values. If this is done in Example 5.2 it is found that the one-tail p-values exceed 0.05. However if any values outside the end-points are hypothesized as median the associated one-tail p is less than or equal to 3/64. If values just outside are taken, e.g. 1.649 and 3.041, in each case $p=3/64$ exactly.

In practice bootstrap confidence interval described briefly in section 2.6 and in more detail by Efron and Tibshirani (1993, Chapters 12–14) may be easier to compute using standard statistical software.

Since, when testing for a zero median (or mean, if it exists), assuming symmetry, for all x_i, $E(x_i) = 0$ and $\text{var}(x_i) = x_i^2$ and since the x_i are independent, it follows that $S = \sum x_i$ has mean zero and variance $\sum x_i^2$. For large m, S is distributed asymptotically $N(0, \sum x_i^2)$ under H_0. In Exercise 5.3 I ask you to show that S_+ and S_- each have mean $\frac{1}{2}\sum|x_i|$ and variance $\frac{1}{4}\sum x_i^2$. Asymptotically each is distributed $N(\frac{1}{2}\sum|x_i|, \frac{1}{4}\sum x_i^2)$. This is the basis for asymptotic tests. The statistic for testing S is

$$Z = S/(\sqrt{\sum x_i^2}). \tag{5.1}$$

and since the denominator of (5.1) is invariant under permutation of the signs of the x_i it follows that Z may be used as an alternative test statistic in an exact test, but it is only asymptotically that Z has a standard normal distribution under H_0.

Statistics that have zero mean and unit variance are often referred to as *standardized* statistics even when they are not normally distributed.

Because, like the parametric *t*-test, the Fisher–Pitman permutation test is not robust against departures from symmetry it is not widely used. Indeed, it and the *t*-test often give broadly similar results not only when each is valid, but even when each is used on inappropriate data. The problem is overcome by methods similar to those developed in Chapter 4 for the two independent sample situation using rank-based methods or by bootstrap sampling using the median or by using trimmed or Windsorized means.

Example 5.3. Compare the parametric *t*-test and the one sample Fisher–Pitman test for the data in Example 4.1 for a sample of numbers of references in papers in *Biometrics* in 1956–7. Test first the hypothesis that the mean/median number of references is 24 and then obtain 95% confidence intervals for the median. The data are:

$$2 \quad 3 \quad 6 \quad 6 \quad 6 \quad 9 \quad 9 \quad 10 \quad 14 \quad 15 \quad 16 \quad 72$$

Clearly the observation 72 indicates a lack of symmetry (or an outlier if one were prepared to assume symmetry apart from any such value).

Any standard program for the *t*-test indicates that a hypothesized mean of 24 is acceptable in a two-tail test at any significance level 5% or less. Indeed, the computed *t*-statistic is $t = 1.84$ and with 11 degrees of freedom in a two-tail test this implies $p = 0.093$. The 95% confidence interval for the mean is the unconvincing (2.05, 25.95). The unrealistically high upper limit is a consequence of the extreme observation 72. StatXact and TESTIMATE provide programs for an exact (Fisher–Pitman) permutation test of the hypothesis $H_0: \theta = 24$ although some care is needed in entering the data correctly as the programs are designed basically for dealing with the paired comparison situation described in section 5.6 below. The procedure in StatXact is to feed the above data after subtracting 24 from all observations, i.e. $-22, -21, -18, -18, -18, -15, -15, -14, -10, -9, -8, 48$, into the appropriate permutation test programme as a *difference variate*. For a two tail-test the program gives $p = 0.0928$ for the exact permutation test, almost identical to that for the *t*-test. The asymptotic approximation gives a two-tail $p = 0.0926$, again close to the exact value. Approximate confidence limits may be obtained from a modification of (5.1) using the asymptotic result. While (5.1) is relevant to testing the hypothesis $\theta = 0$ it is easy to see that the modified statistic for testing $H_0: \theta = \theta_0$, given a sample of *m* observations may be written

$$Z = (S - m\theta_0)/(\sqrt{\textstyle\sum x_i^2}).$$

For an approximate 95% interval one sets $Z = \pm 1.96$ and for this example $m = 12$, $S = 168$ and $\sum x_i^2 = 6244$, and on substituting one finds the confidence limits 1.09 and 26.91, close to those based on the *t*-distribution. Programs like StatXact make feasible a trial and error process for getting exact limits such that the associated one-sided *p* value is close to, but not greater than 0.025in each tail. The *t*-distribution limits or the above asymptotic limits provide a sensible starting point. With some half dozen sweeps making commonsense adjustments based on linear interpolation for these data, I obtained an approximate 95% two-tail confidence interval (5.9, 26.1), again not very different from the *t*-based interval. Because of discontinuities in possible *p*-values this is not an exact 95% interval but in fact is approximately a 95.9% interval.

As in the two sample case, one might hope quantile-based confidence intervals using the bootstrap median would be robust against the outlier, while if one used the bootstrap mean the results might be like those given by the Fisher–Pitman approach, because the outlier would still be influential.

Example 5.4   For the data in Example 5.3 I computed $B=1000$ bootstrap samples and for each I computed the bootstrap mean and the bootstrap median. The quantile-based 95% confidence intervals were (6.67, 25.5) for the mean and (6, 14.5) for the median. The former accords with expectations, being marginally shorter than the exact permutation test interval. The difference may be attributed to the tendency of the uncorrected bootstrap quantile-based interval to undercover. The median interval (6, 14.5) reflects the limited influence of the outlier on the median estimator in line with our finding in section 3.4 and 3.5.

With such heavily tied data unease may be felt because this results in the bootstrap estimates of the median themselves being heavily tied. Indeed for several sets of 1000 bootstrap samples I found virtually no difference between the quantile based 90%interval and the corresponding 95% interval. This problem may be eased by less severe trimming . Trimming by removing the two most extreme observations in each tail gave a bootstrap confidence interval based on $B = 1000$ bootstrap samples (6.125, 19.125). This reflect the fact that, unlike the median, where the outlier has little influence, for the trimmed mean it still has an appreciable influence on the top quantile through those bootstrap samples that include the value 72 three or more times when consequently that value is included in calculation of the trimmed mean.

5.3 The signed-rank test

Transformation to ranks should introduce robustness by reducing the influence of outliers with the added advantage of providing an unconditional test because ranks depend only on the order of the data and not the data values. A relevant test was proposed by Wilcoxon (1945) and is known as **Wilcoxon's signed rank test**. To test $H_0 : \theta = \theta_0$ against a one- or two-sided alternative one first subtracts θ_0 from all sample values to form a new set of m values $y_i = x_i - \theta_0$.

The y_i are ranked in increasing order of unsigned magnitudes, $|y_i|$, and each rank is then given the sign of the corresponding y_i. Ties are assigned mid-ranks. An analogue of the Fisher–Pitman permutation test is now performed on the signed ranks. The sum of all signed ranks is a possible test statistic but the sum of the positive or negative ranks is often used in published tables or in computer programmes. Be careful to check which statistic is used. StatXact or TEST-IMATE both provide programs for the Wilcoxon test which give exact probabilities for small or moderate sample sizes and an asymptotic result for larger samples. StatXact also provides a programme for a Monte Carlo approximation applicable to any sample size. Transformation to ranks is performed automatically by these programs. However, any program for the one (or matched pairs) Fisher–Pitman permutation test may be used providing the signed ranks are entered as data.

Example 5.5 For the data in Example 5.3, test the hypothesis $H_0:\theta = 24$ against $H_1:\theta \neq 24$ using the Wilcoxon signed-rank test.
   Subtracting 24 from each datum gives

$$-22, \ -21, \ -18, \ -18, \ -18, \ -15, \ -15, \ -14, \ -10, \ -9, \ -8, \ 48.$$

The signed ranks using the usual rule for midranks are easily obtained as

$$-11, \ -10, \ -8, \ -8, \ -8, \ -5.5, \ -5.5, \ -4, \ -3, \ -2, \ -1, \ 12.$$

   The extreme value 72 is ranked 12 and this rank would be allocated to *any* count of numbers of references exceeding, in this case, $24 + 22 = 46$. For a two-tail test StatXact gives an exact $p = 0.0327$, appreciably lower than that given by the Fisher–Pitman test for the original data. I show in Example 5.7 that if ranks are used there is also a dramatic difference in confidence intervals compared to those obtained by permuting the raw data.

   A complication occurs if one or more sample values equals the θ postulated under H_0, because any corresponding y_i are zero. Omitting such observations and performing the test on a reduced sample causes loss of power. A widely preferred recommendation is to rank all observations and temporarily giving a zero a rank of 1 or any tied zeros the corresponding lowest mid-rank. After ranks are assigned those corresponding to the zero y_i are replaced by zeros, the others being unaltered. Programmes such as StatXact or TESTIMATE allow either approach.
   Asymptotic results for the Wilcoxon signed-rank test take different but equivalent forms depending on whether the statistic used is the sum of all signed ranks, the sum of the positive signed ranks or the lesser of the sums of the positive or negative signed ranks. If there are no ties and if S is the sum of all signed ranks then under H_0 the mean $E(S) = 0$ and $Var(S) = m(m + 1)(2m + 1)/6$. The latter result follows by replacing the data by ranks in the formula for the variance, $\sum x_i^2$, of the analogous Fisher–Pitman test statistic (5.1). For large n

$$Z = \frac{S}{\sqrt{[m(m + 1)(2m + 1)/6]}} \tag{5.2}$$

has approximately a standard normal distribution. Since S is discontinuous and takes values that increase in steps of two the approximation is improved by using a unit continuity correction and replace S by $S' = S + 1$ when computing the lower tail p-value.
   If there are ties, again $E(S) = 0$, but $Var(S)$ is altered. Some textbooks give specific formulae for $Var(S)$ for particular patterns of ties, but it is best computed by the formula $Var(S) = \sum s_i^2$, where the summation is over all m signed ranks and

tied ranks, s_i. This is simply a special case of the variance result for the Fisher–Pitman permutation test where signed ranks and tied ranks replace the original data, and indeed the principle carries over to any 'scores' used as transformations of the original data.

If S_+ or S_- are used as the test statistic, if we denote either by S^*; then $E(S^*) = \frac{1}{4}m(m + 1)$ and $\mathrm{Var}(S^*) = m(m + 1)(2m + 1)/24$ if there are no ties, whence asymptotically

$$Z = \frac{S^* - \frac{1}{4}m(m + 1)}{\sqrt{[m(m + 1)(2m + 1)/24]}} \qquad (5.3)$$

has approximately a standard normal distribution. Since S^* is discontinuous but increases by steps of 1, it is usual to add a continuity correct of $\frac{1}{2}$ to S^* when computing a lower-tail p-value, when (5.3) becomes:

$$Z = \frac{S^* + \frac{1}{2} - \frac{1}{4}m(m + 1)}{\sqrt{[m(m + 1)(2m + 1)/24]}} \qquad (5.4)$$

For an upper tail p-value [i.e. when $S^* > \frac{1}{4}m(m + 1)$] the continuity correction is subtracted. With ties denote, as above, the ranks and tied ranks by s_i; then the numerator is unaltered but the denominator is replaced by $\sqrt{(\frac{1}{4}\sum s_i^2)}$. With ties it is difficult to justify a continuity correction on theoretical grounds, but in practice using one often gives a better approximation.

The procedure for confidence intervals for a location parameter using Wilcoxon signed ranks is reminiscent of that for WMW confidence intervals for a location shift between two independent samples.

The basic theory assumes no ties; however, for ties, similar methods may be used but without an appropriate computer program it is difficult to assign confidence limits. Version 3.0 of StatXact provides a program to give confidence intervals having at least a specified confidence level (e.g. 95%) although in view of discontinuities in p-values the exact confidence level may be greater.

I outline the theory for the case of no tied ranks. Without loss of generality I assume that the sample of m values $x_1, x_2, \ldots, x_m$ are arranged in ascending order. If the location parameter $\theta = \theta_0$ then all x_i less than θ_0 give a deviation $y_i = x_i - \theta_0$ with a negative signed rank. Among all such negative ranks that associated with x_1, will have the greatest magnitude, that associated x_2 the next greatest magnitude and so on. Similarly, all x_j greater than θ_0 will have positive signed ranks for the deviation y_i and that for x_m will have the highest rank, x_{m-1}, the next highest and so on. Consider now all $\frac{1}{2}m(m + 1)$ paired averages of the form $\frac{1}{2}(x_i + x_j)$, $i=1$, $2, \ldots, m; j=i, i+1, \ldots, m$. Suppose x_1 gives rise to a deviation $y_1 = x_1 - \theta_0$ with signed rank $-p$, where p is positive, then because of the ascending order of observations each of the averages $\frac{1}{2}(x_1 + x_q)$ will be less than θ_0 providing the deviation $x_q - \theta_0$ has either a negative rank or a positive rank less than p. When there are no ties among the ranks no deviation has the rank $+p$, since p is already assigned to x_1. If x_q is the smallest observation whose deviation has a positive rank exceeding p, then $|x_q - \theta_0| > |x_1 - \theta_0|$, whence it follows that $\frac{1}{2}(x_1 + x_q)$ is greater than θ_0. Also, all $\frac{1}{2}(x_1 + x_r)$ are greater than θ_0 for all $r > q$. Thus the number of averages involving x_1 which are less than θ_0 is equal to the (negative) rank associated with x_1. Similarly, if we form all the paired averages of any x_i less than θ_0 with all x_j where $j \geq i$ the number of these averages less than θ_0 will equal the (negative) rank associated with x_i. Clearly, for any x_i with a positive rank none of the averages with itself or a greater x_j will be less than θ_0. The complete set of $\frac{1}{2}m(m + 1)$ averages defined above are often called the **Walsh averages**, having been proposed by Walsh (1949a, b). Clearly the number of Walsh averages less than θ_0 is S_-, and the number greater than θ_0 is S_+. Thus the Walsh averages enable us both to calculate the statistic S_+ or S_- or if we know the values of these for significance at appropriate specified levels, to determine corresponding relevant one or two-tail confidence limits.

Ties in the original data or among the $x_i - \theta$ for any θ values of interest affect the values of the S_+ or S_- required for significance; values given in tables for the no-tie case are no longer appropriate, but programs such as StatXact now provide the required limits. A point estimator of the parameter θ commonly used is the median of the Walsh averages, proposed by Hodges and Lehmann (1963), and referred to as the **Hodges-Lehmann** estimator.

If no program is available to compute them the Walsh averages may be calculated and set out in a triangular matrix in the way indicated in Example 5.6.

Example 5.6   For a sample of six observations there are (Example 5.2) $2^6 = 64$ possible allocations of signs, so if we form a critical region from the three smallest and three largest values of any of the statistics S, S_+ or S_- we reject H_0 for any result falling in that region in a two tail test at the $(6/64) \times 100 = 9.375\%$ significance level. The interval covering all hypotheses acceptable at this level is a $100 - 9.375 = 90.625\%$ confidence interval. The maximum value of, say, S_+ indicating significance at this level is easily verified to be 2, so the confidence limit is easily determined from the Walsh averages.

Suppose the sample values are

$$1.7 \quad 2.2 \quad 2.4 \quad 2.9 \quad 3.8 \quad 4.5$$

We form the triangular matrix of Walsh averages in Table 5.2. The leading row and column both contain the data in ascending order. The entries in the body of the table are the Walsh averages of the observation at the top of that column and at the left of that row.

Like the paired difference matrix used for confidence intervals for two-sample location differences in Table 4.1, the entries in Table 5.2 have a pattern. The smallest entries are near the top left, the largest at the bottom right. To form the 90.625% two-tail confidence interval

Table 5.2   Walsh averages for data in Example 5.6

	1.7	2.2	2.4	2.9	3.8	4.5
1.7	1.7	1.95	2.05	2.3	2.75	3.1
2.2		2.2	2.3	2.55	3.0	3.35
2.4			2.4	2.65	3.1	3.45
2.9				2.9	3.35	3.7
3.8					3.8	4.15
4.5						4.5

we reject those Walsh averages, namely the 2 smallest and 2 largest, that would indicate S_+ or S_- equal to 2 or less. The relevant limits are given by the third largest and third smallest Walsh average. Inspecting the table the relevant interval is clearly (2.05, 3.8), because if $\theta \leq 2.05$ or $\theta \geq 3.8$ we reject H_0 at the 9.375% significance level. The Hodges–Lehmann point estimator is the median of the Walsh averages and you should verify from Table 5.2 that this is 2.9.

All Walsh averages are computed in Table 5.2 so as to get the Hodges-Lehmann point estimator, but if only confidence limits are wanted, only the three largest and three smallest are needed for this small sample, and the pattern makes it easy to see which these are without complete enumeration.

For tied data, except perhaps for only a few ties, unless the sample is large enough to allow the use of asymptotic results, it is virtually essential to use a program to obtain confidence intervals. StatXact 3 allows this.

Example 5.7   Obtain a two-tail confidence interval with level at least 95% for the data on numbers of references in Example 5.3. The data are

$$2 \quad 3 \quad 6 \quad 6 \quad 6 \quad 9 \quad 9 \quad 10 \quad 14 \quad 15 \quad 16 \quad 72$$

StatXact gives the Hodges–Lehmann estimator of the median to be 9.5 with an interval (6, 15) at a nominal 95% confidence level. With tied data such as these there may be dramatic discontinuities in the exact confidence level. With programs such as StatXact it is easy to verify that we would accept $\theta = 6$ or $\theta = 15$ in a two-tail test at a level about 6.6% in each case, whereas we would reject at almost exactly the 5% level a value of $\theta = 15.001$ and at a level 2.5% a value of $\theta = 5.999$.

The confidence interval (6, 15) for the population median obtained from these data has a much stronger intuitive appeal than the corresponding interval based on the t-test or the Fisher–Pitman procedure obtained in Example 5.3, which were respectively (2.05, 25.95) and (5.9, 26.1).

However, remember that the Wilcoxon signed-rank test presupposes a sample from a symmetric distribution and clearly the value 72 is in that sense an outlier. In section 5.5 I describe a test appropriate to skew distributions. However, as already indicated, the Wilcoxon signed rank test is often robust against a small proportion of outliers. The confidence interval (6, 15) is in close agreement with that obtained in Example 5.4 by bootstrapping the median, i.e. (6, 14.5).

The Hodges–Lehmann estimator used above is another example of an *R*-estimator, a type of robust estimator mentioned in section 3.5.

Asymptotic theory for the Wilcoxon signed rank statistic may be used to determine approximate confidence intervals for large samples. The result is sometimes surprisingly good even for relatively small samples, although caution is needed with ties. For the no-tie situation we may use (5.3) with a continuity correction of $+\frac{1}{2}$ added to the numerator to determine a value of S_- that would just give significance at, say, a 5% level for a particular m. This requires a critical lower-tail value of $Z = -1.96$. Putting this value, together with the continuity correction in (5.3) with $m = 12$ gives $-1.96 = (S + 0.5 - 39)/\{\surd(12\times13\times25/6)\}$, whence $S = 13.51$. Because S can take only integral values we round this down to $S = 13$ to ensure a two-tail approximate *p*-value less than 0.05. Although this is only an asymptotic approximation and ignores possible ties it is interesting to show how this result may be used for data like that in Example 5.7 to obtain an approximate 95% confidence interval. A procedure along these lines is provided in some general computer packages.

Example 5.8 Use the asymptotic result above to determine an approximate 95% confidence interval for the population median based on the reference number data in Example 5.7. We have just shown that the approximate sum of negative differences had critical value $S=13$ for samples of 12 for significance at the 5% level in a two sample test. This implies that to determine an approximate 95% confidence interval we reject the 13 smallest and 13 largest Walsh averages. It is not difficult to compute these using the triangular matrix format given in Table 5.2, or some packages such as MINITAB compute them directly. You should verify that Table 5.3 gives all the relevant averages. It is easily deduced from that table that the relevant interval is (6, 15.5) close to the exact-probability-based interval (6, 15) and much superior to that based on the Fisher–Pitman method which is here inappropriate because of the strong influence of the outlier. However, with heavy tying the asymptotic result should be used with caution and avoided if a program for exact intervals is available, or else an adjustment in the denominator of (5.3) for ties might be made.

5.4 Transformation of ranks

Transformation to van der Waerden or normal scores as in the two independent sample case needs modification for single sample location tests, and with a symmetry requirement for the population, transformation to Savage scores is not appropriate since this transformation is designed for use with skewed data.

A problem with transformation to van der Waerden or expected normal scores is that these are symmetrically distributed about zero and are in effect already signed, so that there is no meaningful way of allocating an equivalent to a 'signed' rank. A transformation used to overcome this problem is based on what

Table 5.3 Walsh averages for asymptotic 95% confidence interval for number of references

	2	3	6	6	6	9	9	10	14	15	16	72
2	2	2.5	4	4	4	5.5	5.5	6				37
3		3	4.5	4.5	4.5	6	6					37.5
6			6	6	6							39
6												39
6												39
9												40.5
9												40.5
10												41
14												43
15											15.5	43.5
16											16	44
72												72

in effect are certain positive quantiles of the standard normal distribution that lie above the median and are therefore positive. The recommended quantiles corresponding to the ranks 1, 2, 3, . . ., m are respectively the $(m + 2)/(2m + 2)$th, $(m + 3)/(2m + 2)$th, $(m + 4)/(2m + 2)$th, . . ., $(2m + 1)/(2m + 2)$th quantiles. Signs corresponding to the Wilcoxon signed ranks are allocated to these quantiles. The resulting scores may be used in a permutation test program in a Fisher–Pitman type procedure with these scores in place of the original data. The fact that these quantiles are not symmetrically distributed about their median detracts from the intuitive appeal of the transformation, but my experience with a few examples is that it generally gives results very similar to a Wilcoxon signed rank test. Maritz (1995, Example 2.15) gives an example using these scores or closely related scores based on expected values and proceeds to discuss hypothesis testing and confidence intervals using these and even more general scores. An *ad hoc* approach using normal scores (but not standard normal scores) that preserves symmetry about the median is given in Sprent (1993, section 3.2.1)

5.5 Location tests for asymmetric distributions

The **sign test** is a simple unconditional distribution-free test for inferences about the median of any population. There is no requirement for symmetry and it is closely related to the median test in section 4.6. The sign test is still valid if we assume symmetry, but often it then has lower efficiency and less power than the Wilcoxon signed rank test. The principal of the sign test is simple. If θ is the median of a distribution and we have a sample of n observations from that distribution, then each observation is equally likely to be greater than or less than θ. We associate a plus (+) sign with any observation above θ and a minus (–) sign with any observation below θ and ignore any further information. If θ is the population median the number of plus signs, r, in a sample of m has a binomial distribution with parameters, m, $\pi = \frac{1}{2}$; i.e. under H_0, r is distributed $B(m, \frac{1}{2})$. There is an abundance of relevant tables, and nearly all general statistical software packages enable one to form the exact distribution of r for given m. We may also establish confidence intervals for θ. An example makes the procedure clear.

Example 5.9   I again use the data in Example 5.7, namely

$$2 \quad 3 \quad 6 \quad 6 \quad 6 \quad 9 \quad 9 \quad 10 \quad 14 \quad 15 \quad 16 \quad 72$$

but now test $H_0: \theta = 4$ against $H_1: \theta \ne 4$. Table 5.4 gives the probability of each number r of plus (or minus) signs for a $B(12, \frac{1}{2})$ distribution. Clearly a two-tail test is appropriate and under H_0 we immediately see that there are 2 minus signs (corresponding to the observations 2, 3) and 10 plus signs.

It is clear from Table 5.4 that the probability of 10 or more + signs (or 2 or less minus signs) is $p = 0 + 0.003 + 0.016 = 0.019$. Thus, in a two-tail test we reject H_0 at the $(2 \times 0.019) \times 100 = 3.8\%$ significance level. It is easy to obtain a two-sided confidence interval. If we require a confidence level of at least 95% clearly we reject any θ specified in H_0 that would leave us with 2 or less or 10 or more plus signs. Inspecting the data we see immediately that if $\theta < 6$ we have 10 or more + signs and if $\theta > 15$ we have 2 or less + signs. Thus the required interval is (6, 15) at an exact $(100 - 3.8) = 96.2\%$ level. An appropriate point estimator of the median is the sample median, here 9.

The confidence interval obtained in Example 5.9 is identical with that given by the Wilcoxon signed-rank procedure, but that is a peculiarity of these data and may surprise you since less information is used for the sign test. However, if the data show skewness the sign test often performs as well or better than the less appropriate signed-rank test, although the latter is helped by some robustness. Even for some symmetric distributions, particularly those with long tails, the sign test is more efficient than the signed-rank test. For the double exponential distribution the sign test has Pitman efficiency 4/3 relative to the signed-rank test.

Special procedures are needed if some data equal the hypothesized median. The resulting zeros are uninformative about any alternative so we might perform the test without them although this results in a loss of power for small samples or if there are many values 'tied' at the hypothesized median. An alternative is to allocate + or – signs at random to such tied values (e.g. by tossing a coin). Yet

Table 5.4 $\Pr(X = r)$ when X has a binomial distribution with $m = 12$, $p = \frac{1}{2}$

r	0	1	2	3	4	5	6	7	8	9	10	11	12
$\Pr(X=r)$	0.000	0.003	0.016	0.054	0.121	0.193	0.226	0.193	0.121	0.054	0.016	0.003	0.000

another proposal is to allocate the same sign to all zeros in the way that makes rejection of H_0 less likely. Random allocation of signs usually makes little difference in practice and lacks intuitive appeal and the latter suggestion is unattractive as it makes the test conservative.

Asymptotic tests are based upon the normal approximation to the binomial distribution. Under H_0 the number of positive signs is, for large m, distributed approximately $N(\frac{1}{2}m, \frac{1}{4}m)$, i.e

$$Z = \frac{X - \frac{1}{2}m}{\frac{1}{2}\sqrt{m}} \qquad (5.5)$$

has approximately a $N(0, 1)$ distribution. The estimate is improved by adding a continuity correction of $\frac{1}{2}$ to X if $X < \frac{1}{2}m$. If $X > \frac{1}{2}m$ the $\frac{1}{2}$ is subtracted.

The sign test in the form given here, or modifications of it, provides a versatile tool with a number of applications to problems that at first sight seem rather different from location estimates. I discuss some in sections 6.2 – 6.4..

Although logically distinct, the sign test is equivalent to a special case of the Wilcoxon signed-rank test in which all observations below the median are regarded as tied at a common value, say, -1, and all above the median are tied at a common value $+1$. Because of the high level of tying, tables of critical values for the no-tie Wilcoxon test are inappropriate and misleading.

5.6   Paired sample data

An important feature of tests in this chapter is that they are also applicable to matched-pair data for tests of differences in location.   These tests are nonparametric or distribution free analogues of the matched pair or single-sample t-test. Several logically distinct situations may be considered.

- Measurement (x_i) of some characteristic is made before treatment and measurement (y_i) of the same characteristic is made after treatment on each of m experimental units, $i = 1, 2, 3, \ldots, m$.
- Two treatments are applied either in random order to each of m experimental units, or are applied to similar or paired parts (e.g. eyes, legs, arms, etc.) of each unit, the part to receive each treatment being chosen at random, and responses x_i, y_i are measured.
- Units are grouped into m pairs such that each member of a pair is as similar to the other as possible. Two treatments are applied, one to each unit randomly selected within a pair, and the responses x_i, y_i are measured for all m pairs.

In each case the analysis is based on the m differences $d_i = y_i - x_i$. The conventional t-test procedure for tests about the mean is optimal if it is assumed that the d_i are independently identically distributed $N(\mu, \sigma^2)$ with parameters unknown. Typically a test for no treatment effect is then based on the null hypothesis $H_0: \mu = 0$ and confidence limits for μ are based on t-distribution theory. As I pointed out in section 1.6, in most experiments the experimental units are in no sense samples from a normal distribution but are typically units selected for convenience or perhaps because they are the only units readily available. In such cases if units are paired and treatment allocation is random **within** each pair then an exact test may be made by considering the relevant associated permutation distribution under H_0. The t-test on the differences is a one-sample test, and intuition suggest the single-sample permutation-based tests developed in this chapter may in certain circumstances be appropriate alternatives to a t-test.

Before considering these I show the basic importance of the matched pair experiment in encapsulating the concept of increased precision resulting from appropriate experimental design. It is the simplest example of the familiar randomized block experiment. A numerical example shows how precision may be increased in practice by devices such as blocking units that are similar apart from any effect of treatments under investigation.

Example 5.10   Times in minutes taken by 6 students to complete a standard computation are 16, 22, 19, 27, 21, 24. A computation claimed to be of equal difficulty is completed by 6 students in 25, 18, 19, 17, 23 and 15 minutes. Students were allocated at random. Is there any evidence to reject the claim that the computations are of equal difficulty?

For these data the two-sample t-test, a Fisher-Pitman test or a WMW test are possible alternatives. With none of these tests (Exercise 5.5) would we accept evidence of a difference in median times. The two-tail p-values are t-test, $p = 0.3844$; Fisher–Pitman, $p = 0.4156$; WMW, $p = 0.4156$.

Suppose, however, that the same 6 students had performed **both** computations and that the times each took were those given above, but the assignment to students gave these results:

Student	A	B	C	D	E	F
First calculation (x)	16	19	21	22	24	27
Second calculation (y)	15	17	18	19	23	25
Difference $d = x - y$	1	2	3	3	1	2

Intuitively, the differences suggest that the second calculation is done more quickly, although the times taken by different individuals on either calculation vary greatly; e.g., student A is much faster than student F for both calculations. But still, with the information so far given, it is not clear whether the improved performance should be attributed to the second calculation being easier, or if the second calculation were always given after the first, whether the improvement may in part be due to a 'learning' effect. This second possibility cannot be ruled out, but we can protect against it if the order in which the computations are performed by each individual is decided randomly (perhaps by tossing a coin) . If this is done we show below that the standard one-sample nonparametric tests performed on the differences are valid under some broad assumptions for detecting certain kinds of treatment effect. For the paired data in this example testing for median difference zero the two-tail p-values are: t-test, $p < 0.004$ and for both the one-sample Fisher–Pitman test and the Wilcoxon signed rank test, $p = 0.0313$. For these last two tests this is the smallest possible p-value attainable in a two-tail test for a sample of 6. I have kept the data sets unrealistically small to simplify understanding of basic ideas.

A key assumption usually implicit, even if not explicit, in a matched pair test or estimation procedure is that under the null hypothesis for the ith pair both x_i and y_i are observed values of variates with the same distribution. This implies, in particular, each has the same location parameters for a given i. This may be the mean, if it exists, or more generally the median; in particular, the latter is more appropriate if the distributions are skewed. However, for different i these common distributions need not all be the same. A common assumption is that for all i they are of the same general form (e.g. both normal, both rectangular) and that they differ, if at all, for any i only in location, often with the stronger assumption that this effect is the same for all i. Readers familiar with randomized block designs and the parametric analysis of variance model will recognise this location difference as the 'treatment' effect. In these circumstances under the null hypothesis of no treatment difference all the d_i are independent and identically symmetrically distributed with zero median. Symmetry follows because if X and Y have the same distribution then clearly $D = X - Y$ and $D' = Y - X$ are identically distributed. Under these circumstances the appropriate permutation reference set when there is randomization within pairs is obtained by reversing labels x and y within pairs, implying that under H_0 for the ith pair the observations d_i and $-d_i$ are equally likely. Thus for a hypothesis test that the median of the differences is zero, the Fisher–Pitman or the Wilcoxon rank sum tests immediately become appropriate.

However, some distribution-free tests may still be appropriate under further relaxation of assumptions. In particular, there are many practical situations where it is evident from the nature of the data, if not from broader theoretical considerations, that under H_0 for different i, the common distributions of X and Y may not only differ in location (the so-called treatment effect) but may also differ in other characteristics such as variance, degree of skewness, etc. It is also possible that although under H_0 for a particular i the medians of X and Y may remain the same while the distribution of X_i differs from that for Y_i in other respects, e.g. one may be symmetric and one skew, or both be symmetric but one have longer tails than the other. The overall effect of such complications is to retain under H_0 the situation that the median of the joint distribution of the D_i is still zero, but that joint distribution of the D_i may no longer be symmetric. If the d_i look to be skew the sign test may be appropriate.

Confidence intervals based on any relevant test for the median difference are calculated as for the single-sample case, but in general one cannot assume that the median of the difference is equal to the difference between the medians of X and Y in the general matched-pair situation. A counter example is given by Gibbons and Chakraborti (1992, pp.145–7). Indeed, the very generality of the situations where matched paired samples lead to hypothesis tests for zero differences calls for extreme care in making inferences about the nature of potential treatment differences. When using the same subjects for both treatments a further complication is that carry-over effects may influence experimental responses. For example, if two drugs are to be compared for pain relief it may happen that if drug A is administered before drug B (even if several days elapse between administrations and no clinical effect of drug A persists), that drug A proves effective but drug B does not, whereas if drug B is given first it proves to be highly effective and so also does drug A if it is given later. This is a simple example of a carry-over effect that results in what is called an interaction, topics we return to in later chapters, but the simple models considered here are no longer appropriate.

Maritz (1995, section 3.5) highlights some pitfalls in the analysis of paired-sample data. For example, if it is thought that a treatment is likely to have a proportional, or multiplicative effect, rather than an additive effect, then transformation to logarithms before taking differences should produce a more symmetric distribution of differences, resulting in a situation where the Wilcoxon signed rank test or even a Fisher–Pitman permutation test will have increased power. These tests are not invariant under a transformation of the original data to logarithms before taking differences, whereas the sign test is invariant. However, simply because one happens to obtain, in general, more marked treatment effects for high valued responses than for low valued responses, it does not follow that a transformation to logarithms will be appropriate. Maritz (1995, Example 3.5) considers a set of 97 matched pair measurements for plaque regrowth after two dental treatments. The regrowth measurements for some subjects are very low, of the order of 10 or 20 in the units used, while for others both measurements are over 100. Such variation is not unusual in bacteriological studies. Two of the more extreme results in Maritz's data were the pairs 688, 280 and 32, 20 with a wide range inbetween. Maritz gives asymptotic results which should be close to exact test results for a sample size 97. With $Z = 2.45$, the Wilcoxon signed-rank test suggests rejection of H_0. The sign test however falls well below the value required for significance. For this particular data set transformation to logarithms before differencing reduces the Z value for the asymptotic Wilcoxon test to 1.45 whereas the result of the sign test is unchanged because the ordering of x_i, y_i within pairs is unchanged. Complexity arises with this example because sometimes one treatment did markedly better than the other in reducing plaque regrowth but there were many cases, not obviously related to the extent of regrowth, where there was little treatment difference and some where it went in the other direction. This would suggest that individual patients were responding differently to the two treatments and that it might be worth taking other factors into consideration. Just what these might be is not obvious from the data on plaque regrowth alone, but other factors such as general level of dental care, initial dental status or even genetic variability might influence patient responses to a particular treatment. In this example one should probably be considering tests for more than a simple location shift, no matter whether that, considered apart from other such factors, appears to be either additive or multiplicative.

The non-invariance of the signed rank test under transformation to logarithms (or for other appropriate monotone nonlinear transformations) before differencing is easily demonstrated by a simple example.

Example 5.11   Although the data in this example are fictitious is not untypical of that obtained in many biological situations such as bacterial counts, counts of plant or animal species per unit areas in various localities, counts of cell abnormalities, etc. in appropriate units.

Group	A	B	C	D	E	F	G	H	I	J
Treatment A	60	380	1	211	8	83	47	172	45	17
Treatment B	64	346	3	43	15	27	31	103	22	10
Difference	−4	34	−2	168	−7	56	16	69	23	7

Clearly positive differences dominate, all except one positive difference being greater than any negative differences. For this small sample it is not difficult to compute S for the signed rank test manually but using StatXact or TESTIMATE or some other appropriate software indicates that the two-tail test $p = 0.0293$. The corresponding sign-test gives $p = 0.3438$. As indicated above, transformation to logarithms before differencing gives the same sign-test p-value but that for the signed rank test is $p = 0.2207$. This (see Exercise 5.6) is explicable in terms of the negative differences attaining higher rankings for the logarithmic differences than they did for the raw data differences. Indeed, the evidence of skewness in the original differences casts doubt on the appropriateness of a signed-rank test.

When 'before' and 'after' treatment measurements are made on the same units of, e.g. systolic blood pressure, it is often simplistic to assume that differences are only ones of location. If the before and after measurements for unit i are x_i and y_i a simple location difference model might be $y_i = x_i + d + e_i$ where d is constant for all i and the e_i are random errors perhaps symmetrically distributed about a mean of zero, but a more suitable model might be a regression-type model

(Chapter 11) of the form $y_i = \alpha + \beta x_i + e_i$, for which the methods of this chapter are not appropriate unless $\beta = 1$. An interesting discussion of the possible models for paired data from the same units is given by Cox and Snell (1981, Example E).

5.7 Observational data

Data obtained from an experiment without random treatment allocation, or from units not randomly selected from a larger population, or that are culled from existing records are observational data. Thousands, perhaps millions of analyses using statistical methods that require or assume a relevant form of randomization are carried out on observational data and I indicated pitfalls in doing so in section 1.4 and elsewhere. Nevertheless, in many cases it is reasonable to assume that the ill consequences of not randomizing may be small. For example, if one wants to check whether two quality control inspectors are consistent and equally efficient at spotting faults in a product a sensible way to do this would be to ask each to examine the same set of items and to score each item in the set for faults and then to examine differences between the scores given by each inspector. Selecting the items for test at random from a large batch would ensure that any inferences made from the results would have a population validity. However, we might get better information about how well the inspectors were doing their job if a batch of items were chosen that exhibited a wide range of flaw patterns; some with only a few, others with many faults, to see if inspectors coped equally well at detecting faults over a range of conditions. Since each specimen is to be submitted to each inspector, if it may reasonably be assumed that examination by one inspector does not alter the physical condition of the specimen or leave any indication of faults that would make them either clearer or more obscure to the second inspector, it would not seem to matter in which order the tests were done by the inspectors. On the other hand if the first inspection left clues that could influence the second inspection findings, randomization of the order of inspection would be needed to validate inferences from any analysis of differences. Even that may be insufficient if there are interactions effects like that suggested for drugs in section 5.6. There might also be spurious differences if one inspector carried out the tests on a Monday and the other on Friday after a particularly stressful week.

However, if in the light of common sense and past experience one can be confident that the order of inspection will not influence the inspectors' findings, randomization of test order might be discarded as unnecessary, and if it were convenient to do so, one inspector might look at all sheets first and the second inspector at the same batch of sheets at a later time. This implies an assumption that faults do not become either more or less evident and that no new ones appear between inspections. If such possibilities can be ruled out it seems not unreasonable to analyse the data collected as though there had been a random order of presentation of specimens.

A major pitfall in not randomizing is that one may not be measuring the effect of interest. For example, if, contrary to beliefs, new faults do occur between the first and second inspector's examinations, finding that the second inspector detects more faults need not imply that the latter is better at spotting faults, but rather that the time lag results in there being more faults to spot. Again, if the time lag does lead to more faults but the numbers detected by both inspectors are the same this may give a false impression of equal fault-spotting ability, whereas the second inspector is really doing less well.

Example 5.12 Kauman, Gottstein & Lantican (1956) as part of a larger study on methods of scoring blemishes in wood veneer presented 20 sheets of a particular type to each of two inspectors and asked them to score each sheet according to well defined rules. The higher the score the more serious were the faults. The scoring system required each person to measure the physical extent of each of four kinds of blemish. Essentially these consisted of measurements of the total lengths or areas, as appropriate, of various kinds of flaw (cracks, warping, splits, etc.). Each of the four measurements was then given a score between 0 (non-existent) and 5 (serious flaw) according to defined rules. Since the four different types of faults were considered to be not all equally serious the score (out of 5) for each type was weighted with weights between 1 and 4 and added to give a composite fault score between 0 and 50. Details are given in the original paper. Scores were allocated to each of the 20 sheets for inspections by a well-trained observed (A) and by a newly-trained observer (B).
The scores allocated are given in Table 5.5.

If the data are used to test for a median difference zero a one-tail test is justified if one believes that the newly trained observer cannot do better than the experienced observer, but if one makes the less strong *a priori* assumption that does not preclude the possibility of the newly trained observer doing better a two-tail test is appropriate. Taking the latter view the potentially relevant tests give the following two-tail probabilities:

Test	Two-tail p
t-test	0.0099
Fisher–Pitman	0.0101
Signed rank	0.0147
Sign test	0.0118

All tests are in reasonable agreement, and the corresponding 95% confidence intervals do not differ greatly. However, differences for individual sheets suggest departures from normality (I discuss tests for departures from normality in section 6.7). No matter whether we accept or reject a hypothesis all data should be closely inspected for evidence of features other than those under test. From Table 5.5 it is clear that four sheets, namely numbers 1, 7, 8 and 17 yield large differences, whereas the remaining 16 sheets all give differences in the range –6, 6. Indeed, if we confine attention to these 16 remaining sheets, all the tests considered above give *p*-values well above those indicating significance even in a one-tail test, so it appears that only the four high values are influential in suggesting lower scoring by observer B.

Table 5.5 Fault scores by two observers for 20 sheets of wood veneer

Sheet number	Observer A	Observer B	Difference A–B
1	22	9	13
2	31	29	2
3	34	40	–6
4	44	42	2
5	10	8	2
6	31	34	–3
7	42	25	17
8	28	13	15
9	29	28	1
10	39	34	5
11	21	16	5
12	42	41	1
13	40	34	6
14	11	7	4
15	38	44	–6
16	9	11	–2
17	20	6	14
18	28	24	4
19	28	25	3
20	15	9	6

A good experimenter would look for reasons. Score were based on severity of several different faults. For 16 from 20 sheets there is little indication of major differences between scores awarded by the two observers (differences being between +6 and –6), for the remaining 4 sheets the difference exceeds 13 points and each is in the same direction. Is observer B underscoring or even missing some type of fault that is prevalent on these 4 sheets? It is clear from the original paper that some of the faults were more easily spotted than others, and perhaps their extent more easily measured. A reasonable and sensible approach would be to look at scores awarded for each fault in detail if these were available, and to see whether any were being overlooked or under-scored by observer B and whether these were peculiar to sheets showing large scoring discrepancies. If so, further training would be indicated in these areas.

Exercises

5.1 Show that the usual t-statistic may be used in Fisher–Pitman test, i.e. that this statistic is a monotonic function of the statistic S proposed in section 5.2. Explain why the usual t-

distribution tables are not appropriate for determining whether a result is significant when the Fisher–Pitman test is used.

5.2 Verify that the procedure described in Section 5.2 for obtaining one-sample permutation (Fisher–Pitman) test confidence limits produces the requisite 2^m-1 changes in the order of the test statistic needed to give values covering all possible sign permutations.

5.3   Establish the means and variances of S_+ and S_- used as statistics in the one-sample Fisher–Pitman permutation test and hence derive the appropriate asymptotic statistics corresponding to these.

5.4 Verify the 95% confidence limits computed in Example 5.3 using asymptotic results.

5.5 Confirm the assertion in Example 5.10 that, without pairing, the data show no indication that the two tests are not of equal difficulty and confirm the p-values quoted in that example.

5.6 By computing the relevant rankings confirm the reasoning suggested for the behaviour of the signed rank test under transformation in Example 5.11.

5.7 Show that the permutation test statistic distribution for the Wilcoxon signed rank test for the hypothesis $H_0{:}\theta = 7$ where θ is the population median, based on a sample of m observations, r of which take the value 5 and s of which take the value 9, where $r + s = m$ is identical to the sign test distribution for the same hypothesis. Would this equivalence still hold if the null hypothesis were $H_0{:}\theta = 6$?

*5.8 I counted the numbers of pages in a random sample of 12 of my statistics books. The numbers were

126   142   156   228   245   246 370 419 433 454 478 503

Use the Wilcoxon signed rank test to test the hypothesis that the median number of pages in my collection of statistics books is 400. Obtain a 95% confidence interval for the mean number of pages based on the Wilcoxon signed-rank test and compare it with the interval obtained on an assumption of normality. The sample in this example comes from my personal collection of statistics books. Do you think it likely that inferences about the mean number of pages in my collection could be applicable to all published statistical books? Give reasons.

*5.9 Kimura and Chikuni (1987) give data for lengths of Greenland turbot of various ages sampled from commercial catches in the Bering Sea as aged and measured by the Northwest and Alaska Fisheries Center. For 12 year old Turbot the numbers of each length were:

Length (cm)	64	65	66	67	68	69	70	71	72	73	75	77	78	83
No. of fish	1	2	1	1	4	3	4	5	3	3	1	6	1	1

Would you agree with someone who asserted that, on this evidence, the median length of 12-year-old Greenland turbot was almost certainly between 69 and 72 cm.?

*5.10   Hand et al (1994, Set 86) give data provided by the UK electricity council for winter energy consumption in Mwh for 10 houses in Bristol, England before and after introduction of cavity wall insulation. Is there acceptable evidence that insulation saves energy? If so is there evidence that saving is proportional to the amount of energy used? You may not be able to answer this latter question satisfactorily until you have studied chapters 9 and 10, but plotting energy savings against preinsulation energy use should give a good indication of the answer. Do you think there are outliers in the data that require action?

Before insulation	12.1	11.0	14.1	13.8	15.5	12.2	12.8	9.9	10.8	12.7
After insulation	12.0	10.6	13.4	11.2	15.3	13.6	12.6	8.8	9.6	12.4

6

More one- and two-sample tests

6.1 Some further considerations

Chapters 4 and 5 dealt mainly with test and estimation procedures for location or location differences when the data were often from continuous, or potentially continuous, distributions. Many procedures in those chapters extend to other problems. In section 4.2 and elsewhere I pointed out that some could be used to test for dominance of one distribution over another. There may also be differences in, for example, dispersion (as measured by variance, inter-quartile range, etc.). More generally one may ask whether data are consistent with samples being from some completely specified distribution; e.g., an exponential distribution with unit mean, or a uniform distribution over the interval (0, 6) or a normal distribution with mean 7 and variance 6.5. Less specifically, one could ask whether it is reasonable to suppose data are a sample from **some** normal distribution or **some** exponential distribution, or **some** gamma distribution without specifying parameter values. For two samples we might wish to test whether they are from identical but otherwise unspecified distributions against an alternative that they are from different distributions.

Many data are not continuous – for example, they may be counts of items with specific characteristics. A simple case is binary data, where observations take only one of two values, often conveniently denoted by 0 and 1. Analyses are then often based on counts of the number of zeros or ones in m observations. If the probability of observing a 1 is π and outcomes are independent, then X, the number of ones in m observations has a binomial distribution with parameters m, π and $\Pr(X = r) = {}^mC_r \, \pi^r(1 - \pi)^{m-r}$. For the sign test (section 5.5) I converted measurements into a dichotomy of counts of numbers above and numbers below a hypothesized median. The sign test uses only the information condensed into those counts. This immediately implies that if we have binary data (counts of 0s or 1s) and want to test $H_0: \pi = \frac{1}{2}$ against one- or two-tail alternatives we are in a situation equivalent to that for a sign-test. In the next three sections I discuss some problems where a sign test 'equivalent' is appropriate, and closely related tests where we relax the restriction $\pi = \frac{1}{2}$.

6.2 Generalizations of the sign test

It may be difficult or impossible to make a physical measurement, yet to be able to decide in a matched pair situation which of a pair shows the more favourable response; in effect giving a sign to the difference between units.

Example 6.1 A drug counsellor interviews 18 parents of children known to have been experimenting with drugs, recording in each case whether the mother (M) or the father (F) shows the better understanding of the dangers of potential addiction. If both show equal understanding this is recorded as (E). The results for the 18 parents are 9 F, 6 M and 3 E. If the parents may be regarded as a random sample of a large population of parents of children who have drug problems, is this sufficient evidence to reject the hypothesis that it is equally likely that either parent may have the better understanding?

This is essentially a sign test problem. If we ignore the 3 E as uninformative about the alternatives we may count the 9 F preferences as 9 minus, scoring each as 0 and the 6 M as 6 plus scoring each as 1. The plus total of 6 is the value r of the statistic X used to test if the sample is acceptable as one from a binomial B(15, ½) distribution. Using appropriate software, or tables, it is easily established that if $H_0: \pi = $ ½ is true, then $Pr(r \leq 6) = 0.304$, so clearly one would not reject H_0. If we want to know for what values π_0 we would accept $H_0: \pi = \pi_0$ we need a confidence interval for π when we observe r successes in m observations. This is a standard exercise in elementary statistics and a number of software packages give either exact results or an asymptotic approximation for large m. Details are given in many elementary texts or in the StatXact manual. For this example the nominal 95% confidence interval is (0.1634, 0.6771). Thus for this simple example there is considerable doubt about the precise probability that the father (or mother) will have the better understanding. If one scores F as plus and M as a minus the conclusions are unaltered.

A variation of the sign test applies to any quantile. For example, candidates in examinations are often classified into one of several grades usually on the basis of total marks awarded. In the simplest situation there are two grades *pass* or *fail*. If, for a large examination entry, it is known that 80 per cent of all pupils obtain a pass grade, the parent–teacher association at a school may be alarmed if only 19 out of 30 candidates from that school obtain passes. The headmaster might argue that if 30 candidates are selected randomly from all candidates one should not be surprised if that group contains only 19 passes despite the fact that the expected number (80% of 30) would be 24. Calling a pass a plus and a fail a minus, the test is based on a binomial distribution with $m = 30$, $\pi = 0.8$ with observed $r = 19$, or equivalently, using failures, with $m = 30$, $\pi = 0.2$, $r = 11$.

Example 6.2 Is 19 passes among 30 candidates consistent with a population pass rate of 80%? One may use binomial distribution probability tables with $m = 30$ and $\pi = 0.8$ or the StatXact or a similar program to determine whether a 95% confidence interval for π includes 0.8. If it does, this is equivalent to accepting $H_0: \pi = 0.8$ in a two-tail test (Exercise 6.1 asks for an appropriate one-tail test). An approximation is to use an asymptotic test. Under H_0 the number of passes, X, has mean $m\pi = 30\times0.8 = 24$ and variance $m\pi(1 - \pi) = 30\times0.8\times0.2 = 4.8$, or standard deviation $\sqrt{4.8} = 2.19$ Asymptotically $Z = (X - 24)/2.19$ has a standard normal distribution. Adding the usual continuity correction ½ when X is below expectation we find $Pr(X \le 19) = Pr\{Z \le (19.5 - 24)/2.19)\} = Pr(Z \le 2.05) = 0.0202$. On this basis we reject H_0 in a one-tail test at the 2.02% significance level and in a two-tail test at the 4.04% level. However, if binomial tables are available it is easily verified that the corresponding exact levels are 2.56% and 5.12%. StatXact gives the 95% confidence interval for π under H_0 to be (0.4386, 0.8007) suggesting again that we should not reject at the 5% level in a two-tail test. The value $\pi = 0.8$ only just falls within the confidence interval, as we would expect because the exact test only just fails to indicate significance at the 5% level.

With the exact binomial distribution result available it is unlikely that one would use bootstrap sampling to estimate confidence intervals or to test the hypothesis $\pi = 0.8$ in this situation. However this may be done as an exercise by resampling with replacement from a set of observations consisting of 19 *ones* (representing passes) and 11 *zeros* (representing failures) to generate the bootstrap samples. I generated 1000 such samples and for each computed $\pi* = $ *number of ones in sample*/30. An estimated *p*-value for a one-tail test of $H_0:\pi = 0.8$ against $H_1:\pi < 0.8$ is given by (*number of* $\pi* \ge 0.8$ *divided by 1000*). For my samples this gave $p = 0.031$, close to the binomial probability of 0.0256. The 95% bootstrap quantile based two-sided confidence interval for π was (0.467, 0.800) again in good agreement with the binomial interval (0.4386, 0.8007) bearing in mind the tendency for quantile-based intervals (section 2.6) to undercover. Heavy tying among the $\pi*$ here introduces large discontinuities in possible *p*-values.

In Example 6.2, as is generally good practice, the decision whether a one- or two-tail test is appropriate should be made before carrying out the analysis. If there is no *a priori* reason to expect that pupils from the given school might do either worse or better a two-tail test is suitable. However, if it were known that pupils from this school had considerably less than the average amount of teaching in all schools a one-tail test could be justified on the grounds that it would be irrational to expect under-prepared pupils to do better than those receiving average or better tuition. A similar argument applies if a large portion of the syllabus had not been covered at this school. The two-sided confidence interval above suggests pupils might indeed be below-average performers, although this is not firmly established. If we had 90 pupils and the same expected pass rate (0.8), and the same proportion passed, i.e. 57 passed, it is easily to show (Exercise 6.2) that asymptotically $Pr(X \le 57) = 0.00007$ (using a continuity correction). The exact probability is $p = 0.0002$. Note the clear indication that power of the test (and by implication here the precision of ones findings) is increased by taking a larger sample when all other relevant factors are unchanged. The asymptotic approximation is not good for this small p, even for a large sample. The asymptotic result is more reliable even for somewhat smaller values of m when π takes values close to 0.5 than it is when π is close to 0 or 1 (Exercise 6.3).

6.3 McNemar's test

McNemar (1947) proposed a test that is a special case of the sign test in a matched pair situation when the pairs consist of units, typically humans or animals, where a binary response is recorded before and after some event which may be regarded as an applied treatment and interest centres on whether there are changes in the response in one direction, as opposed to the other, before and after treatment. I denote change in one direction as a plus and that in the opposite direction as a minus. No change is a tie and is ignored in testing whether or not there is evidence of a preponderance of changes in one direction or the other. The appropriate test is a sign test in which there is often a large number of ties.

Example 6.3 To test whether a broadcast on behalf of the 'yes' campaign before a referendum in which there is to be a yes/no vote swings more voters from 'no' to 'yes' (a plus) than *vice versa* (a minus), a sample of 200 voters might be asked to state their voting intentions both before and after the broadcast with the following outcome:

		Pre-broadcast intention	
		Yes	No
Post-broadcast intention	Yes	93	12
	No	6	89

Those on the diagonal (93, 89) have not changed allegiance. It is the off-diagonal numbers that measure the effectiveness (if any) of the broadcast. The number switching from *no* to *yes* (a plus) is 12 and the number switching the other way (a minus) is 6. For a binomial distribution with $m = 18$, $\pi = \frac{1}{2}$ it is easily verified that $\Pr(X \geq 12) = 0.199$ so in practice one would not reject the hypothesis that switches are equally likely to be either way even in a one-tail test.

The 2×2 contingency table above is a useful data summary, but that format does not make it obvious that the appropriate test is a sign-test. A special feature of these data is that only the off-diagonal elements are relevant to the test. For large m the usual asymptotic normal approximation (Exercise 6.4) for the binomial and a chi-squared equivalent (but not the traditional chi-squared test for independence in an $r \times c$ contingency table!) is available.

The McNemar test seems wasteful because it only uses outcomes for 18 units from 200, but in Chapters 12 and 13 I discuss other situations where many observations are required to detect what are, as here, likely to be small real differences. Typically in the situation in Example 6.3 a propaganda broadcast will only affect the opinion of a few wavering voters. The McNemar binary response sign test arises in other situations. In Sprent (1993, Example 6.4, pp. 95–6) I gave an example that is not a 'before' and 'after' situation. In it members of a mountaineering club argued about which of two climbs was the harder. To help resolve their arguments they examined club records which stated for all members who had attempted both climbs whether they succeeded or failed at each. They counted the numbers who had successfully climbed the first but failed to climb second and those who had climbed the second successfully but failed at the first. The numbers in these categories were small compared to those who had succeeded at both or failed at both, but these last two categories provide no information on relative difficulty.

I discuss in section 13.4 extensions of McNemar-type tests to cases with several possible categorizations before, and the same possible categorizations after an event, but where individuals may move between categories before and after.

6.4 A Sign test for trend

Tests for monotonic trends with time, increasing or decreasing, are often of interest and at their simplest take the form of testing H_0:*no trend* against H_1:*a monotonic trend*. Cox and Stuart (1955) proposed a simple test for increasing or decreasing trend in a set of independent observations $x_1, x_2, \ldots, x_m$ ordered in time. If m is even, say $m = 2n$, then the test is based on the signs of the n differences $x_{n+1} - x_1, x_{n+2} - x_2, \ldots, x_m - x_n$. For odd $m = 2n + 1$ the procedure is similar after deleting x_{n+1} and using the signs of $x_{n+2} - x_1, \ldots, x_m - x_n$. If there is an increasing (decreasing) trend one expects most of these differences to be positive (negative). Under H_0, the plus or minus signs have a binomial distribution with parameters n, $\pi = \frac{1}{2}$.

Unfortunately the independence assumption does not hold for much data in time series. In Sprent (1993, Example 2.6) I gave an example concerning mileages travelled by cars in the USA based upon estimates using different samples of vehicles in each of the years 1970–83 and because different samples were used in each year the estimates were independent. Had the same samples being used each year there may well have been evidence of a trend but the range of inference would be confined to cars actually sampled because these might not reflect the over-all usage of such vehicles in the USA. With observational data in particular it is important to consider the effects of lack of independence. Lack of independence in many time series analysis is characterized by seasonal or other periodic effects and by serial correlations. Long term trends often overlie seasonal trends, so if seasonal trends can be removed one might still go through the mechanics of performing a sign test and get meaningful results even though the independence assumption may be questionable insofar as the irregular departures from a trend may still be serially correlated.

For example, at any chosen meterological station average daily maximum temperature per day varies from month to month, generally being higher in summer and lower in winter. However, if at one station, we took the average daily maximum temperature per year then data for, say, 30 years (i.e. 30 data values) might be used to detect an increasing trend (as would be hypothesized by those who believe in global warming). However, this would be a crude test, because global warming, if indeed it is taking place, is doing so at a rate that might easily be masked by annual irregularities over a short to moderate period. It is also questionable whether evidence from such a simple test even over a 30 year period would be acceptable evidence of a truly new trend, since many weather cycles extending over several decades, or even centuries, have been reported, most of them discovered or confirmed by more sophisticated analyses.

If there are known seasonal or annual trends, or data show clear evidence of such trends, it is tempting to take a series of 24 monthly observations to see if there is some additional superimposed longer-term trend and to form a pseudo Cox-Stuart test comparing the first 12 with the second 12 observations, meaning that the differences compared are those for consecutive Januaries, consecutive Februaries, and so on. In fact this is essentially a matched pair sign test with the pairs being provided by the months and the data for consecutive years for each month regarded as a 'before' and 'after' measurement. This, however, may not be justified and may be misleading if there are correlations between observations in consecutive months. For instance, ozone measurements at ground stations, like atmospheric temperature, vary by a factor of 2 or more between January and July at many points on the earth's surface but there are also correlations between successive months – for instance, an exceptionally high ozone reading in March in any year is likely to be followed by high readings in April and perhaps also in May due to persistence of the gas in the atmosphere, so the test in such circumstances should be avoided, or used only with caution. When there are seasonal trends the Cox–Stuart sign test should certainly not be applied to monthly records over, say, a 16-month period as it would almost certainly not detect any trend because seasonal variation, for example, in ozone concentrations if that were the subject under study, would mask any longer underlying trend.

Example 6.4 The data below gives the mean value of the $US exchange rate per £UK computed on an annual average basis for each of the years 1979–93 in that order.

1.92 2.12 2.33 1.75 1.52 1.34 1.30 1.47 1.64 1.78 1.64 1.78 1.77 1.76 1.50

To apply the Cox–Stuart test, since there are 15 observations we ignore the 8th observation of 1.47 and note the signs of the differences $1.64 - 1.92$, $1.78 - 2.12$, $1.64 - 2.33$, . . ., $1.50 - 1.30$. The relevant signs are $-, -, -, +, +, +, +$ the almost equal numbers of $+$ and $-$ giving the strongest possible evidence of no consistent increasing or decreasing trend. The pattern of signs suggests there is at first a descreasing trend, followed by an increasing one, although this does not bring the rate back to the initial level.

This example is of little practical importance and of dubious validity because of a lack of independence. For exchange rates it is likely that rates in nearby years are highly correlated and influenced by events (such as a recession or implementation of a deflationary policy) with effects lasting several years.

6.5 Tests for dispersion

Historically, statistical testing and estimation developed with emphasis on the effect on *location* of some response to stimuli such as treatments in an agricultural, biological or medical contexts or to changes in materials used or to production methods in an industrial context, or to different teaching methods in education. There was sometimes a secondary interest in the effect on dispersion as measured by a variance, standard deviation or range. Increasingly, particularly in the field of quality control, dispersion differences have assumed more impor- tance because consumers demand that a product performs well not only on average, but that it is consistently reliable.

In the two independent sample situation if we assume normality the appropriate statistic to test equality of population variances given samples $x_1, x_2, \ldots x_m$ and $y_1, y_2, \ldots, y_n$ from the respective populations is

$$F = \frac{\sum(x_i - \overline{x})^2/(m - 1)}{\sum(y_j - \overline{y})^2/(n - 1)}$$

which has and F-distribution with $m - 1$, $n - 1$ degrees of freedom under the hypothesis that the population variances are equal, irrespective of whether the means are equal.

In Chapter 4 I covered tests basically designed for shifts only in location without making specific population assumptions. I pointed out, however, that most of these were also relevant to tests of dominance where the alternative hypothesis was of the form $H_1:G(x) \le F(x)$ with strict inequality for at least some x. This is important, because for some distributions if we are comparing samples from distributions in the same family it is impossible to make a location shift without also altering other distribution characteristics such as the variance. A simple example is the exponential distribution with parameter λ. The probability density function is $f(x) = \lambda e^{-\lambda x}$ and $E(X) = 1/\lambda$, $Var(X) = 1/\lambda^2$ and $med(X) = \ln(2/\lambda)$. If λ is changed not only are the mean and median changed, but so is the variance and we have a dominance situation. The parameter λ, or sometimes its reciprocal, is called a scale parameter. This effect on both mean and variance leads to difficulties in developing completely distribution-free tests for differences in variances or measures of dispersion generally. In most situations a test for a dispersion differences alone requires either an assumption that the medians or means are equal or that we know the difference between the medians or means and can, by subtracting that difference from all values in one sample, align our samples locationwise. This matter is discussed in detail by Gibbons and Chakraborti (1992, Chapter 10). Most tests for a difference in dispersion either implicitly or explicitly assume that the samples are from populations with identical medians, or that the difference in population medians is known and an adjustment has been made to align the samples by subtraction or addition of that difference to one set of sample values as appropriate. If it is clear from the sample values, or if a test indicates that equality of medians is not tenable, it is often suggested that one should adjust sample values to align the sample medians. This has an intuitive appeal, but examples exist that show this may sometimes be counterproductive and the resulting tests are not in general truly distribution-free.

Several methods that **assume** the population means or medians are identical have been proposed for testing for equality of population variances. Siegel and Tukey (1960) proposed a simple test that bears their name. It is analogous to the Wilcoxon rank sum test, but ranks are assigned differently. The motivation is that if there is a difference in variance, then if the observations in the combined samples are ordered the more extreme observations will tend to be from the population with the greater variance. Siegel and Tukey rank the smallest observation in the combined samples 1, the largest observation 2, the next largest 3, the next two smallest 4 and 5, the next two largest 6 and 7 and so on working inwards until all $N = m + n$ ranks are allocated. Mid-ranks are used for ties. The test statistic is now the sum of the ranks associated with the sample of m (or any equivalent Wilcoxon rank sum statistic). Sufficiently small values of this sum indicate that the associated sample comes from a population with larger variance – sufficiently large values that it comes from a population with smaller variance relative to that for the other sample. Any program for the WMW test may be used providing ranks are allocated by the rule given above. A disappointing feature of this and some other rank-based tests for dispersion, is the low Pitman efficiency, 0.61, compared to the F-test if samples are from normal populations.

Example 6.5 Sugar is packed into bags each with a nominal weight of 1 kg. Because weight varies from bag to bag, to ensure that bags do not fall below 1 kg weight, the packing machine is set to deliver bags with a median weight of 1002 g, it being known from experience that weights are seldom more than 2 g above or below the median. For a sample of 10 bags from this machine the weights are

    1001.3 1002.7 1003.1 1003.8 1000.9 1001.4 1002.2 1002.4 1002.9 1000.5

A cheaper machine set to deliver the same median weight of 1002 g is offered to the packaging company but they are concerned that the dispersion of weights may be greater, with a potential for some to fall below 1000 g. A sample of 12 bags filled by this machine have weights

1002.7 1000.1 1003.4 1001.2 1000.3 1003.6 1002.4 1004.1 1001.1 1000.2 1003.7 1003.9

These values for the combined samples are arranged in ascending order below, those in the first sample being printed in bold and the corresponding Seigel–Tukey ranks given by the rule above appear below each observation

  1000.1 1000.2 1000.3 **1000.5 1000.9** 1001.1 1001.2 **1001.3 1001.4 1002.2** 1002.4
    1   4   5   **8**   **9**   12   13   **16**   **17**   **20**   21.5

  1002.4 1002.7 **1002.7 1002.9 1003.1** 1003.4 1003.6 1003.7 **1003.8** 1003.9 1004.1
   21.5   18.5   **18.5**   **15**   **14**   11   10   7   **6**   3   2

The sum of the ranks associated with the 10 first sample values (in bold) is 145.

    StatXact gives an exact $p = 0.0239$ for a one-tail test, indicating that we would reject at the 2.39% significance level the hypothesis of equal variance in favour of the hypothesis that the second machine produces weights with a greater variance. A one-tail test is justified on the grounds that it is not unreasonable to expect the cheaper machine to be less consistent.

The Siegel–Tukey test is inefficient if the samples show evidence of misalignment, i.e. differences in location. Lehmann (1975, p.33) proposes alignment using the Hodges-Lehmann estimator of the median difference, subtracting this from the appropriate sample to align the medians. This is intuitively reasonable and unlikely to be seriously misleading although the test is no longer strictly distribution-free or exact.

    Many tests use other scoring systems, but these often have no higher Pitman efficiency against, for example, the F-test for normal distributions. Several essentially equivalent tests were formulated at or about the same time by more than one statistician so there is often confusion in nomenclature, particularly for a group of tests that form a relevant statistic based upon dispersion of ranks about the combined samples mean ranks.

    If there are $N = m + n$ observations in the combined samples and these are ranked from 1 to N in ascending order the mean rank is $\frac{1}{2}(N + 1)$ and the deviation of any rank i from this mean is $d_i = i - \frac{1}{2}(N + 1)$. Clearly the sum of these deviations over all N ranks is zero. If the dispersion in one sample is greater than that in the other, low and high ranks tend to dominate in that sample. Since the d_i corresponding to low and high ranks tend to have opposite signs the magnitudes of the d_i are relevant and these are reflected by taking either absolute values of the d_i, or the squares of the d_i, as scores. The sum of these for one sample is an appropriate score to use in a Fisher–Pitman type permutation test for a difference in dispersion (under the assumption that the population medians are the same). Mood (1954) uses the sum of the d_i^2 for one of the samples in what is commonly called the Mood test. Several tests that are essentially equivalent use the sum of the $|d_i|$ for one of the samples have been proposed with minor variations by Freund and Ansari (1957), David and Barton (1958) and by Ansari and Bradley (1960). I follow Gibbons and Chakraborti (1992) and refer to these tests collectively as the Freund–Ansari–Bradley–David–Barton test (or FABDB test for short). Any program for the Fisher–Pitman two independent sample permutation test may be used with appropriate scores, although StatXact provide programs that compute the scores given the raw data. Midranks are widely used for ties. This does not alter the mean rank but affects some of the scores.

Example 6.6 For the data in Example 6.5 test for equality of variance using the Mood test. I give below the combined sample data, the corresponding ranks and the corresponding scores for the Mood test. Data for the first sample are given in bold.

Data	1000.1	1000.2	1000.3	**1000.5**	**1000.9**	1001.1	1001.2	**1001.3**	1001.4	1002.2	1002.4
Ranks	1.	2	3	4	5	6	7	8	9	10	11.5
Score	$(10.5)^2$	$(9.5)^2$	$(8.5)^2$	$\mathbf{(7.5)^2}$	$\mathbf{(6.5)^2}$	$(5.5)^2$	$(4.5)^2$	$\mathbf{(3.5)^2}$	$(2.5)^2$	$(1.5)^2$	0

Data	**1002.4**	1002.7	**1002.7**	1002.9	1003.1	1003.4	1003.6	1003.7	**1003.8**	1003.9	1004.1
Ranks	**11.5**	13.5	**13.5**	15	16	17	18	19	**20**	21	22
Score	**0**	2^2	$\mathbf{2^2}$	$\mathbf{(3.5)^2}$	$\mathbf{(4.5)^2}$	$(5.5)^2$	$(6.5)^2$	$(7.5)^2$	$\mathbf{(8.5)^2}$	$(9.5)^2$	$(10.5)^2$

The sum of the first sample scores is $S = (7.5)^2 + (6.5)^2 + \ldots + (8.5)^2 = 228$

StatXact gives the exact one-tail $p = 0.0200$, slightly less than that obtained for the Siegel–Tukey test in Example 6.5. In passing, note that StatXact gives $S = 228.5$. The small discrepancy is due to the way it deals with ties. Consistent with the practice in that package for many other tests, rather than using the mid-rank, the Mood score is calculated on the assumption that the ties are separated by a small amount and the scores calculated for two such observations are then averaged. For example, the tied values at 1002.7 are looked upon as separated and assigned ranks 13 and 14. The corresponding Mood scores are $(1.5)^2$ and $(2.5)^2$ giving a mean score for the tie of $\frac{1}{2}(2.25 + 6.25) = 4.25$ compared to the score 4 allocated above. It is easily verified that if we proceed in this way the entries I have scored as 0 are each replaced by 0.25 leading to the value of S given by StatXact. In practice the difference is usually small unless there is considerable tying.

The FABDB test uses the magnitude of the deviations rather than their squares as scores.

Example 6.7 Apply the FABDB test for equality of variance to the data in Examples 6.5 and 6.6. The combined sample data and the relevant scores are given below with the scores for the first sample in bold.

Data	1000.1	1000.2	1000.3	**1000.5**	**1000.9**	1001.1	1001.2	**1001.3**	**1001.4**	**1002.2**	1002.4
Ranks	1	2	3	**4**	**5**	6	7	**8**	**9**	**10**	11.5
Score	10.5	9.5	8.5	**7.5**	**6.5**	5.5	4.5	**3.5**	**2.5**	**1.5**	0

Data	**1002.4**	1002.7	**1002.7**	1002.9	1003.1	1003.4	1003.6	1003.7	**1003.8**	1003.9	1004.1
Ranks	**11.5**	13.5	**13.5**	15	16	17	18	19	20	21	22
Score	**0**	2	**2**	**3.5**	**4.5**	5.5	6.5	7.5	**8.5**	9.5	10.5

The sum of the first sample scores is $S = 7.5 + 6.5 + \ldots + 8.5 = 40$. StatXact gives a one-tail $p = 0.0289$. StatXact deals with ties in the way described in Example 6.6, and also uses as test statistic a linear function of S which is effectively $S' = \frac{1}{2}m(N + 1) - S$.

The test has Pitman efficiency only 0.61 for samples from normal distributions. Klotz (1962) proposed using squares of van der Waerden scores in a test analogous to the Mood test and Capon (1961) suggested a similar test using expected normal scores. StatXact provides a program for the former and for the data in Examples 6.7 this gives a one-tail $p = 0.0183$

Sukhatme (1957) proposed an ingenious test having analogies with the Mann–Whitney formulation of the WMW test, but the scoring is slightly more complicated and there are some practical limitations to its use. It is described fully by Gibbons and Chakraborti (1992, section 10.7) who point out that it can be adapted to construct confidence intervals for a scale parameter.

A different approach is taken by Conover (1980) on the assumption that the means of the two populations are known. If the population random variables X, Y have means μ_X, μ_Y respectively, then equality of variance implies $E[(X - \mu_X)^2] = E[(Y - \mu_Y)^2]$. Conover proposed a test for variance equality based on the square of the ranks of the squared deviations $(x_i - \mu_X)^2$, $(y_j - \mu_Y)^2$ in the combined sample. In the common situation where the population means are unknown Conover suggests replacing them by the sample means m_x and m_y stating that the test he proposes is then approximately valid. My experience has been that doing this is unlikely to be misleading in practice though the test is no longer completely distribution-free. The test is carried out by first computing deviations from the population means (or more usually the sample means when the former are unknown) for each sample and arranging these in order of increasing magnitude for the combined samples and ranking the resulting data. A valid test statistic analogous to a WMW statistic could be based on these ranks but a more powerful test is obtained by using squares of these ranks in what is then essentially a Fisher–Pitman type permutation test.

Example 6.8 I illustrate the Conover squared rank test for the data in Example 6.5. The first sample values were

1001.3 1002.7 1003.1 1003.8 1000.9 1001.4 1002.2 1002.4 1002.9 1000.5

The sample mean is 1002.12. The **magnitudes** of the deviations from the mean are

0.82   0.58   0.98   1.68   1.22   0.72   0.08   0.28   0.78   1.62

The second sample values were

1002.7 1000.1 1003.4 1001.2 1000.3 1003.6 1002.4 1004.1 1001.1 1000.2 1003.7 1003.9

and the sample mean is 1002.225 and the magnitudes of deviations from this mean are

0.475 2.125 1.175 1.025 1.925 1.375 0.175 1.875 1.125 2.025 1.475 1.675

Arranging these deviations in ascending order for the combined sample and allocating ranks we obtain the squared ranks given below. First sample values are printed in bold.

Deviation	**0.08**	0.175	**0.28**	0.475	**0.58**	**0.72**	**0.78**	**0.82**	**0.98**	1.025	1.125
Squared rank	**1**	4	**9**	16	**25**	**36**	**49**	**64**	**81**	100	121

Deviation	1.175	**1.22**	1.375	1.475	**1.62**	1.675	**1.68**	1.875	1.925	3.025	2.125
Squared rank	144	**169**	196	225	**256**	289	**324**	361	400	441	484

The sum of the first sample squared ranks is 1014. An exact test may be performed with these scores in a Pitman–Fisher test program but StatXact provides a program specifically for this test which computes the squared ranks given the data. For the relevant one-tail test $p = 0.0223$ in good agreement with values obtained above by other methods.

In this example I lined up on the sample means despite an indication when I first introduced these data in Example 6.5 that the population medians were in each case 1002. If we further assume symmetry we would be justified in lining up the samples on this common median. See Exercise 6.15.

There were no ties in the above example. Where ties occur it is better to use the mean of adjacent squared ranks involved in ties rather than squaring mid-ranks, although the difference is likely to be small with only a few ties.

Asymptotic results for all the above tests may be based on (1.4)–(1.6) using the appropriate scores as 'data'. Simplifications are available for specific tests and may prove useful especially when there are no tied scores. In particular the result for the WMW test may be used for the Siegel–Tukey test and for the Mood test it is easily shown (though only after tedious algebra, Exercise 6.5) that for the scores used and taking the sum of the first sample scores as the statistic S, the relevant mean and variance given by (1.4) and (1.5) are

$$E(S) = m(N^2 - 1)/12$$

and

$$\mathrm{Var}(S) = mn(N + 1)(N^2 - 4)/180,$$

while for the FABDB test using again the sum of the first sample scores for the statistic S, the corresponding values, which differ depending upon whether N is even or odd (see Exercise 6.6), are

For N even: $E(S) = \frac{1}{4}m(N+2)$ and $\mathrm{Var}(S) = [mn(N^2 - 4)]/[48(N-1)]$,

For N odd: $E(S) = \frac{1}{4}[m(N^2 - 1)/N]$ and $\mathrm{Var}(S) = [mn(N+1)(N^2+3)]/[48N^2]$.

Many equivalent statistics are used and if these are linear transformations of S the mean is often different but the variance terms above are either unaltered or multiplied by a constant.

For the Conover test, for the statistic S that I used, the mean and variance for the no-tie case are

$$E(S) = m(N+1)(2N+1)/6$$

and

$$\mathrm{Var}(S) = mn(N+1)(2N+1)(8N+11)/180.$$

6.6 Gastwirth scores

Gastwirth (1965) proposed statistics for location and scale differences based on simple scores assigned to the top sth and bottom rth non-overlapping fractions of the combined samples ordered data where $0 < s, r < 1$, with zero scores given to any data with intermediate ranks. The motivation for this approach was the fact that tests for location based on transformation of ranks such as that using van der Waerden scores that have high Pitman efficiency relative to optimal normal theory tests when the latter are appropriate give greater weight to the scores associated with extreme ranks. Gastwirth was especially interested in tests that were more efficient for dispersion than tests such as the Siegel–Tukey test or the FABDB test.

His scoring system is simple but differs slightly depending upon whether the combined sample size N is odd or even. To illustrate his proposal and some of its implications I consider only the case N odd; the reader should refer to Gastwirth's paper for a detailed account of the complete procedure and its implications. Optimum choices of r and s were shown to be strongly dependent on the type of populations envisaged. For symmetric distributions a choice $r = s$ is generally appropriate and the same common value is often appropriate whether one is interested in location or dispersion, but an optimum or near optimum choice depends on factors like whether the distributions have long tails or are compact like a uniform distribution. For skew or truncated distributions it is sometimes appropriate to select $r \neq s$ and even sometimes to take one of these as zero. If N is odd all non-zero scores are integral and if the bottom fraction is r and the top fraction s we set $R=[Nr]+1$ and $S=[Ns]+1$ where $[u]$ denotes the largest integer less than or equal to u. The ranks are then transformed to the scores below:

Rank	1	2	3	...	$R-1$	R	$R+1$	...	$N-S$	$N-S+1$	$N-S+2$	...	$N-1$	N
Score	R	$R-1$	$R-2$	...	2	1	0	...	0	1	2	...	$S-1$	S

Setting T_R equal to the sum of the scores corresponding to Sample I values in the bottom fraction and T_S equal the sum of the scores corresponding to Sample I values in the top fraction, Gastwirth considers two statistics

$$U = T_S + T_R \text{ and } V = T_S - T_R.$$

It is easy to verify (see Exercise 6.16) that if we choose $s = r$ large enough to ensure that $S = R = \frac{1}{2}(N - 1)$ then V is a linear transformation of, and hence equivalent to, the Wilcoxon rank-sum statistic and U is similarly equivalent to the FABDB statistic.

In general, for any choice of s, r, Gastwirth therefore suggested using U as a statistic for differences in dispersion and V for differences in location. Clearly these scores may be used to compute exact p-values if they are inserted in a Fisher– Pitman permutation test program such as that in StatXact or a program specifically designed for this test like one in TESTIMATE, version 5.2 or later. Asymptotic tests are got by using these scores as z_i in (1.4), (1.5) and (1.6).

Gastwirth obtained the Pitman efficiency of tests using U relative to the F-test and of V relative to the t-test for samples from normal distributions and for various other symmetric and asymmetric distributions for various choices of s and r. For samples from normal distributions the FABDB test has Pitman efficiency only 0.61, but Gastwirth found that this rises to 0.85 for U if we set $s = r = 1/8$. For the uniform distribution it rises from 0.60 for FABDB to a spectacular 5.85 when $s = r = 1/20$. For the gamma distribution, as the index n increases optimal Pitman efficiency for U is obtained by allowing r to tend to zero and s to tend to 1. In this situation it is easily verified that both U and V are equivalent to the Wilcoxon rank sum statistic and the situation reflects the blurring between the effects of changes in the scale parameter on both location and dispersion. As Gastwirth points out the scale parameter of a gamma distribution with index n determines the mean and as n increases the gamma distribution approaches the normal and the Wilcoxon test is known to have good efficiency for shifts in the mean of the normal distribution.

Remember that Pitman efficiency is an asymptotic concept, and although often efficiency for small samples may be of the same order as the Pitman efficiency, if s, r are both small the number of nonzero scores that are assigned may also be small, resulting in marked discreteness in possible p-values. In some instances all p-values may then exceed 0.05 or any other acceptable 'significance' level. For example, if we suspect we have samples from uniform distributions and $N=19$

and we take $s = r = 1/20$ as seems desirable in the light of Gastwirth's Pitman efficiency finding, only the two extreme observations would receive nonzero scores and U could only take 3 possible values. However for large data sets (where there is often a clear indication of the likely broad form of the population distributions) these special yet simple scoring techniques may be useful. Even if optimum choice of s, r lead to too few scored observations, taking larger fractions in one or both tails may still give increased efficiency relative to the more orthodox tests for location or scale differences.

Another approach to situations with potential differences in location and dispersion is the bootstrap. Efron and Tibshirani (1993, section 16.2) discuss a bootstrap solution to the Behrens–Fisher problem using only the assumption that under H_0 the distributions for X and Y have the same mean. Their method draws attention to an important aspect of the bootstrap methods in any hypothesis testing context, namely the need to exercise care in the choice of the statistic to be bootstrapped. This is closely linked to the relevant H_0, a point that is also stressed and illustrated for several examples by Hall and Wilson (1991).

Given two samples $x_1, x_2, \ldots x_m$ and $y_1, y_2, \ldots, y_n$ an appropriate Studentized statistic for testing equality of means under the assumption that the unknown population variances are σ_1^2 and σ_2^2 is

$$t_{FB} = \frac{\overline{x} - \overline{y}}{\sqrt{[(s_1/m + s_2/n)]}}$$

where $s_1^2 = \sum(x_i - \overline{x})^2/(m-1)$ and $s_2^2 = \sum(y_i - \overline{y})^2/(n-1)$ are the usual unbiased estimators of σ_1^2 and σ_2^2. This estimator is widely used with the additional assumption of normality in the Behrens–Fisher test. Without the normality condition it is tempting to base a bootstrap test for equality of means computing t_{FB} for separate bootstrap samples of the x and y along the lines of Example 2.9. However, this is not appropriate for testing the hypothesis H_0 of zero difference in means, because although we want to test this hypothesis on the basis of the given data we cannot be sure that it actually holds for the data. However if we denote the mean of the combined sample of $m + n$ observations by $\overline{z}$ it is immediately clear that if we adjust all first sample values by subtracting $\overline{x} - \overline{z}$ and all second sample values by subtracting $\overline{y} - \overline{z}$ the derived samples will both have mean $\overline{z}$. Denoting the derived sample values by $\tilde{x}_i, \tilde{y}_j$ we may calculate, say, $B = 1000$, bootstrap samples by sampling with replacement separately within each sample and computing for each the statistic $t_{FB}{}^{*b}$ where a fresh computation of the denominator is made for each bootstrap sample. An approximate p-value for a test of H_0 against a one-sided alternative is given by counting in the appropriate tail the number, r, of $t_{FB}{}^{*b}$ greater/less than the value of t_{FB} for the given data, the estimate being $p = r/B$.

Example 6.9 For the data in Exercise 2.9, i.e.

Unvaccinated	22	21.5	30	23				
Vaccinated	21.5	0.75	3.8	29	2	27	11	23.5

a standard *t*-test gives a one-tail $p = 0.08$ for testing for a zero mean difference against the alternative that vaccination reduces infestation. Clearly there is a difference in variance between samples. I carried out the bootstrap procedure suggested above where here it is easily verified that for the given data $t_{FB} = 2.02$ and that the combined sample mean is $\bar{z} = 17.92$ and that with slight but sensible rounding the derived data with equal means becomes

Unvaccinated	15.8	15.3	23.8	16.8				
Vaccinated	24.6	3.9	6.9	32.1	5.1	30.1	14.1	26.6

This transformation does not alter the denominator of t_{FB}. It simply makes the numerator zero. One thousand bootstrap samples of these data with each treatment resampled independently gave 29 samples with $t_{FB}{}^* \geq 2.02$, giving an estimated one-tail $p = 0.029$.

This indicates that vaccination reduces infestation, where this small data set is certainly nonnormal in appearance. Chatfield (1988) discusses these data and points out that a reasonable deduction from them is that vaccination is ineffective for some sheep and partly to almost wholly effective for others. He wisely suggests that additional data rather than more sophisticated statistical analysis is the path to more definitive conclusions.

6.7 Goodness of fit tests

In practice deviations from a hypothesis may involve factors more complex than location, dominance, or differences in dispersion. We might ask whether data are consistent with their being a sample from some specified (or partially specified) distribution or whether two samples may reasonably be supposed to come from the same unspecified distribution. Two widely used tests for the first problem are the *chi-squared goodness-of-fit test* and the *Kolmogorov goodness-of-fit test*. The former was proposed originally for data in the form of counts of items categorized by some qualitative or quantitative measure which might or might not be ordinal. The latter was designed basically for continuous, or measurement, data.

To decide if it is reasonable to suppose two samples (usually of measurements) come from the same population, a well-known test is the Smirnov test considered in section 6.8 and which is in some aspects an extension of the one-sample Kolmogorov test.

In their basic forms both the chi-squared test and the Kolmogorov test are distribution-free. A major difference is that the chi-squared test in the asymptotic form, in which it is best known, is an approximation to an exact permutation test. The Kolmogorov test has a known and exact distribution for samples from any completely specified continuous distribution and in that sense differs from most exact tests that I have discussed so far, all of which could be regarded as permutation tests, even though that might not always be the easiest way to view them.

The chi-squared goodness-of-fit test. In its simplest form the test assumes that we have counts f_i of the numbers among m items occurring in each of k categories and that the probability of an item falling into category i is π_i, $i = 1,2, \ldots, k$ where $\sum \pi_i = 1$ and $\sum f_i = m$. If $k = 2$ under an independence and homogeneity assumption the distribution of the numbers falling into the categories is binomial with parameters m, π. For $k > 2$ it is multinomial and denoting the random variable specifying the number in category i by X_i then a standard results gives the probability mass function:

$$\Pr(X_1 = f_1, X_2 = f_2, \ldots, X_k = f_k) = \frac{m!}{\prod(f_i!)} \prod[\pi_i^{f_i}] \qquad (6.1)$$

The test is distribution-free in the sense that the counts may be derived from data following any completely specified distribution for which we are able to calculate the probabilities π_i. For example, if our sample is m counts from a Poisson distribution with known mean λ and we group the data into $k = 6$ classes corresponding to the events $X = 0, 1\ 2, 3, 4, 5$ or more, then

$$\pi_i = e^{-\lambda} \lambda^i / i!, \ i = 0, 1, 2, 3, 4$$

and

$$\pi_5 = \sum_{i \geq 5} e^{-\lambda} \lambda^i / (i!),$$

If we have only counts of the number of sample values from a completely specified continuous distribution with cdf $F(x)$ for each of k groups covering the complete range of possible sample values and the ith group is the count of all sample values in the interval $x_{i-1} < x \leq x_i$ then $\pi_i = F(x_i) - F(x_{i-1})$. However, if the choice of the x_{i-1}, x_i to determine the k groups is arbitrary, the test outcome will depend upon the grouping, a weakness I discuss further below.

In the general notation given above for any multinomial distribution the chi-squared statistic takes the well-known form

$$\chi^2 = \sum_i (f_i - m\pi_i)^2 / (m\pi_i) \qquad (6.2)$$

which is often expressed verbally as the sum over all k categories of

(observed number – expected number)²/(expected number).

Standard distribution theory may be used to show that asymptotically χ^2 has a chi-squared distribution with $k - 1$ degrees of freedom. For computational purposes a more convenient equivalent to (6.2) is

$$\chi^2 = \sum[f_i^2/(m\pi_i)] - m. \qquad (6.3)$$

The literature abounds in conflicting advice on how large the expected number in each group should be for inferences based upon the asymptotic χ^2-test to be reliable. The most conservative advice is that none should be less than 5, and the most optimistic is that providing each group has expectation 1 or more, not more than 20% of the groups should have expectations less than 5. Data sets have been produced both to support and also to cast doubt upon nearly all such recommendations. Clearly, for small counts there is an attraction in basing tests of goodness-of-fit on a permutation test working out the probabilities of all possible groupings of m items for the given multinomial probabilities and computing the probability of getting that grouping or any less likely grouping under the null hypothesis.

Because the χ^2 statistic is so intuitively reasonable it might be hoped that this permutation test would be equivalent to an ordering on the basis of computed χ^2. Unfortunately the two are not exactly equivalent in small samples as the next example shows.

Example 6.10 Flowers of a certain species may be red, yellow or pink. In random sample of 4 flowers there are 2 red and 2 pink (and none yellow); is this consistent with H_0: Pr(red) = ¼, Pr(pink) = ¼, Pr(yellow) = ½? It is easy to verify that there are 15 ways of allocating 4 items to the 3 categories and with each we may associate a probability under H_0 calculated from (6.1). For the given outcome this probability is $[4!/(2!\times2!\times0!)] \times (¼)^2 \times (¼)^2 \times (½)^0 = 0.0234$. From (6.2) we get $\chi^2 = 4.0$. Table 6.1 lists all 15 permutations of the data and the corresponding multinomial probabilities under H_0 and also the corresponding value of χ^2. For convenience the data are arranged in order of increasing multinomial probabilities.

It is easily verified from the above multinomial probabilities that the probability of the observed configuration, 2 2 0, or any less likely configuration, is

$$p = 0.0234 + 0.0156 + 0.0156 + 0.0039 + 0.0039 = 0.0624$$

so we would not reject H_0 at the 5% significance level. However, if we use an exact test based on the χ^2 statistic the value of this statistic for the observed configuration, 2 2 0, is $\chi^2 = 4.0$ and using the probabilities in the above table associated with $\chi^2 \geq 4.0$ it is easily shown that $p = 0.1873$, almost exactly three times the exact permutation test value. The asymptotic chi-squared distribution probability is approximately $p = 0.14$, but little notice should be taken of this for such small samples.

Traditionally, the asymptotic form of the chi-squared statistic has been used with the proviso that it should only be used if the expected values associated with each group are sufficiently large, but as I have indicated opinion differs on what is meant by 'sufficiently large'. StatXact provides a facility for calculating exact values corresponding to any observed χ^2 using multinomial probabilities, but as Example 6.10 shows these values may be different from those given by the exact

Table 6.1 Multinomial probabilities and χ^2 values for samples of 4 in 3 groups.

Sample configuration			Multinomial probability	χ^2
Red	Pink	Yellow		
4	0	0	0.0039	12
0	4	0	0.0039	12
3	1	0	0.0156	6
1	3	0	0.0156	6
2	2	0	0.0234	4
3	0	1	0.0312	5.5
0	3	1	0.0312	5.5
0	0	4	0.0625	4
2	1	1	0.0937	1.5
1	2	1	0.0937	1.5
2	0	2	0.0937	2
0	2	2	0.0937	2
1	0	3	0.1250	1.5
0	1	3	0.1250	1.5
1	1	2	0.1875	0

permutation test. An asymptotically equivalent alternative to the χ^2 statistic is the likelihood ratio test statistic, which is in the notation of this section uses

$$T = 2\sum f_i \ln[f_i/(m\pi_i)] \tag{6.4}$$

For the data in Example 6.10 this statistic gives an exact $p = 0.1249$ (see Exercise 6.8), slightly nearer the exact permutation probability than that given by χ^2, but there is no general rule to determine which is the closer to the exact multinomial permutation test for any particular data set.

An important practical situation is that where the multinomial probabilities π_i are not specified *a priori* but must be estimated from the data. For example, if we know how many employees in a factory have sustained various numbers of accidents during one year we may wish to test whether the data are consistent with a Poisson distribution with some unspecified mean. The procedure for an asymptotic test is to estimate the mean from the data and calculate the Poisson probabilities of a person having each number of accidents on this assumption. These probabilities are then used as estimated multinomial probabilities in calculating a χ^2 statistic. Because the probabilities have been estimated, the degrees of freedom in an asymptotic test are reduced by 1 to allow for the computation of one parameter, the mean. The permutation test based on the multinomial distribution is no longer valid because for every permutation of the data one would obtain a different estimate of the multinomial probabilities. I do not know whether an exact test has been developed for this case.

If expected numbers in any category are small a common procedure, used also when the multinomial probabilities are known, is to combine adjacent categories until expected numbers are reasonable. This is illustrated in Example 6.11.

Example 6.11 In a factory with 120 employees the numbers of these having 0, 1, 2, 3, . . . accidents in a given year are recorded below. Are these data consistent with the number of accidents having a Poisson distribution?

Number of accidents	0	1	2	3	4	5	6	7
Numbers of employees	72	34	10	1	0	1	0	2

The calculated mean number of accidents per employee gives an estimate of the parameter λ i.e. $\lambda' = (0 \times 72 + 1 \times 34 + 2 \times 10 + 1 \times 3 + 1 \times 5 + 2 \times 7)/120 = 76/120 = 0.6333$. Expected numbers are then estimated by the Poisson formula $Pr(X = r) = e^{-\lambda}\lambda^r/(r!)$, where $\lambda = 0.6333$ and multiplying each probability by $m = 120$ giving

Number of accidents	0	1	2	3	4	≥ 5
Expected frequency	63.7	40.3	12.8	2.7	0.4	0.1

The expected combined frequencies of 3 or more accidents totals only $2.7 + 0.4 + 0.1 = 3.2$. We may take the conservative view that for an asymptotic test we should combine classes to obtain all frequencies at least 5, or the more liberal view that most should be above 5, but that one of these lying between 1 and 5 is acceptable. Taking the latter view we are left with 4 groups having observed and expected numbers:

Numbers of accidents	0	1	2	≥ 3
Observed	72	34	10	4
Expected	63.7	40.3	12.8	3.2

Using (6.2) or the alternative form (6.3) for computation, we easily find $\chi^2 = 2.88$. The degrees of freedom are 2 and not 3 because a parameter, λ, has been estimated. This is well below the value of chi-squared required for significance and asymptotically $Pr(\chi^2 \geq 2.88) = 0.237$ with 2 degrees of freedom. The statistic T in (6.4) provides an alternative test.

I mentioned above that if one has counts of the number of units in specified intervals for a sample from any continuous distribution – so-called grouped data – a chi-squared test may be used and indeed if one only has grouped data this may be the best one can do. For a complete sample of values from a continuous distribution the test is sometimes applied by subjective grouping of the data and then performing the test on the resulting grouped data. Such applications are justifiably viewed with suspicion because the result of the test is not unique, but will depend upon the grouping that is chosen.

If the χ^2 test is used with grouped data from a continuous distribution (perhaps because these are the only data available) and parameters have to be estimated a degree of freedom is deducted as usual for each parameter estimated. Even if the complete data are available, but parameters have to be estimated, it is better to base the estimates on the grouped data rather than the complete data or anomalies may arise. These are discussed by Kimber (1987). For continuous data the Kolmogorov test is generally preferable.

The Kolmogorov test. This test, proposed by Kolmogorov (1933, 1941) assumes that the data are a random sample from a continuous distribution and tests whether they are consistent with that distribution having some completely specified form. The test is distribution-free as the exact distribution of the statistic does not depend upon the specified distribution. The complete theory is beyond the scope of this book and detailed computation of the relevant distribution requires considerable care. A full treatment is given by Gibbons and Chakraborti (1992, section 6.3).

The test is based on the fact that the sample distribution function (which also plays the same key role in bootstrapping as shown in Chapter 2) is an appropriate estimator of the population distribution function. For a sample of n observations, which may be regarded with no loss of generality as being arranged in ascending order $x_1, x_2, \ldots, x_n$, the sample cumulative distribution function $S(x)$ introduced in section 2.1 is a step function with a step of height $1/n$ at each distinct sample value x_i, or a step of height r/n if there are r tied values each equal to x_i.

If the true continuous cdf is $F(x)$ the differences $|S(x_i) - F(x_i)|$ should all be small. How small depends upon the sample size. Kolmogorov proposed a two-tail test based on the statistic $K = \sup|S(x) - F(x)|$. The statistic has these points in its favour:

- It is distribution-free for any completely specified continuous cdf $F(x)$. This implies that a calculated value of K gives rise to the same p-value (or indicates significance at the same level), irrespective of the precise form of $F(x)$.
- For any given n one may compute p-values corresponding to each specified value of K or alternatively compute the value of K indicating significance at a preassigned level, α, irrespective of the selected $F(x)$.
- One can obtain a confidence region for $F(x)$ that determines with a given confidence level the maximum amount that the true cdf value may differ from that indicated by $S(x)$ for any value of x.
- The test may be adapted to a one-tail test for alternatives of the form that for all x the true cdf is always above or always below the function $F(x)$ specified in H_0.

The distribution-free nature of the test holds because the statistic, K, depends only on the distribution of $F(x)$ at values corresponding to the order statistics $x_{(1)}, x_{(2)}, \ldots, x_{(n)}$ and it is well-known that for any continuously distributed variable

X with cdf $F(X)$ the variable $Y = F(x)$ has a continuous uniform distribution over $(0,1)$. A formal proof of the distribution-free property is given by Gibbons and Chakraborti (1992, section 4.3) and also in many standard statistics textbooks.

Computation of exact p-values, or of the values of K indicating significance at specified levels α, is in principle relatively straightforward but requires care in specifying the limits of integration. Details are again given by Gibbons and Chakraborti. Extensive tables of various quantiles of the distribution of K for small to moderate values of n have been published. For larger n an asymptotic result is often quoted. This takes a general form requiring appreciable computation for accurate results and this again is given by Gibbons and Chakraborti, but if the values for significance at only the 5% and 1% level in a two-tail test are required for $n > 40$ useful approximate values of K are $1.36/\sqrt{n}$ and $1.63/\sqrt{n}$ respectively. StatXact, TESTIMATE and other packages provide similar or more accurate approximations.

Availability of exact values of K for significance at level α makes it easy to obtain $100(1 - \alpha)\%$ confidence limits for $F(x)$ for all x by the usual inversion argument. If K is the value of the statistic that just signifies significance at the specified value of α then the required confidence band for any x is $S(x) \pm K$ subject to the constraint that if this gives values less than zero or greater than 1, such values are replaced by 0, 1 respectively.

One-tail tests are less often appropriate than two-tail tests. Approximate one-tail tests are derived by determining the two-tail K that gives a significance level twice that required for the one tail test, or for a given K, the appropriate one-tail p-value is approximately half the two-tail value. An interesting sidelight is that it is easier to compute exact one-tail p-values than it is to compute exact two-tail p-values. Doubling these exact one-tail values gives an approximation to (but not the exact) two-tail p-value. This method is used by some software to compute p-values relevant to both one and two-tail tests because of the considerable computational simplification.

The theory of one-tail tests is again covered comprehensively by Gibbons and Chakraborti, Section 6.3. I quote below one form of the general result due to Birnbaum and Tingey (1951). The one-tail statistics are modifications of the two tail statistics taking respectively the forms

$$K^{+} = S(x) - F(x) \text{ and } K^{-} = F(x) - S(x)$$

depending whether the alternative hypothesis specifies a cdf greater than or less than $F(x)$. Because of symmetry K^{+} and K^{-} are identically distributed. Denoting either by K' Birnbaum and Tingey show that

$$\Pr(K' > k) = (1 - k)^{n} + k\sum {}^{n}C_{j}(1 - k - j/n)^{n-j}(k + j/n) \qquad (6.5)$$

where the summation is over $j = 1$ to $[n(1 - k)]$ where the notation $[u]$ denotes the greatest integer less than or equal to u. A simple asymptotic approximation is

$$\Pr(K' \geq k/\sqrt{n}) = \exp(-2k^2)$$

In particular, if we set $k = 1.36$, then $\Pr(K' \geq 1.36/\sqrt{n}) = 0.0247$. Doubling this probability for an approximate two tail test gives $p = 0.0494$, very close to 0.05, and this is the basis of the asymptotic value for a significance level given on p.162 (see also Exercise 6.9). It is not difficult (see Gibbons and Chakraborti, p.113) to show that if we set $V = 4nK'^2$ then asymptotically V has a chi-squared distribution with 2 degrees of freedom.

Conover (1980) gives an iterative formula for computing (6.5), although his prime aim was to show how these computations may be extended to provide approximately exact tests of the Kolmogorov form for grouped data. StatXact effectively calculates exact p-values using (6.5) or its asymptotic analogue and extends computations to cover the grouped data situation discussed by Conover.

Example 6.12 A parking attendant notes the times recorded on meters in excess of permitted parking time for each offending car he observes. For a random sample of eight cars the excess times in minutes are

$$57 \quad 29 \quad 1 \quad 32 \quad 129 \quad 43 \quad 27 \quad 3$$

Are these consistent with the hypothesis that the excess times are distributed exponentially with mean 20? To compute $K = \sup|F(x_i) - S(x_i)|$ it is convenient, if not using software that does so automatically, to set out the data and values of differences between sample and population distribution functions in a table. Under H_0, $F(x) = 1 - e^{-x/20}$, and from the graph in Figure 6.1 it is easily seen that the difference $F(x) - S(x)$ has greatest magnitude either at one of the given x_i or at a value of x arbitrarily close to but less than an x_i at which point $F(x) = F(x_i)$ for a continuous distribution, but because of its stepwise nature $S(x) = S(x_{i-1})$, with the convention $S(x_0) = 0$. Thus we need the differences in both the fourth and fifth columns in Table 6.2.

A scan of columns 4 and 5 in Table 6.2 immediately shows that $K = 0.491$ just before the step in $S(x)$ at $x = 27$. Tables for the Kolmogorov test show that this lies between the values required for significance at the 5% and at the 1% levels in a two tail test. The exact two tail p-value given by a program such as that in StatXact is $p = 0.0267$. This implies that the data are not consistent with an exponential distribution with mean 20 as their source. They would, however, be consistent with a variety of other distributions; some of these will certainly not be exponential distributions; indeed it is possible the data would not be consistent with any one-parameter exponential distribution.

Tables indicate that the value of K required for significance at the 5% level with a sample of 8 is 0.454. It follows that if, as in Figure 6.2, we draw two further step functions everywhere at 0.454 units above and below the step function representing $S(x)$, subject to the constraints that the functions must not fall below 0 or above 1, these would provide a 95% confidence region for $F(x)$ in

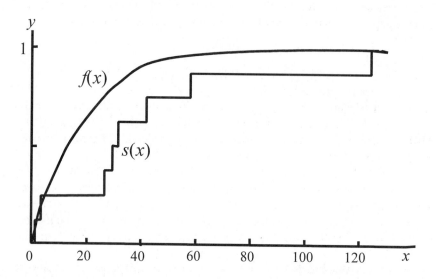

Figure 6.1 Population cdf and sample cdf for the data in Example 6.12 based on Table 6.2.

Table 6.2 Computations to determine the value of the Kolmogorov Statistic K

x	$F(x_i)$	$S(x_i)$	$F(x_i)-S(x_i)$	$F(x_i)-S(x_{i-1})$
1	0.049	0.125	−0.076	0.049
3	0.139	0.250	−0.111	0.014
27	0.741	0.375	0.366	0.491
29	0.765	0.500	0.265	0.390
32	0.798	0.625	0.173	0.298
43	0.884	0.750	0.134	0.359
57	0.942	0.875	0.067	0.192
129	0.998	1.000	−0.002	0.123

the sense that *any* $F(x)$ lying entirely inside the region bounded by these new step functions would be accepted by the Kolmogorov test when testing at a 5% significance level in a two-tail test.

A common practical situation is one where it is reasonable to suppose a sample is from one of an (often small) number of completely specified distributions. We might then use a Kolmogorov test for each. In each test the relevant $S(x_i)$ will be the same for a given sample. For such tests it is not unusual to reject some of these hypotheses but to accept others. It is then of some interest to get at least an

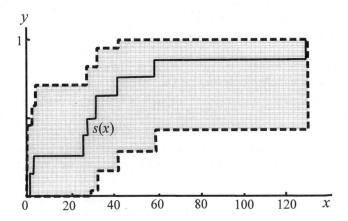

Figure 6.2   Confidence limits (95%) for the cdf giving rise to the data in Example 6.12 The confidence region is the shaded area between the broken lines.

over-all eye impression of how well the match of $S(x)$ is to each of the acceptable $F(x)$. Ross (1990, p.85) gives a useful procedure for such eye comparisons. Basically it consists of plots of the maximum of the two deviations given in the last two columns of *each* row in tables like Table 6.2 against the $S(x_i)$ for that row. Plots may easily be done on the same graph for the several different acceptable hypothesized distributions. An example is given in Sprent (1993, section 3.3.2).

Lilliefors Test. A statistician who has to decide whether parametric tests are justified may want to know whether it is reasonable to suppose data come from a normal population. Unless the parameters μ, σ^2 are known the Kolmogorov test is not immediately applicable. An obvious approach is to estimate these parameters from the data, and to compute an analogous statistic K", say, on this basis. This approach was suggested by Lilliefors (1967) who gave approximate tables for significance at the conventional levels based on a Monte Carlo procedure. He also provided a similar procedure for exponential distributions where the parameter or parameters had to be estimated. There is no closed expression giving exact p-values as there is for the Kolmogorov test, hence the resort to Monte Carlo methods to obtain tables at specific levels. Modern computing methods make it possible to use a Monte Carlo method to obtain approximate p-values without recourse to tables. StatXact provides a program for this, together with an asymptotic result which is essentially equivalent to using the Kolmogorov asymptotic result. The difficulty with using the Kolmogorov test with estimated parameters is that the test becomes conservative in the sense that it is less likely to detect departures. The intuitive explanation for this is that if parameters are estimated from the sample data there is a tendency to get a rather better fit to that sample than would be obtained if one took another sample from the same population because we are effectively, by the estimation process, tailoring our population to the observed sample. What programs like StatXact do is to use a Monte Carlo approach that generates repeated samples from a distribution with the estimated mean and variance and to form in this way an approximate distribution of the resulting K'' and use this to estimate the probability of getting a value as or more extreme than that observed. This procedure is closely related to the parametric bootstrap mentioned in section 2.8.

Example 6.13   Sprent (1993) gave the ages at death recorded on 30 tombstones in the Badenscallie Burial Ground in Wester Ross, Scotland for a random sample of 30 male members of 4 Scottish clans i.e. all were members of a clan bearing one of 4 surnames:

$$11 \quad 13 \quad 14 \quad 22 \quad 29 \quad 30 \quad 41 \quad 41 \quad 52 \quad 55 \quad 56 \quad 59 \quad 65 \quad 65 \quad 66$$
$$74 \quad 74 \quad 75 \quad 77 \quad 81 \quad 82 \quad 82 \quad 82 \quad 82 \quad 83 \quad 85 \quad 85 \quad 87 \quad 87 \quad 88$$

These data are also used in the StatXact 3 manual, Chapter 7, to illustrate Lilliefors' test and also the Shapiro–Wilks test given below. To test whether it is reasonable to accept normality, it is easiest to make comparisons between the standard normal cdf $\Phi(z)$ and $S(z)$, the sample standardized cumulative distribution function based on the transformation $z_i = (x_i - m)/s$ where m is the sample mean and s is the usual estimate of the sample standard deviation, i.e.

$$s = \sqrt{[\{\sum(x_i^2) - (\sum x_i)^2/n\}/(n-1)]}.$$

For the given data $m = 61.43$ and $s = 25.04$. The z_i corresponding to each x_i is calculated, i.e. for $x_i = 52$, $z_i = (52 - 61.43)/25.04 = -0.377$ and from tables of the standard normal distribution we find $\Phi(-0.337) = 0.353$. The complete calculations need for K'' are given in Table 6.3.

The magnitude of the largest difference (final column) is 0.192 when $x = 74$. Tables giving approximate critical values for Lilliefors test (e.g. Sprent (1993, Table IV) indicate significance at the 1% level since it exceeds the critical value 0.187. StatXact output gave a Monte Carlo estimated $p = 0.0055$ with 99% confidence limits for p of 0.0012, 0.0098. Further runs with the same data would give similar, but not quite identical results. StatXact also gives the asymptotic probability based on the Kolmogorov test as 0.0062. Because the sample size is large this agreement is not surprising. One further run with StatXact using the same data gave an estimated exact $p=0.0062$ with a 99% confidence interval for p of (0.0042, 0.0082).

There are a number of other tests for normality when parameters are estimated from the data. One that is widely used and that has been shown to have good power against a number of alternatives is due to Shapiro and Wilk (1965). Special tables of constants, given by Conover (1980, Table A17) are required to implement it. StatXact has a program that computes these constants and carries out the required test. It is illustrated in the StatXact 3 manual using the data in

Table 6.3 Lilliefor's test for normality. Badenscallie death data

x	z	$\Phi(z)$	$S(z)$	$\Phi(z_i)-S(z_i)$	$\Phi(z_i)-S(z_{i-1})$
11	−2.014	0.022	0.033	−0.111	0.022
13	−1.934	0.026	0.067	−0.044	−0.007
14	−1.894	0.029	0.100	−0.071	−0.038
22	−1.575	0.058	0.133	−0.075	−0.042
29	−1.295	0.098	0.167	−0.069	−0.035
30	−1.255	0.105	0.200	−0.095	−0.062
41*	−0.816	0.207	0.267	−0.060	−0.007
52	−0.377	0.353	0.300	0.053	0.086
55	−0.257	0.399	0.333	0.066	0.099
56	−0.217	0.414	0.367	0.047	0.081
59	−0.097	0.461	0.400	0.061	0.094
65*	0.142	0.556	0.467	0.089	0.156
66	0.183	0.572	0.500	0.072	0.105
74*	0.502	0.692	0.567	0.125	0.192
75	0.542	0.706	0.600	0.106	0.139
77	0.622	0.733	0.633	0.100	0.133
81	0.781	0.782	0.667	0.115	0.149
82*	0.821	0.794	0.800	−0.006	0.127
83	0.861	0.805	0.833	−0.028	−0.005
85*	0.942	0.827	0.900	−0.073	−0.006
87*	1.021	0.846	0.967	−0.121	−0.054
88	1.061	0.856	1.000	−0.144	−0.111

*denotes a repeated value.

Example 6.13 above and gives $p = 0.0007$. Conover (1980, section 6.2) discusses possible applications in some detail but does not give the full theory of the test which involves regression of the order statistics on normal scores.

Other tests for goodness of fit to normal distributions have been proposed by, among others, Bowman and Shenton (1975), and are based on estimates of skewness and kurtosis and one of their tests has been shown by Jarque and Bera (1987) to be powerful against a range of alternatives. However departures from normality do not always reflect themselves directly in skewness or kurtosis, especially if there is bimodality, a situation that often occurs if data are generated from a mixture of distributions.

6.8 Comparing two distributions

The WMW test (sections 4.2, 4.4) provides a robust distribution-free test for shifts in location or dominance of one distribution, and several tests for dispersion differences are given in section 6.5. However, one may be interested in more general differences between population continuous cdfs $F(x)$, $G(x)$ of the form $H_0:F(x) = G(x)$ against $H_1:F(x) \neq G(x)$. While one-sided alternatives such as $H_1':F(x) < G(x)$ are also testable by the method I now give these are of secondary interest because the WMW test (as a test for dominance) is then often more efficient and more powerful. The test below was proposed by Smirnov (1939) and the computations and test statistic are closely related to those for the Kolmogorov test in section 6.7, but now an exact permutation test is available. The two tests are sometimes jointly described as the Kolmogorov–Smirnov tests. The Smirnov test is distribution-free as it depends only on the ordering of the observations in the two samples and may be applied either to the original data or the ranks, since the latter retain the order. For the two-tail test, the statistic is the maximum value of $|S_1(x) - S_2(x)|$ where $S_1(x)$, $S_2(x)$ are the sample cdfs, the appropriate point estimators of the corresponding population cdfs. For a one-tail test with the alternative hypothesis $G(x) > F(x)$ the appropriate statistic is the maximum of $S_2(x) - S_1(x)$ while for the alternative $G(x) < F(x)$ the appropriate statistic is the maximum of $S_1(x) - S_2(x)$.

The distribution of each of the statistics is given by forming all $^{m+n}C_n$ allocations of the combined data to samples of the given size m, n as we did for the Fisher–Pitman test in section 1.6 and for the WMW test in section 4.2. The Smirnov statistic given above is calculated for each of the $M = (m + n)!/(m!n!)$ permutations. The exact p-value corresponding to the relevant statistic, S, is

$$p = (\textit{number of permutations giving the observed or a greater value of } S)/M.$$

For small samples it is tedious, but not difficult, to compute the distribution of the Smirnov statistic S.

Example 6.14   I illustrate the computation of the Smirnov one- and two-tail test statistics for a simple example and outline the procedure for determining the exact distribution. StatXact gives a program for the exact test for small samples and an asymptotic test for larger samples.

We test whether the following samples of 2 and 4 observations may be from the same population against the two-tail alternative that they are from different populations.

Sample I	3.2	5.9		
Sample II	3.1	6.2	6.3	6.4

The computations in Table 6.4 below are more detailed than are needed for the present test, but they provide also the statistics relevant to one-tail tests. The first column gives the ordered sample I values and the second column the ordered sample II values so arranged that there is only an entry in one column or the other and the combined entries corresponding to the overall ordering of the samples in ascending order. Columns 3 and 4 give values of $S_1(x)$ and $S_2(y)$ computed at each sample point where $x = y$ because the sample cdfs are constant between sample values. The remaining columns give the differences shown at the head of each column.

Table 6.4 Computations for the Smirnov one- and two-tail test statistics

x	y	$S_1(x)$	$S_2(y)$	$S_1(x)-S_2(y)$	$S_2(y)-S_1(x)$	$|S_1(x)-S_2(y)|$
	3.1	0	0.25	−0.25	0.25	0.25
3.2		0.5	0.25	0.25	−0.25	0.25
5.9		1.0	0.25	0.75	−0.75	0.75
	6.2	1.0	0.5	0.5	−0.5	0.5
	6.3	1.0	0.75	0.25	−0.25	0.25
	6.4	1.0	1.0	0	0	0

Table 6.5 Number of times each Smirnov statistic is observed in permutations of samples of 2 and 4

Value observed	Number of times observed		
	$S=\max|S_1(x)-S_2(y)|$	$S'=\max\{S_1(x)-S_2(y)\}$	$S''=\max\{S_2(y)-S_1(x)\}$
0.0	0	3	3
0.25	1	4	4
0.5	8	5	5
0.75	4	2	2
1.0	2	1	1

The final column 7 is relevant to the two-tail test for the alternative $H_1:F(x) \neq G(x)$ and the maximum entry 0.75 implies $S = 0.75$. To test the alternative $H_1':F(x) > G(x)$ column 5 provides the statistic $S' = 0.75$. For the alternative $H_1'':F(x) < G(x)$ column 6 shows $S'' = 0.25$.

Inspecting Table 6.4 we see that the sample x, y values do not enter directly into the calculation of S, S' or S'' because we only need to know the ranks of these to compute sample cdfs and the statistics. This, as pointed out above, indicates the distribution-free nature of the test. To obtain the complete distribution of the statistics we need to calculate their values for each permutation giving samples of 2, 4. There are $^6C_2$ such permutations, one of which corresponds to the observed sample. It is a tedious but simple exercise to compute the statistics for each of the 15 permutations. Table 6.5 summarizes the results for the 15 permutations, indicating the number of permutations in which each possible value of S, S', S'' is obtained (Exercise 6.10).

In the current example we found $S = 0.75$. The probability of getting a permutation with this or a higher value (the only possible higher value is 1) is seen from column 2 of Table 6.5 to be $(4+2)/15 = 0.4$ so in this case no serious consideration would be given to rejecting H_0. Indeed for such small samples the smallest value of p is $p = 1/15 = 0.06667$, and this probability is only associated with one-tail test statistics.

For larger samples the range of possible p-values increases rapidly. For example, in a two-tail test with samples of 9 and 12 the minimum possible p-value is $p = 0.0001$.

Table 6.6 Computation of the Smirnov statistic S for the data on numbers of references.

Sample I	Sample II	$S_1(x)$	$S_2(y)$	$\lvert S_1(x)-S_2(y)\rvert$
2		0.0833	0	0.0833
3		0.1667	0	0.1667
6*(3)	6	0.4167	0.0714	0.3453
	8	0.4167	0.1429	0.2738
9*(2)	9	0.5833	0.2143	0.3690
10	10	0.6667	0.2857	0.3810
14		0.7500	0.2857	0.4643
15		0.8333	0.2857	0.5476
16	16	0.9166	0.3571	0.5595
	18	0.9166	0.4286	0.4880
	20	0.9166	0.5000	0.4166
	22*(2)	0.9166	0.6429	0.2737
	23*(2)	0.9166	0.7857	0.1309
	26	0.9166	0.8571	0.0595
	28	0.9166	0.9286	0.0012
	59	0.9166	1.0000	0.0834
72		1.0000	1.0000	0

Example 6.15 For the data on numbers of reference in papers in *Biometrics* in Example 4.1 apply the Smirnov test for the null hypothesis that the samples are from identical populations against the two-tail alternative that they come from different populations. Table 6.6 shows the necessary computations. There are a number of repeated sample values and care is needed in computing the sample cdfs at such points although the procedure is straightforward. Repeated values in each sample are indicated by an asterisk with the number of repetitions in that sample in braces after the asterisk. The sample sizes were 12 and 14 respectively.

Inspecting the final column shows $S = 0.5595$. StatXact confirms this and gives $p = 0.0157$. This compares with $p = 0.0065 \times 2 = 0.0130$ for the WMW test, which, when it is appropriate, is generally more powerful. The Smirnov test will, however, pick up more general differences.

Asymptotic results are available, but they are not very satisfactory unless both samples are large. One is given by StatXact, but for the data in Example 6.15 the asymptotic estimate is $p=0.0350$, almost double the exact value.

Alternative tests for comparison of population cdfs are available. One that is often powerful and has the attraction that a good asymptotic approximation exists and is easy to use and valid for moderate sample size is the Cramér–von Mises test stemming from work by Cramér (1928) and von Mises (1931). Instead of using only the maximum difference between sample cdfs the statistic is based on the sum of squares of all differences. With ties as in Example 6.15 values of the difference corresponding to ties in either sample should be counted r times where r is the number of ties in the combined samples. For samples of size m, n if S_d denotes the sum of squares of the differences $S_1(x) - S_2(y)$ then the statistic used is $T = mnS_d/(m + n)^2$ and for moderate or large m and n values of T exceeding 0.461 indicate significance at approximately the 5% level and values of T exceeding 0.743 indicate significance at approximately the 1% level. This virtual independence of the significance level on sample size for moderate to large samples is useful. It is theoretically possible to work out the permutation distribution for this statistic as was done for the Smirnov statistic, and although programs to do this may exist, none has come to my attention. For the data in Example 6.15 it may be verified that $T = 0.754$, just indicating significance at the 1% level. If programs become readily available for an exact test this test might be preferred to the Smirnov test.

6.9 Run tests

We often want to know if a sample or samples may be regarded as random in some context.   This may often be answered in part by what are called run-tests (sometimes runs-tests). In the two-sample problem one of these tests provides an alternative to the Smirnov test to decide whether the samples may be from the same population.

For a single sample the simplest form of run test is applicable to an ordered sequence of binary digits, e.g.

$$0, \ 1, \ 1, 0, 0, 1, 1, 1, 1, 0, 1, 0, 0, 1, 1$$

Such a sequence may characterize any binary variable, eg. success (1) or failure (0); male (1) or female (0); alive (1) or dead (0) and so on. The notion extends to derived variates from samples of continuous measurements, e.g., 1 may denote a sample value at or above the sample median, 0 a value below, or if the data consist of counts, 1 may correspond to an even number and 0 to an odd number.

Methods for obtaining a random sample, especially in simulation studies, often depend upon sophisticated mechanisms called pseudo-random number generators purporting to be equivalent to repeated tossing of a coin or to repeated independent selection of one of the ten digits 0, 1,2, 3, 4, 5, 6, 7, 8, 9 each having a probability 0.1 of selection.

A characteristic of the coin tossing situation is that in the long run there should be approximately equal numbers of heads and tails. A sign test equivalent is clearly appropriate to see if this criterion is being met, but acceptance is no guarantee of randomness.   One should be highly suspicious of a computer program that allegedly produced the digits 0, 1 in random order if, in a sequence of 20, it produced digits in the order

$$0 \ 0 \ 0 \ 0 \ 0 \ 1 \ 1 \ 1 \ 1 \ 1 \ 1 \ 1 \ 1 \ 1 \ 0 \ 0 \ 0 \ 0 \ 0 \ 0$$

and equally suspicious of the orders

$$0\ 1\ 0\ 1\ 0\ 1\ 0\ 1\ 0\ 1\ 0\ 1\ 0\ 1\ 0\ 1\ 0\ 1\ 0\ 1$$

or

$$0\ 0\ 0\ 0\ 0\ 0\ 0\ 0\ 0\ 0\ 1\ 1\ 1\ 1\ 1\ 1\ 1\ 1\ 1\ 1.$$

Although each contains 10 zeros and 10 ones, the characteristic that reflects our reservations about randomness in these outcomes is the **number of runs** or **lengths of runs** where a run is a sequence of one or more ones or zeros followed either by an outcome of the opposite kind or of no further observation.   It is easily verified that in the three examples above the numbers of runs are respectively 3, 20 and 2. Intuitively we should be happier about the functioning of the random digit generator if we got an outcome such as

$$0\ 0\ 0\ 1\ 0\ 1\ 1\ 0\ 0\ 1\ 0\ 0\ 1\ 0\ 1\ 1\ 0\ 1\ 0\ 1$$

where the number of runs is 14. We feel happier about this sequence reflecting the coin tossing situation (with, say, H = 1, T = 0) because there is a less discernable pattern in the sequence and the number of runs, 14, is intermediate between the extremes of 2 (corresponding to only one alternation of heads to tails or vice versa) and 20 (corresponding to an alternating sequence of heads and tails or vice versa). Although the numbers of 1s and 0s in this last example are not quite equal, being 9 and 11, they are well within the limits acceptable in a sign test for the hypothesis that each occurs with probability $\pi = \frac{1}{2}$.

Both the numbers of and the lengths of runs can be used to test for randomness. The two are clearly inter-related, long runs tending to go with few runs and very short runs with many runs, in a sequence of given length.

A permutation test may be based upon the distribution of all possible numbers of runs in a sequence of a fixed number of binary digits, extremely large or small values indicating nonrandomness. A detailed treatment of this approach is given in Bradley (1968, Chapters 11 and 12) and a concise but complete account in Gibbons and Chakraborti (1992, Chapter 3).   Tests based on length of the longest run are also described by these writers.

The distribution of numbers of runs is obtained using certain standard results in permutation algebra. I omit the derivation here and simply quote result:

Let the random variable R denote the number of runs in a sequence of $N=m+n$ binary observations 0, 1 composed of m 0s and n 1s. If r, the number of runs observed is odd, say $r = 2s + 1$, then

$$\Pr(R = 2s+1) = (^{m-1}C_{s-1} \times {}^{n-1}C_s + {}^{m-1}C_s \times {}^{n-1}C_{s-1})/{}^{N}C_m. \qquad (6.6)$$

If the number of runs is even, say $r = 2s$, then

$$\Pr(R = 2s) = 2 \times {}^{m-1}C_{s-1} \times {}^{n-1}C_{s-1}/{}^{N}C_{m}. \tag{6.7}$$

For an odd number of runs the first term in the numerator is set at zero if $s \geq n$ and the second term is set at zero if $s \geq m$.

The tail-probabilities associated with the distribution of R given above have been tabulated and StatXact provides a program that gives exact probabilities and also an asymptotic test. The latter uses results obtained after tedious algebra (see Gibbons and Chakraborti, 1992, section 3.2), namely

$$E(R) = 1 + 2nm/N$$

and

$$\text{Var}(R) = 2nm(2nm - N)/[N^2(N - 1)].$$

Asymptotically,

$$Z = [R - E(R)]/[\sqrt{\text{var}(R)}],$$

has a standard normal distribution. The approximation is improved by a numerator continuity correction, adding $+\frac{1}{2}$ if $R < E(R)$ and subtracting $\frac{1}{2}$ if $R > E(R)$.

In practice a two-tail test is most often relevant, a significant result implying non-randomness, but a one-tail test may be meaningful. If there are few runs this implies clustering of like values as opposed to randomness, and many runs implies a pattern of alternation rather than randomness.

Example 6.16 The tail probabilities, or if required, the complete distribution of the number of runs under the hypothesis of randomness, may be computed directly for small samples. Suppose $m = 4$ and $n = 3$. There must be at least two runs if both 0s and 1s occur and there cannot be more than 7. Substitution in (6.6) and (6.7) (Exercise 6.13) gives

$$
\begin{aligned}
\Pr(R = 2) &= 2/35 = 0.0571 \\
\Pr(R = 3) &= 5/35 = 0.1429 \\
\Pr(R = 4) &= 12/35 = 0.3429 \\
\Pr(R = 5) &= 9/35 = 0.2571 \\
\Pr(R = 6) &= 6/35 = 0.1714 \\
\Pr(R = 7) &= 1/35 = 0.0286
\end{aligned}
$$

For such small samples only a one-tail test is meaningful and if we adhere to conventional significance levels we reject randomness only if $R = 7$, when $p = 0.0286$, implying alternation as opposed to randomness.

Example 6.17 The time to failure (in days) of 20 machine components is recorded in the order in which the machines were installed. If there is a monotone trend either increasing (or decreasing) in times to failure this would mean the later observations would tend to be greater (or less) than early observations. A simple test to decide whether there is evidence of this as opposed to random fluctuation is to record for each observation whether it is above or below the sample median and to apply a runs test to the resultant sequence of zeros (for below median) and 1s (for above median). The data are:

72 65 81 56 68 71 92 86 69 73 59 82 83 95 93 112 84 122 101 85

It is easily verified that the sample median is 82.5 and thus the sequence of below and above the median values is

0 0 0 0 0 0 1 1 0 0 0 0 1 1 1 1 1 1 1 1

There are clearly 4 runs (compared with a minimum of 2 and a maximum of 20). StatXact gives the probability of observing 4 or less runs as 0.001. A one-tail test is appropriate here as the alternative we are interested in is a clustering of like symbols (0 or 1) at the beginning or end of the series, and this would be indicated by a reduced total number of runs. The asymptotic result given by StatXact is $p=0.0014$ for one tail.

An interesting application of the two-tail test is the Wald–Wolfowitz runs test (Wald and Wolfowitz, 1940) for a population difference given two independent samples. Like the Smirnov test, it is an overall test for differences. The test is carried out by arranging the combined data in ascending order and coding each first sample value as zero and each second sample value as 1. The number of runs will be distributed as for a single random sample under the hypothesis of population identity. The alternative is $H_1:F(x) \neq G(x)$ and a moment's reflection shows that only a reduced total number of runs in the combined sample supports H_1, because if the samples are from the same population a large number of runs does not indicate a population difference, though it might suggest that the sampling mechanism has some tendency to produce related values, i.e. that the samples are not independent. A difficulty arises with tied values in different samples. One suggestion, which is conservative, in that it makes it less likely to reject H_0, is that these be scored in an order that maximises the number of runs, and the p-value be calculated on the basis of that ordering. This avoids false conclusions about population differences. On the other hand, a rearrangement among ties to get the minimum number of runs may over-estimate significance. This feature makes the test unsatisfactory with many ties. StatXact gives p-values for both possibilities.

Example 6.18 For the data on numbers of references in Example 4.1 use the Wald–Wolfowitz runs test for population identity. We first arrange the combined sample in ascending order. This is done below with the sample 1 values in **bold**. Below each value is the indicator variable 0 or 1 arranged in two different orders where ties occur. The first gives the greatest possible, the second the least possible total number of runs. Such orderings may not be unique (Exercise 6.14).

2 3 6 6 6 6 8 9 9 9 10 **10 14 15** 16 **16** 18 20 22 22 23 23 26 28 59 **72**
0 0 1 0 0 0 1 0 1 0   1   0 0 0   1 0   1 1 1 1 1 1 1 1 1 0
0 0 0 0 0 1 1 1 0 0   1   0 0 0   0 1   1 1 1 1 1 1 1 1 1 0

You should verify that when $m = 12$ and $n = 14$ and with the given pattern of ties, no other ordering of the cross-sample ties gives a greater number of runs, here $R = 13$, than the first set of 0s and 1s , or less than the second set where $R = 7$.   This dramatic difference is reflected in the corresponding p-values obtained for the exact test using StatXact, namely $p = 0.0040$ when $R = 7$ and $p = 0.4296$ when $R = 13$.

This indicates that for heavy tying across samples the runs test is less satisfactory than the Smirnov test.

6.10   A case study

Here is a simple data set for the analysis of which many of the test and estimation procedures covered so far in this book may be used and their performance compared to that of classical procedures. I summarize some approaches and suggest others that might be explored. The example shows how easy it is to generate simple data sets that raise a number of pertinent questions, and I encourage readers to produce such sets from their own fields of interest.

Example 6.19   These data are given in Hand *et al* (1994, Set 2) and were collected by Professor T. Lewis in Perth, Western Australia, shortly after metric units of length were officially introduced in Australia to replace imperial units. One group of 44 students were asked to guess, to the nearest metre, the width of the lecture hall in which they were sitting and another group of 69 students were asked to guess the width of the same room to the nearest foot. The room was in fact 13.1 metres (43.0 feet) wide. The data were collected to answer such questions as

● Are students better at guessing in units with which they are more familiar (in this case feet)?
● Do student guesses tend to over or underestimate the width of the room when using one or other unit? If so, is any bias similar in both units?
● Are the guesses more variable when using one unit than for the other?
● Is there evidence of skewness or of outliers?

The data obtained were:

Guesses in metres
8   9   10   10   10   10   10   10   11   11   11   11   12   12   13   13   13   14   14   14   15   15
15   15   15   15   15   15   16   16   16   17   17   17   17   18   18   20   22   25   27   35   38   40

Guesses in feet
24   25   27   30   30   30   30   30   30   32   32   33   34   34   34   35   35   36   36   36   37   37
40   40   40   40   40   40   40   40   40   41   41   42   42   42   42   43   43   44   44   44   45   45
45   45   45   45   46   46   47   48   48   50   50   50   51   54   54   54   55   55   60   60   63   70
75   80   84

The data are observational data so care is needed in considering the range of validity of any inferences. The units in each set are different, so for comparative purposes one needs to transform the units in one set. Most statistical software will easily convert feet to metres or *vice versa* introducing spurious accuracy by giving results to many decimal places unless some control is applied. It is good statistical practice first to compare small data sets like these visually and to do this a sensible step is to retain one decimal place in converted units to avoid spurious accuracy while maintaining order in the data without losing relevant information as would happen, for example, if transformations from feet to metres were rounded to the nearest metre (then e.g. 42, 43 and 44 ft would all round to 13 m). The *guesses in feet* converted to metres using this convention are:

```
7.3   7.6   8.2   9.1   9.1   9.1   9.1   9.1   9.1   9.8   9.8  10.1  10.4  10.4  10.4
10.7  10.7  11.0  11.0  11.0  11.3  11.3  12.2  12.2  12.2  12.2  12.2  12.2  12.2  12.2
12.2  12.5  12.5  12.8  12.8  12.8  12.8  13.1  13.1  13.4  13.4  13.4  13.7  13.7  13.7
13.7  13.7  13.7  14.0  14.0  14.3  14.6  14.6  15.2  15.2  15.2  15.5  16.5  16.5  16.5
16.8  16.8  18.3  18.3  19.2  21.3  33.9  24.3  28.7
```

Many of the questions of interest are answered by looking at each data set separately in their original units, while others require comparison of the two sets, thus we need both one-sample and two independent sample methods.

A useful preliminary step is to inspect and summarize the data invoking tools such as bar charts, five number summaries (least value, first quartile, median, third quartile, greatest value) perhaps representing these on adjacent box plots [see, e.g. Chatfield (1988)]. It is immediately clear from such summaries that the data are skewed with long upper tails, this being particularly marked for guesses in metres. It is also clear that the sample median (15 m) for those who guess in metres overestimates the true width by 1.9 m (6.2 ft) but that the median (42 ft) for those who guess in feet slightly underestimates the true width by just 1 ft (0.3 m). The modal, or most common estimates, for each group are respectively 15 m and 40 ft, indicative of a well known tendency in situations like this to round estimates to the nearest 5 units. The Lilliefors test confirms non-normality in both cases ($p \approx 0.0005$ for data in metres and $p \approx 0.008$ for data in feet using Monte Carlo sampling in StatXact). This is not surprising as one might reasonably expect a complex mixture of distributions reflecting different estimation techniques used by individuals. Some may be experienced in judging lengths of the order of the width of the room and be able to do so fairly precisely while others may make little more than vague guesses. In each set there are a few guesses that over-estimate the true width by a factor of two or more, while at the lower limit there are none as low as half the true width. A few very large estimates show indications of being outliers in the sense that they cause surprise. The statistic proposed in (3.8) provides a robust test for outliers and using it the three top values in each set are designated outliers. Although removal of these values leaves the data still somewhat skewed the Lilliefors test no longer rejects normality ($p \approx 0.0865$ for data in metres and $p \approx 0.1116$ for data in feet).

To test whether one would accept that the data (including the outliers) are consistent with a mean or median equal to the true width, available one-sample tests include the *t*-test (suspect in view of non-normality although the central limit theorem implies some robustness) the Wilcoxon signed rank test (suspect in view of the skewness) and the sign-test (which ignores a lot of information in these moderately large samples). A two-tail test is appropriate as we have no prior reason to specify a particular tail in our alternative hypothesis. The relevant *p*-values for each of these tests are:

	Metres	*Feet*
t-test	0.009	0.65
Wilcoxon	0.047	0.52
Sign test	0.174	0.46

Despite the evident nonnormality the *t*-test performs well because for the somewhat larger samples here relative to many met earlier the influence of the central limit theorem is stronger. Although the Wilcoxon test does not reflect so strongly the upward bias in guesses in metres it is of some interest to compare the 95% confidence intervals with those for the *t*-test. These are:

	Metres	*Feet*
t-test	(13.85, 18.20)	(40.70, 46.70)
Wilcoxon	(13.7, 16.0)	(40.0, 45.0)

The shorter intervals for the Wilcoxon test reflects release from the restriction of symmetry about the sample mean for the limits explicit in those based on the *t*-test.

Had I decided to reject the three highest data from each sample as outliers, different results would be obtained. The reader should explore the consequences of their removal. In particular, will it remove the tendency for guesses in metres to produce an over-estimate?

Because the guesses in metres tend to over-estimate, it does not follow that these samples of guesses in the different units necessarily provides evidence that the two populations differ in location. This and other questions such as whether guesses in the two units show differences in dispersion lead to two-sample tests. A range of tests are possible here. We might use the Smirnov test for overall unspecified population differences, or more specifically the WMW or normal scores test for a median shift or dominance. To test for dispersion differences an appropriate test depends upon whether or not we accept a possible difference in location. If the WMW or other tests indicate a location difference, a test such as the modified Conover test which allows for adjustment for a location difference would be appropriate. One might also use the standard two-sample test for difference between means relevant to normal distributions (not necessarily having the same variances) in the light of our large sample sizes, hoping that the central limit theorem will once again overcome the problems of nonnormality. This last test might be expected to give similar results to a Fisher–Pitman test. The skewness suggests that the log rank test may not be unreasonable for location differences. These tests led to the following two-tail *p*-values using the data after transforming the measurements in feet to metres Those marked * are Monte Carlo approximations to exact values obtained using StatXact with 50 000 samples:

Smirnov test	0.004*	*Fisher–Pitman*	0.011
WMW	0.029	*log-rank*	0.010*
Conover Test	0.014*	*Normal scores*	0.023*
Normal test for means	0.020		

These results suggest strong evidence that the samples indicate populations with differences in location and dispersion and perhaps also differences in other respects, e.g. skewness. The fact that all tests indicate differences despite the fact that for some of them strict conditions for validity do not hold indicates that increasing sample sizes do bring some robustness to hypothesis testing. It is interesting to look at how 95% confidence intervals (using metres) for the difference between mean guesses in metres and feet compare. Normal distribution theory 95% confidence limits are (0.7, 4.7) and those based on the WMW test are (0.2, 2.9) again indicating the stretching in normal theory confidence intervals to give symmetry about the sample mean difference that occurs when samples are from skew distributions. Thus confidence intervals do not always exhibit the robustness displayed by some hypothesis tests.

I leave the reader to explore the use of techniques such as the bootstrap, the jackknife and cross validation, all of which are potentially useful in the analysis of these data. An alternative approach using classic methods would be to use a transformation of data prior to analysis. A transformation to logarithms should reduce the skewness and the Lilliefors test may be used to test for normality of the transformed data. The reader might like to try this, and also to consider the problem of interpreting results from transformed data to answer questions of interest about the original data.

Finally, whatever method of analysis is used one must consider the range of validity of any inferences. I have pointed out that the data are observational data. How far should one be prepared to accept these are typical of data one might get had we taken a random sample of observers from a large population, or allocated all available observers to the two groups at random? It is likely that the observers were students or pupils attending two different lecture courses. Had the two groups differed significantly in age or sex composition might we not be measuring in part at any rate the influence of age or sex differences in guessing width? Such effects are then described as being **confounded** with the difference of interest and our experiment gives us no way of untangling one from the other. Our decision on the range of validity of any inferences about differences between group data, and indeed on the best method of making them, must inevitably be influenced by how much background information we have about factors such as, in this case, potential age or sex differences and perhaps other factors (e.g., whether one group were engineering students and the others social science students). It would be interesting to know what information Professor Lewis had about such points. One might also consider whether the tendency to round guesses to the nearest 5 units might not account for overestimation of guesses in metres and to slight underestimation in feet, since the true width of the room rounded to the nearest 5 units is 15 m or 40 ft.

Exercises

6.1 How would you modify the procedure in Example 6.2 to test the hypothesis $H_0 : p = 0.8$ against the alternative $H_1 : p < 0.8$?

6.2 If 57 out of 90 pupils from one school pass an examination for which the national pass rate is 80% is this consistent with a hypothesis that the school's performance is not below the national average?

6.3 The asymptotic approximation for the exact p-value relevant in Exercise 6.2 was shown on p. 143 to be poor. By working out corresponding probabilities, both asymptotic and exact for several values between $p = 0.01$ and $p = 0.1$ for the given m, π values in Exercise 6.2 determine whether this approximation is better for these greater values of p.

6.4 Determine a normal and also a chi-squared approximation for the McNemar test. (Hint: these are in essence the approximations for the relevant binomial distribution).

6.5 Verify the mean and variance (p. 152) for the Mood test statistic for equal dispersion.

6.6 Verify the mean and variance (p. 153) for FABDB test statistic for equal dispersion.

6.7 Verify the mean and variance (p.153) for the Conover test statistic for equal dispersion.

6.8 Calculate the exact probability associated with the likelihood ratio test for goodness of fit with the data in Example 6.10.

6.9 Verify the 1% asymptotic result for Kolmogorov test given on p.162 using the Birnbaum and Tingey approximation given on p.162-3.

6.10   Verify the entries in Table 6.5 for the Smirnov test for samples of size 2 and 4 by considering all possible permutations of the rank order.

*6.11 Perform the Smirnov one-tail test for the number of references data for Biometrics used in Example 6.15 for the alternative of stochastic dominance for the later year.

*6.12   Use Lilliefors test for each sample of Biometrics references in Example 6.15 to determine whether or not there is evidence of non-normality.

6.13 Confirm the probabilities for each number of runs given in Example 6.16.

6.14 Confirm by considering other possible arrangements that the orderings given in Example 6.18 for least and greatest number of runs are not unique. Do you think the Wald–Wolfowitz test would be suitable for the data in Example 6.19? If you have software for carrying out this test try it for these data.

*6.15 For the data used in Example 6.8 perform the Conover squared rank sum test on the assumption that each sample is from some symmetric population with known median 1002.

6.16 Confirm that the Gastwirth statistics U, V defined on p. 154 correspond respectively to the FABDB statistic and the Wilcoxon rank-sum statistics when $P = R = \frac{1}{2}(N - 1)$.

*6.17 Use the Lilliefors test on the data in Exercise 5.8 to decide whether it is reasonable to assume that the numbers of pages in my statistics books are normally distributed.

*6.18 Use the Smirnov test to decide whether it is reasonable to suppose the samples in Exercise 4.10 come from different populations. Do you consider this test preferable to any tests you used in Exercise 4.10?

*6.19 It is often argued that the probability a horse wins a race on a circular track is influenced by its starting position? Does the following data for the starting position of the winners of 160 nine-horse races support this argument. Position 1 is on the inside of the track.

Position	1	2	3	4	5	6	7	8	9
No. of wins	28	21	30	14	12	16	19	8	12

*6.20 Rutherford and Geiger (1910) give the following data for counts of the numbers of scintillations in 72 second intervals caused by radioactive decay of a specimen of polonium. Are the data consistent with a Poisson distribution?

Count	0	1	2	3	4	5	6	7	8	9	10	11	12	13	14
Frequency	57	203	383	525	532	408	273	139	45	27	10	4	0	1	1

*6.21 A lot of data has been published about eruptions of the Old Faithful geyser in Yellowstone National Park, USA. The data below is a subset of some given by Azzali and Bowman (1990) for times between the starts of successive eruptions. Geological theory suggests it is unlikely that these are uniformly randomly distributed. Perform appropriate tests of the validity of a hypothesis of uniform random distribution (read across columns).

```
80 71 57 80 75 77 60 86 77 56 81 50 89 54 90  73 60 83 65 82 84 54
85 58 79 57 88 68 76 78 74 85 75 65 76 58 91  50 87 48 93 54 86 53
78 52 83 60 87 49 80 60 92 43 89 60 84 69 74  71 108 50 77 57 80 61
```

An extended table of these data is given in Hand *et al* (1994, Set 280).

7

Three or more independent samples

7.1 Some extensions and some differences

When extending classic normal theory location tests from 2 to k independent samples the t-test gives way to the analysis of variance (ANOVA) in overall tests for location differences, the link being that if a statistic t has a t-distribution with v degrees of freedom then t^2 has an F-distribution with 1, v degrees of freedom. Thus, inference based on a two-sample t-test has an equivalent analysis of variance format which generalizes to k independent sample problems. This leads to an omnibus two-tail test of the hypothesis that the k independent samples are all from the same normal population, against an alternative that there is at least one difference in location among the k populations.

Even if we consider only location differences, extension from 2 to $k > 2$ samples often calls for testing more detailed hypotheses about the pattern of such differences rather than just testing for the existence of some otherwise unspecified location difference. Typically, if an overall test suggests location differences, we want to know precisely which samples or groups of samples indicate population locations different from those of others or whether there is some ordered pattern in the differences. For example, if I have k samples from normal populations with means $\theta_1, \theta_2, \ldots, \theta_n$, but which are otherwise identical, I may wish to test H_0:*all θ_i are equal*, against an *ordered* alternative, e.g., H_1:$\theta_1 \le \theta_2 \le \ldots \le \theta_k$, where at least one of the inequalities is strict. Other well-known problems dealt with in parametric ANOVA include tests for *treatment patterns* like those of main effects and interactions in factorial treatment structures. I defer to the next chapter nonparametric equivalents to ANOVA and related tests for designed experiments involving the grouping of treatments into blocks, etc., many of which are basically extensions of the two dependent or paired sample concept introduced in section 5.6

Extensions of nonparametric location tests developed in Chapter 4 may show parallels to, but also some divergences from, their parametric counterparts. The theory is often straightforward, but if relevant computer software for calculating exact p-values is not available there are appreciable practical difficulties in application, particularly, in situations where asymptotic results are inappropriate. In this chapter I consider these practical details as they arise, often in specific examples.

7.2 Permutation analysis for *k*-samples or treatments

The two-sample Fisher–Pitman permutation test extends to k independent samples. If the original data are used this test provides the randomization test appropriate to a situation where k treatments are applied to randomly selected subsets of N units, the ith treatment being applied to n_i units, where $\sum n_i = N$. We write the observed samples:

$$
\begin{array}{ll}
\text{Sample 1} & x_{11}, x_{12}, \ldots, x_{1n_1} \\
\text{Sample 2} & x_{21}, x_{22}, \ldots, x_{2n_2} \\
\ldots\ldots \\
\text{Sample } k & x_{k1}, x_{k2}, \ldots, x_{kn_k}
\end{array}
$$

I pointed out in Section 1.6 that although the two sample Fisher–Pitman test is usually based on a simple statistic like the sum or the mean of the smaller sample values, an alternative equivalent statistic is the usual t-statistic as this is a monotonic transformation of the simpler statistic, but that the usual t-tables were no longer appropriate. Similarly, for the general permutation test based on the original data for k samples one may use the standard variance ratio

$$F = (Treatment\ mean\ square)/(Residual\ mean\ square).$$

However, the usual normal theory F-tables or programs for determining p-values for F under normality and homogeneity of variance assumptions are no longer relevant. For exact p-values one must determine the number of permutations giving as great or a greater F than that observed, or some equivalent statistic, where the relevant reference set is all permutations of the x_{ij} subject to the constraint that exactly n_i of these are allocated to sample i, $i = 1, 2, \ldots k$

If there are r permutations giving a value as great or greater than the observed F then the relevant p-value for a two-tail test for location difference is $p = r/M$ where M is the number of possible permutations. Standard permutation theory gives:

$$M = \frac{N!}{n_1! n_2! \ldots n_k!}$$

For even moderate N, the number, M, is large and highly efficient algorithms may take a long time, or be unable to cope with the computation of exact p-values. For $N = 20$, and 3 samples of sizes 7, 7 and 6 then $M = 133\ 024\ 320$. If exact p-values cannot be calculated, StatXact gives good Monte-Carlo approximations together with a confidence interval (at a confidence level chosen by the user, but typically 99%) for the true p.

Just as we could use a simpler statistic than t in the two independent sample case we may use one here. The total sum of squares and the correction for the mean are invariant under permutation, whence it follows (Exercise 7.1) that we may use the conventional uncorrected treatment (or between samples) sum of squares as our statistic. If the total for treatment (or sample) i is denoted by S_i, i.e.

$$S_i = x_{i1} + x_{i2} + \ldots + x_{in_i} \tag{7.1}$$

then the (uncorrected) between samples sum of squares is $S = \sum(S_i^2/n_i)$

In practice the permutation test for continuous data, which for brevity I call the permutation ANOVA, although it is not technically an analysis of variance, often behaves in a manner similar to its parametric analogue and certainly will perform little differently when basic assumptions about normality and homogeneity of variance hold. Permutation, or nonparametric, ANOVA results and those for parametric ANOVA are often similar when there is marked heterogeneity of variance, but then both may fail to detect fairly obvious location differences.

Data like that in Example 7.1 occur in trials of insecticides where the variate measured is the number of insects per plant on plants of the same variety subjected to three treatments, those in Samples I and II being sprayed with two different levels of an insecticide and Sample III being an unsprayed control. In the controls it is common to find some but not all plants heavily infested. An experienced statistician would transform the data before carrying out a parametric ANOVA, so I emphasize that I use these data only to illustrate the relationship between certain procedures. The samples are unrealistically small to make it easier to explain basics while avoiding computational problems.

Example 7.1 For the 3 samples below I use a permutation ANOVA to test for differences in location and compare results with those for a parametric ANOVA without a transformation.

Sample I	0	1	3	4
Sample II	6	8	12	
Sample III	10	11	61	65

I assume familiarity with the one-way classification ANOVA, so I omit details, but the variance ratio for testing for a location difference is $F = 3.81$ with 2 and 8 degrees of freedom and this implies an exact 7.3% significance level, so H_0 is not rejected at a conventional 5% level.

For the permutation ANOVA it is easily verified that there are 11 550 possible permutations. StatXact computes the exact $p=0.0632$, implying a significance level of 6.32%, not very different from the parametric ANOVA result.   There are, in fact, 730 permutations giving the same or greater values of the statistic S defined above (since 730/11550=0.0632).

It is easy to see the clear evidence of heterogeneity of variance even in these small samples. A transformation to logarithms of the form $y = \log(1 + x)$ would be appropriate [$\log(1 + x)$ avoids the difficulty with the zero observation that arises with $\log x$]. This transformation leads to a parametric ANOVA F-value indicating significance at an exact 0.8% level while the permutation ANOVA on the transformed data indicates significance at a 0.24% level. In this particular case the latter performs better than the former, but such differences at small p-values are common, and for not greatly different data sets sometimes one, or sometimes the other test, performs better.

The logarithmic transformation effectively eliminates variance heterogeneity but with small samples only glaring departures from normality can be detected.

Because the number of permutations increases rapidly with sample size exact tests are impossible for larger samples. As indicated above, for moderate sample size good Monte Carlo approximations to exact p-values are available and for large samples asymptotic results may be used. The best known asymptotic result is based not directly on the statistic S, but uses a statistic proposed by Hajék and Sidak (1967) which is a hybrid between that used in parametric ANOVA and S. It is, T, the ratio of the *treatment sum of squares* to the *total mean square*, both after subtracting a correction for the mean. Symbolically this may be written

$$T = \frac{(N-1)[S - (\sum x_{ij})^2/N]}{\sum x_{ij}^2 - (\sum x_{ij})^2/N} \tag{7.2}$$

For large N this statistic has approximately a χ^2 distribution with $k-1$ degrees of freedom.   Special cases of it are used freely in this and the next chapter.

The permutation ANOVA has Pitman efficiency 1 when samples are from normal distributions that differ, if at all, only in mean. The F-test is reasonably robust for testing when there are moderate departures from normality or when there is moderate heterogeneity of variance, although related estimation procedures then sometimes give bizarre results.

Example 7.2   For reasonably large and fairly well-behaved data sets (i.e. not too far from normality and homogeneity of variance) the Pitman efficiency of 1 suggests that the permutation ANOVA may be expected to give similar results to a parametric ANOVA.   The data below are counts of mean numbers of words in samples of five consecutive sentences chosen in paragraphs (or where necessary adjacent paragraphs) selected at random from sectors of continuous text (i.e., text not including data or mathematical formulae) from each of 5 books that deal with distribution-free and related methods. Compare the performance of parametric and permutation ANOVA to test for a difference in mean sentence lengths between authors.

The choice of the average length of five consecutive sentences as the sample unit helps to smooth out tendencies for short sentences to be followed by long sentences in many types of argument and so gives a better impression of overall average sentence length. This is known as a data-smoothing ploy.

Sprent (1993)	29.4	16.2	23.0	21.8	20.5	25.0	26.0	34.0	
Gibbons & Chakraborti (1992)	27.0	30.8	26.0	40.8	28.8	27.2	27.6	26.8	21.8
Maritz (1995)	24.4	24.4	18.6	23.8	18.8	29.4	17.6		
Good (1994)	36.2	30.2	23.2	24.4	19.6	23.2	25.2		
Edgington (1995)	28.4	31.6	30.4	21.6	27.6	25.6	29.2	31.4	

Most statisticians would happily perform a classic ANOVA for these data despite a tendency for a few values in some samples to suggest dichotomies into groups of short and long sentences rather than a smoother spread with clustering around the mean that one expects in samples from normal distributions.

I omit details, but the parametric ANOVA for a one-way classification gives the variance ratio $F = 2.18$ with 4 and 34 degrees of freedom, corresponding to an exact significance level of 9.22%. The problem is too large for StatXact to give an exact p-value for the permutation ANOVA, but a Monte Carlo approximation gave $p = 0.0915$, with 99% confidence interval for the true p of (0.0881, 0.0948), indicating approximately a 9.15% significance level, in close agreement with the parametric ANOVA result.

Example 7.3   If the data in Example 7.2 are supplemented with a sample from R.A. Fisher's classic *Statistical Methods for Research Workers* by adding the set

Fisher (1946)   20.2   41.0   44.4   35.0   45.6   49.2   53.0

the situation changes. Eye inspection suggests that Fisher tends to use longer sentences than the other authors and there is also an indication of greater variability in sentence length, or perhaps that the datum 20.2 is an outlier. However the parametric and the permutation ANOVA again give similar results, both now indicating significance at levels below 0.1%. Clearly, arguing from the results in Example 7.2, it is reasonable to conclude that most of the 'significance' reflects a difference between Fisherian sentence lengths and those for other authors. In parametric analysis this might be confirmed by comparing the Fisher mean with the pooled mean for all other authors, using an appropriate t-test, perhaps followed by more specific tests to see if there remain any differences after this particular contrast is eliminated. A natural question, considered briefly in section 7.8, is whether similar procedures are available in a nonparametric context.

In ANOVA the bootstrap provides an alternative to permutation test procedures. For straightforward hypothesis testing this has little to commend it, the main difference being that sampling is with, rather than without, replacement. For some interval estimation problems where the above permutation approach does not lend itself readily to establishing confidence intervals for differences a bootstrap approach using resampling within treatments may be preferable. The basic concept is not difficult but if accurate results are wanted correction of quantile based intervals may be needed.

7.3 Rank related location tests

If θ_i is the location parameter (usually the median or mean) for the ith sample, the WMW test is easily extended using the Wilcoxon rank sum format to test H_0:*all* θ_i *are equal* against H_1:$\theta_i \neq \theta_j$ *for at least one $i \neq j$*, valid under the assumption that all populations have identical distributions apart from a possible location difference. To carry out the test all N sample values are arranged in ascending order and ranked, using mid-ranks for ties. An exact analysis of ranks may be carried out using any program for a k-sample permutation ANOVA where ranks replace the original observations. This test, called the Kruskal–Wallis test (KW), was proposed by Kruskal and Wallis (1952). An asymptotic test is based on (7.2) using ranks or mid-ranks as data in that formula. In the special case of no ties this reduces (Exercise 7.2) to

$$T = 12S/[N(N+1)] - 3(N+1) \qquad (7.3)$$

If N is large T has, like (7.2), a χ^2 distribution with $k-1$ degrees of freedom.

The test extends immediately to rank transformations like those to normal scores or to Savage scores. Again the appropriate scores may be used in any program for exact, or Monte Carlo estimates of, exact permutation test p-values and an asymptotic test may be based on (7.2) with all terms computed for the appropriate scores. The performance of some of these tests for the data sets used in Examples 7.1–7.3 is studied in Examples 7.4–7.6.

Example 7.4 For the small data set in Example 7.1, Sample III shows greater variability than Samples I or II. Nevertheless the Kruskal–Wallis test may be expected to perform reasonably well as there is a suggestion that any shift in location for that population is associated with dominance or a general shift in the distribution to the right for that sample relative to the pattern in the other samples, although that implies a more general H_1 than that proposed above. As the WMW is reasonably powerful in such circumstances it is not unduly optimistic to expect KW to share this property. For these small samples StatXact gives an exact $p = 0.0029$ and also an asymptotic approximation to p using (7.2) applied to ranks, but as one might anticipate, this is misleading for these small samples; giving $p = 0.0198$, which is still significant but more than six times the exact p.

The above exact-test result is more satisfying than that for the permutation ANOVA test on the raw data (Example 7.1) where $p = 0.0632$.   If normal (van der Waerden) scores are used instead of ranks we again find the exact $p = 0.0029$, but the asymptotic value is $p = 0.0236$.

Although there is little evidence that the data come from populations that have anything remotely like an exponential distribution, one might regard sample III as long-tailed relative to the other samples and on those, admittedly slim, grounds try using Savage (log-rank) scores. StatXact gives an exact $p=0.0258$ for these scores. It is not surprising that Savage scores are less satisfactory than ranks or normal scores as there seems little rationale for using the former here so far as Sample I and Sample II are concerned.

Example 7.5. For the larger samples in Example 7.2 there is little obvious evidence of heterogeneity of variance and as I indicated in that example only a small suggestion of non-normality. Here using the KW or normal scores tests seems unlikely to improve performance. Monte-Carlo tests only are available in StatXact as the samples are too large for exact tests, and for these larger samples one also hopes that asymptotic approximations would be adequate. Table 7.1 gives Monte Carlo estimated exact *p*-values together with a 99% confidence interval for these estimated *p*, and also the asymptotic *p*-value for each of these tests.

Although the asymptotic results are conservative (i.e. show less strong evidence of 'significance') they are broadly in line with the exact results, which in turn are in reasonable agreement with those for the parametric and permutation ANOVAs for the raw data. There is no case for using Savage scores with these data.

The Pitman efficiencies of the KW test and the normal scores tests relative to normal theory tests when conditions for the latter hold are respectively $3/\pi$ and 1, so these results are not surprising.

Example 7.6. It is worth repeating the calculations in Example 7.5 after adding the data in Example 7.3 for Fisher sentence lengths. These data introduce some heterogeneity of variance which one might expect to have less influence in the KW test than in ANOVA with raw-data although, as already indicated, both the parametric and permutation ANOVA do not seem unduly diluted by this possible heterogeneity. The results for the KW and normal scores test are given in Table 7.2

The result for the KW test is not as sharp as that for the parametric or permutation ANOVA, while for the normal scores the *p*-value is similar. The asymptotic test results are conservative.

Sweeping conclusions cannot be drawn from just two examples, but Examples 7.5 and 7.6 suggest that when conditions that validate parametric ANOVA hold the use of the above rank-based or related distribution-free methods should result in only a small loss in efficiency, while they may do better if parametric assumptions do not hold.

Table 7.1 Exact *p*-value estimates and asymptotic approximations for data in Example 7.2

Test	Estimated exact p-value (99% confidence interval)	Asymptotic p-value
Kruskal–Wallis	0.0583 (0.0571, 0.0598)	0.0660
Normal scores	0.0694 (0.0665, 0.0723)	0.0785

Table 7.2 Exact *p*-value estimates and asymptotic approximations for data in Example 7.3

Test	Estimated exact p-value (99% confidence interval)	Asymptotic p-value
Kruskal–Wallis	0.0024 (0.0018, 0.0030)	0.0063
Normal scores	0.0009 (0.0005, 0.0012)	0.0034

For censored data and proportional hazard models like those discussed in section 4.8 and 4.9, the log-rank test extends in an obvious way and the appropriate scores may be used in a general permutation ANOVA program such as that in StatXact. Computing scores is tedious but StatXact will generate these using a method for ties similar to that used for the two-sample test and differing slightly from that introduced in section 4.9, but this small difference has little practical impact on test results.

7.4 The median test

The median test in section 4.6 uses minimal information. It readily extends to k samples, the test being based on numbers in each sample below, and numbers above, the combined samples median. These numbers are the cell entries in a $k \times 2$ table with each row corresponding to a sample and the first column referring to numbers below, and the second to numbers above the median. Row totals correspond to sample sizes and column totals to the total numbers of observations below and above the median for all samples. These totals are adjusted if necessary to allow for values equal to the overall median, which are usually rejected. The probability P associated with any configuration is given by a generalization of (4.5). Writing B, A for the total numbers (column totals) and b_i, a_i for the numbers in sample i, below and above the median, the expression is

$$P = \frac{A!B![\prod_i(n_i!)]}{N![\prod_i(a_i!)][\prod_i(b_i!)]} \tag{7.4}$$

The exact p-value in a two-tail test is the sum of the P for the data and that for all permutations of the data conditional upon the fixed row and column totals that give the same or smaller values of P. This extension of the Fisher exact test is often referred to as the Fisher–Freeman–Halton test, having been proposed by Freeman and Halton (1951).

An asymptotic test uses the chi-squared test for independence which here has asymptotically a χ^2 distribution with $k-1$ degrees of freedom. The asymptotic test is unreliable if there are many cells with small expectations, i.e. expectations between 1 and 4 and is often unsatisfactory with one or more small samples when there tend to be marked discontinuities in possible p-values. The difficulty discussed in section 4.6 carry over to here and are illustrated in Example 7.7.

For larger samples the test may do less well than other permutation tests such as KW or permutation ANOVA when the conditions for these hold, although as Example 7.8 indicates, it may occasionally do better even in situations where it is felt conditions for one of these fuller information tests hold reasonably well.

Example 7.7 For the data in Example 7.1, the median of all 11 observations is 8. In the median test we therefore ignore the observation 8 in sample II leading to the following assignment of observations above and below the median.

	Below	Above	Row total
Sample I	4	0	4
Sample II	1	1	2
Sample III	0	4	4
Column total	5	5	

For small samples P given by (7.4) is easily calculated as $P = (5!5!4!2!4!)/(10!4!1!0!0!1!4!)$ = 1/126. Further computations indicate that no permutation of cell counts subject to the given marginal totals produces a smaller P, but one permutation gives an identical P. Thus, the appropriate p-value (the minimum possible) is 2/126 = 0.0159 so that one cannot detect significance at a lower level in this case. StatXact gives a slightly different result because, rather than rejecting the value equal to the median it assigns values equal to the median to those below the median. It is easily verified that in this simple case that method gives $p=0.0130$. Despite the low expected numbers in most cells, the asymptotic result is not too misleading, Using the StatXact method of dealing with the value 8 it is 0.0157, very close to the exact value I got neglecting that observation 8, but this similarity is fortuitous.

Example 7.8 Using the median test for the data in Example 7.2 the exact $p = 0.0194$ and the asymptotic value $p = 0.0217$ using StatXact. Somewhat surprisingly, because we are using less information, these are lower than the corresponding results for the permutation ANOVA, KW or normal scores tests, none of which indicated significance at the conventional 5% level. I expressed some reservations about a normality assumption for these data and that may explain the good performance of the median test here. Applying the median test to the extended data in Example 7.3 we find $p = 0.0160$ and the asymptotic equivalent $p = 0.0197$. A significant location shift is again indicated, but here the p-values are greater than either those for the KW or normal scores test, a more common and not surprising result when we are using less information.

7.5 Permutation tests for ordered location shifts

Many experiments compare responses like those to increasing doses of a drug or an insecticide, to increasing levels of a specific fertilizer, or to increasing time of instruction in carrying out a task, and so on. It is reasonable to expect that if the treatment is effective there will be ordered responses, these either increasing or decreasing as the treatment level increases. A common parametric model is to assume that the mean response is linearly related to some function of treatment level. In the analysis of variance a single degree of freedom *sum of squares for regression* is computed. This is a component of the usual *treatment sum of squares* and may be used as the numerator in an *F*-test (the denominator being the residual mean square) to determine whether the regression is significant. The null hypothesis is effectively one of zero correlation between treatment level and response.

The equivalent permutation test is based on Pitman's (1937b) test for correlation which I describe in more detail in section 9.1 We are interested in correlations between some function $f(t_i)$ of the treatment levels t_i, $(i=1, 2, \ldots, k)$ and the responses x_{ij} $(j= 1, 2, \ldots, n_i)$ in the notation of section 7.2, of units subjected to each treatment. I assume treatments are allocated to units at random. This may be looked upon as a situation where we have n_i tied assigned values, or scores, $f(t_i)$ corresponding to each of the responses

$$x_{i1}, x_{i2}, \ldots, x_{in_i}$$

Choice of $f(t)$ should reflect prior beliefs about likely response. If the response is expected to be proportional to treatment levels, t_i, then an appropriate choice is $f(t) = t$. This is often the situation with fertilizer experiments, whereas with responses to increasing dose level of a drug a more appropriate choice in some circumstances is $f(t) = \log t$, or if a zero dose is included, $f(t) = \log(1+t)$.

The model is a special case of the linear-by-linear association model that I discuss in detail in section 12.4, but which I introduced in another context in section 1.6. An appropriate test statistic is $S = \sum f(t_i)x_{ij}$ where the summation is over all i, j. Defining S_i by (7.1) this may be written $S = \sum_i f(t_i)S_i$. The relevant permutation set is that for the permutation ANOVA as in section 7.2.

Example 7.9 Consider again the data in Example 7.1, i.e.

Sample I	0	1	3	4
Sample II	6	8	12	
Sample III	10	11	61	65

with the additional information that these are counts of insects on plants where for sample I an insecticide was applied at a high level, for sample II at a medium level and for sample III no insecticide was applied. A reasonable supposition is that any effect of insecticide is likely to reduce the average level of infestation, θ, by an increasing amount as the dose is increased. Dependent on the experimenter's belief about how this increase may be related to dose it may be reasonable to score the highest dose as 2, the intermediate dose as 1 and the lowest (here zero) dose as 0. However, a preferred scoring system (arrived at either from some theoretical model, or empirically on the basis of experience) may be to assign scores $\log(1 + 2) = \log 3$ = 0.477 to the higher dose, a score of $\log(1+1) = \log 2 = 0.301$ to the intermediate dose and a score of $\log(1+0) = 0$ to the zero dose. In experimental jargon this last 'untreated' category is often referred to as a *control*. With either scoring system [or some based on another form of $f(t)$] a linear-by-linear association program may be used to compute exact *p*-values. As indicated in section 1.6 high or low values of the statistic S indicate significance and the *p*-value represents the probability of getting the observed or a more extreme S. If it is assumed

Table 7.3 A contingency table format for testing for higher mean response to increasing doses of insecticide.

Column scores x_{ij}	0	1	3	4	6	8	10	11	12	61	65	
Row scores												*Row totals*
0.477	1	1	1	1	0	0	0	0	0	0	0	4
0.301	0	0	0	0	1	1	0	0	1	0	0	3
0	0	0	0	0	0	0	1	1	0	1	1	4
Column totals	1	1	1	1	1	1	1	1	1	1	1	11

that increasing doses are more effective, a small value of S is appropriate for acceptance of the one-tail alternative to the null hypothesis of no effect. In this example a one-tail test is reasonable if we exclude the possibility that the insecticide would increase level of infestation.

StatXact allows a choice of forms for $f(t_i)$ which are introduced as row scores in the linear-by-linear association program, so scores other than the two sets given above may be used if appropriate. The observed x_{ij} are entered as column scores. The data may be set out as a $k \times N$ contingency table with cell entries zero or 1 and the row and column scores and totals indicated in Table 7.3. The row scores 0.477, 0.301, 0 are the second of those suggested above. The cells entries of 1 in a given row indicate presence of the value at the top of the column in the corresponding sample. All other entries are zero. Permutation is over all possible tables with the given row and column totals.

In this example there are no ties among the x_{ij}. Had there been we only need include one column for each such tied value. The column totals corresponding to a tied value will no longer be unity and for computing purposes the statistic S should be replaced by an equivalent $S = \sum u_i v_j n_{ij}$ as defined in section 1.6 where u_i, v_j, n_{ij} are respectively the row and column scores and the count in cell (i, j).

For these data we easily compute $S = 11.642$ (Exercise 7.3). The StatXact program for linear-by-linear association gives the probability of this or a lesser S as $p = 0.0004$. If the column weights are replaced by the alternative choice of 2, 1, 0 suggested above the corresponding results are $S = 42$ and $p = 0.0003$.

Heterogeneity of variance may limit robustness for a raw data permutation test which is linked to the possible effects of observations that are influential in the sense described in section 3.4. Difficulties associated with such observations are often alleviated by transforming responses to ranks as for the Kruskal–Wallis test. To test for ordered responses, instead of the KW test procedure, we may now use the ranks as column scores in place of the original data in a linear-by-linear association test. It would also be in order to use either of the sets of row scores suggested above if these were thought appropriate, but if there is doubt about such scores a rank or mid-rank type score as described below may be appropriate. The row scores could also be any scores in decreasing order such as 3, 2, 1 or 2, 1, 0. This is a generalization of the scoring system used for the WMW test formulated as a linear-by-linear association test. A scoring system with some appeal is suggested by collapsing adjacent rows corresponding to ties, thus in effect regarding all observations belonging to the same sample as ties with respect to a variable allocating observations to samples. If this variable takes the form of ranks all four entries allocated to Sample III are allocated the mean of the ranks 1 to 4, i.e 2.5. The three allocated to Sample II are allocated the mean of ranks 5 to 7, i.e. 6 and those allocated to Sample I the mean of ranks 8 to 11, i.e 9.5. In this particular example since the ranks 2.5, 6 and 9.5 are linear functions of scores 1, 2, 3 it makes no difference which scores are used. If there is no other theoretical basis for allocating scores there is logic in using mid-rank allocations for both row and column scores. I show in section 9.2 that this procedure is a special case of the Spearman rank correlation model which in turn is a special case of the general correlation model and the linear-by-linear association model. An objection to using mid-ranks for row scores is that these dependent on the number of data in each sample and their relative values may be changed by obtaining more data for a particular treatment or sample. Examples can be constructed when this has a substantial influence on the resulting p-value that is not justified by the logic of the problem. An illustration is given in Example 12.7.

Asymptotic results are available in StatXact for linear-by-linear association models. If the row and column scores are u_i, v_j and the fixed row and column totals are r_i, c_j respectively, then for the statistic S defined in Example 7.9 it may be shown that

$$E(S) = (\textstyle\sum u_i r_i)(\textstyle\sum v_j c_j)/N$$

and

$$\text{Var}(S) = [\textstyle\sum u_i^2 r_i - (\textstyle\sum u_i r_i)^2)/N][\textstyle\sum v_j^2 c_j - (\textstyle\sum v_j c_j)^2)/N]/(N-1)$$

and for large N and the sample sizes not too small $Z = [S - E(S)]/\sqrt{\text{Var}(S)}$ has approximately a standard normal distribution.

I showed in section 7.3 that the Kruskal–Wallis test was basically a generalization of the WMW test using an extended Wilcoxon rank-sum formulation. For ordered location shifts an alternative to the tests already given in this section is the Jonckheere–Terpstra test proposed independently by Jonckheere (1954) and Terpstra (1952). This tests H_0: *all* θ_i *equal* against H_1:$\theta_1 \le \theta_2 \le \ldots \le \theta_k$ *with at least one strict inequality*. The inequality signs may be reversed throughout H_1 if that is the alternative of interest. To carry out the test samples are arranged in the order implied in H_1 and we count for each value in Sample I the number of values in Samples II to k that are greater than that value (counting a tie as ½). We now repeat the process starting with each value in Sample II counting the numbers of higher values in all succeeding samples (i.e. Samples III to k) and add these to the previous result, continuing in this way starting at succeeding samples up to Sample $(k - 1)$ and counting the number of Sample k values that exceed each value in that sample. This gives a statistic U, say, which is used in an exact permutation test, the reference set again being all permutations of the data subject to the given sample sizes. Extremely large (or small) values of U indicate significance. StatXact, TESTIMATE and some other software provide a program for the exact test and many general packages provide at least an asymptotic test. I show in section 9.2 that this test is a special case of the well-known Kendall rank correlation coefficient. If the statistic U is computed as described above then

$$E(U) = \tfrac{1}{4}(N^2 - \textstyle\sum n_i^2)$$

and

$$\mathrm{Var}(U) = \{N^2(2N + 3) - \textstyle\sum[n_i^2(2n_i + 3)]\}/72.$$

For large N and the individual n_i not too small the distribution of

$$Z = [U - E(U)]/\sqrt{[\mathrm{Var}(U)]}$$

is approximately standard normal. Clearly this test is an extension of the WMW test.

Example 7.10 For the data in Example 7.9 we easily calculate $U = 38$, because the entry 0 in Sample I is exceeded by all 7 entries in Samples II and III and the same applies to the remaining three entries in that sample so that all four contribute a total of 28. The first two entries in Sample II are exceeded by all four entries in Sample III contributing a further 8 to U and the last entry in Sample II is exceeded by two in Sample III, contributing a further 2, whence $U = 38$. The StatXact program gives this value and the exact $p = 0.0007$ for the relevant one-tail test. The sample size is too small to rely upon the asymptotic result.

Example 7.11 Chris Theobald supplied the following data from a study of 49 patients suffering from a form of cirrhosis of the liver. The study is described by Bassendine *et al.* (1985). One purpose was to examine whether there was evidence of association between spleen size and blood platelet count. Blood platelets form in bone marrow and are destroyed in the spleen, so it was thought that an enlarged spleen might lead to more platelets being eliminated and hence to a lower platelet count. The spleen size of each patient was found using a scan and scored from 0 to 3 on an arbitrary scale, 0 representing an ordinary spleen and 3 a grossly enlarged spleen. The platelet count is the number in a fixed volume of blood. I use the Jonckheere–Terpstra test to decide whether these data indicate an association between spleen size and platelet count in the direction anticipated by the experimenter.

Size	Platelet count
0	156 181 209 220 222 238 295 325 334 342 348 359 365 374 391 395 481
1	65 105 121 150 150 158 170 214 235 238 255 265 322 390
2	33 70 87 98 100 109 114 132 150 179 184 241 281 323
3	79 84 94 259

Arranging all values in ascending order facilitates the counting by inspection for U and direct counting gives $U = 166$. The samples were slightly too large to allow computation of exact p on my computer, but a StatXact Monte Carlo estimate gave $p < 0.0001$. The asymptotic value of $Z = -4.574$ is consistent with this and for samples of this size, despite the relatively small sample in size grade 3, the asymptotic result might be expected to be reasonable. These findings are backed up using the linear-by-linear association model (i) with raw data scores for columns and rank scores for row and (ii) with rank scores for both rows and columns, the values of Z for the asymptotic tests being just slightly smaller in magnitude at -4.262 and -4.306 respectively. The data might make one reluctant to accept normality assumptions for all populations, especially that for spleen size 3, so most experimenters would be reasonably happy to analyse these data by any of the above methods or to use them as backing for a result obtained by using a regression test in a standard parametric ANOVA.

*7.6 Factorial treatment structures

If you are not familiar with factorial experiments you may wish to read an introduction to the main aspects of factorial experimentation in a general statistical text or in one on experimental design that deals with parametric ANOVA for factorial treatment structures before proceeding with this section. I assume only a rudimentary knowledge of the basics of these structures, in particular, an understanding of the ideas of main effects of and interactions between factors and some familiarity with $2\times2\times \ldots \times2$ factorial experiments and the rudiments of experiments with several factors each at 2 or more levels. I do not cover features like confounding or partial replication.

A factorial treatment structure is one where two or more factors are represented each at several levels. Factors may be quantitative or qualitative. For example, in studying a chemical process, yield may be measured for experimental units each operating at one of 4 temperatures (quantitative levels 1, 2, 3, 4 of factor A) at one of 3 pressures (quantitative levels 1, 2, 3 of factor B), in each of 2 factories (qualitative levels 1, 2 of factor C), providing $4\times3\times2 = 24$ possible 'treatment' combinations. I use the notation $a_ib_jc_k$ to indicate the treatment combination of factor A at level i, $(i = 1, 2, 3$ or $4)$, factor B at level j, $(j = 1, 2$ or $3)$ and factor C at level k, $(k = 1$ or $2)$. I later use the same notation for the expected yield of units receiving that treatment combination, the context usually making it clear in which sense the notation is used. When experimental units are allocated at random to each factor combination (treatment), this constitutes a factorial experiment which might be analysed initially either by the conventional normal theory parametric ANOVA or by a distribution-free method such as a permutation ANOVA or a KW analysis to determine whether there are any treatment

*This section gives a slightly more advanced treatment of a specialist topic and may be omitted at first reading.

differences, but the factorial structure makes possible, in the case of a parametric ANOVA, a further analysis to test for or estimate what are called *main effects* and *interactions*.

In parametric ANOVA, providing certain constraints are imposed on the numbers of units receiving each treatment to ensure what is called orthogonality, straightforward analyses enable one to make inferences about how the different levels of each factor influence response and also whether or not the effects of two or more factors are additive. If they are not additive, there is said to be an **interaction** between the factors. The term interaction is also used in wider statistical contexts and has important implications for both the choice of a suitable method of analysis of specific data sets and for the interpretation of any analysis. Interaction in general contexts is examined in detail by Cox (1984) where the complexity of the concept and the implications of that complexity are fully explored. Distribution-free procedures for factorial analyses may be appropriate if the usual assumptions for parametric ANOVA do not hold. Much of the work on these deals with asymptotic approximations although some covers exact tests for main effects and interactions. The effects of the breakdown of assumptions appropriate to a parametric analysis are often far-reaching, and there are no nonparametric or distribution-free analogues of some parametric tests. This is especially true where interactions are involved. These topics are treated more fully by Good (1994, Chapter 4) and by Edgington (1995, Chapter 6) and a summary of the main alternative approaches is given in Manly (1997, Chapter 7, where section 7.4 is of special relevance in the context of factorial experiments).

The treatment that follows is introductory and far from comprehensive. A difficulty in choosing and interpreting nonparametric analogues of ANOVA for factorial treatment structures is that the conventional interpretation of main effects and interactions depends heavily upon the concept of a linear model with additive effects. It is when those assumptions break down that distribution-free analyses are most appropriate. Nonadditivity is often associated with heterogeneity of variance and many of the classic transformations of data such as using logarithms of, or square roots of, the original data in parametric ANOVA aim to impose variance homogeneity or induce additivity. In doing so, they often have a radical influence on the pattern of main effects and interactions; in particular, they may reduce or eliminate interactions. Transformation to ranks, used so widely in distribution-free analyses sometimes have similar effects, but unfortunately this transformation also sometimes introduces interactions that are not evident in the original data. Thus care is needed in interpreting results of analyses based on ranks. In Example 7.12 I use a trivial data set to show how different types of interaction are affected by making a transformation to ranks as one might do, for example, as a preliminary to a KW analysis.

Example 7.12 A small 2×2 factorial experiment consists of two replicates of each factor combination. Factors are denoted by A, B and levels by 1, 2.   Treatments are allocated at random to the 8 experimental units subject to the two-replicate constraint. The yields are

B at level 1	*A at level 1*	1.1	2.8
	A at level 2	4.3	7.5
B at level 2	*A at level 1*	6.4	8.5
	A at level 2	121.7	165.4

If you are familiar with the conventional parametric analysis with an additive model you will realise immediately that there is a marked interaction, because of the large difference in response to factor A at level 2 of factor B compared to that at level 1 of factor B. There is also a strong indication of heterogeneity of variance, the response at levels 2 of both factors combined (last line of the above table) appearing, though only on very slender evidence with just two replicates, to be more variable than that for the other factor combinations.

   If I transform the data to ranks (as a preliminary perhaps to a KW test for overall treatment effect) the corresponding ranks are

B at level 1	*A at level 1*	1	2
	A at level 2	3	5
B at level 2	*A at level 1*	4	6
	A at level 2	7	8

and inspection of the data shows that there is no trace of an interaction since the response to A, as measured by the mean rank differences between level 2 and level 1 of A, is the same at each level of B, i.e. 7.5—5 = 4–1.5 = 2.5.

   However, transformation to ranks will not remove some interactions. Had the responses with A, B both at level 2 been 0, 1.5 instead of 121.7, 165.4 there would still be an indication of interaction in the original data, the responses to A at level 2 of B being in the opposite direction to that at level 1 of B. This difference is also evident in the new ranks which are

B at level 1	*A at level 1*	2	4
	A at level 2	5	7
B at level 2	*A at level 1*	6	8
	A at level 2	1	3

so the interaction persists. Cox (1984) calls interactions that persist after transformation to ranks order-based interactions.   In the presence of order-based interactions care is needed to interpret main effects, irrespective of whether parametric or distribution-free methods are used.   Several writers who have discussed rank-based analyses of factorial structures (e.g. Mack and Skillings, 1980, Groggel and Skillings, 1986) have discussed methods where interactions are specifically excluded. This is a major constraint because an advantage of factorial treatment structures is that they let us examine possible interactions.

Edgington (1995) points out that the additive treatment effect model is not essential to randomization tests, indeed all we might be able to say about the effect of a factor is that it tends consistently to increase (or decrease) yield as the factor level changes, but that the magnitude of such changes may vary between experimental units, i.e. there is dominance rather than simple location shift. While many randomization tests do not require effects to be strictly additive, appropriate randomization tests do exist for models that specify additivity.

Theoretically it is relatively straight-forward to devise exact permutation tests for main effects on the assumption that there are no interactions or at any-rate no order-based interactions. There are tests for certain types of interactions and also for the so-called particular effects at preassigned common fixed levels of all factors other than the one of current interest. A key feature of these tests is that the permutation reference set is different for each test. I confine illustrations to the equal replication case for all factor combinations (treatments) although this is not always essential when using distribution-free methods.

The appropriate permutation reference set depends upon the hypothesis spec-ified in H_0. In the case of a 2×2 experiment a different reference set is appropriate for testing for a main effect of A, for a main effect of B, or for an order-based interaction. The following simple example demonstrates some principles.

Example 7.13 In a greenhouse experiment bean plants were grown in the presence or absence of a growth hormone in natural daylight and with extended light (daylength increased by using artificial light). Five plants were allocated randomly to each hormone/light regime. The heights of plants (in cm) after one week were recorded:

no hormone	*normal light*	1.8	2.1	1.7	2.9	3.6
	extended light	2.4	2.9	3.5	3.9	4.1
hormone	*normal light*	7.3	7.5	8.1	8.4	9.2
	extended light	14.3	14.9	15.2	18.1	18.9

Inspecting these data suggests that hormone enhances growth, being dramatically more effective in extended light than in normal light, whereas without hormone there is some, but certainly not a dramatic benefit with extra light. The data suggest a direct benefit both from extra light and from hormone, but the enhanced benefit of both, relative to just one, implies an interaction. Growth differences are unrealistically large in practical terms but this helps to highlight some characteristics. Assigning ranks to these data removes the evidence of interaction, indicating it is not an order-based interaction.

I give an appropriate analysis based on the data without transformation. If we consider the null hypothesis H_0 that there is no effect of light (irrespective of what hormone does) and wish to test this against an hypothesis that additional light is beneficial, the appropriate randomization test uses a reference set consisting of permutation of data done **separately**

within each hormone level, because under H_0 there is no effect of additional light *within* these two 'strata' whereas there may or may not be (and the data strongly suggests that there is) differences between strata (i.e. due to different levels of hormone). Thus for testing for a main effect of light, the reference set consists of all permutations into two groups of 5 of the 10 observations with no hormone, combined with all permutations into two groups of 5 of the 10 observations with hormone, giving in all $^{10}C_5 \times ^{10}C_5 = 63\ 504$ permutations. A suitable test statistic is the sum S of the sample values, both with and without hormone, at the normal light level only, here $S = 52.6$. Low values of this statistic indicate departure from H_0 in the direction of better growth with extended light. An alternative statistic is suggested in Exercise 7.4. StatXact provides a program (under the heading of 2 independent stratified samples) for this test and the indicated p-value for obtaining the same or a lesser sum (appropriate for a one tail-test against the hypothesis that the main effect of extended light is to increase growth) is $p = 0.0002$.

To test the main effect of hormone the data is better rearranged as

Normal light	No hormone	1.8	2.1	1.7	2.9	3.6
	Hormone	7.3	7.5	8.1	8.4	9.2
Extended light	No hormone	2.4	2.9	3.5	3.9	4.1
	Hormone	14.3	14.9	15.2	18.1	18.9

This form again highlights the effect of hormone under normal light and a more marked benefit under extended light. To test the main effect of hormone, permutation is now done separately in the stratum corresponding to normal light and that for extended light. An appropriate test statistic is the sum of the *no hormone* scores over each strata, here 28.90 and the reference set again contains 63 504 permutations, but these are not the same permutations as those for testing the main effect of light. In this case StatXact indicates a $p<0.000\ 05$ for a one-tail test of no effect of hormone against the alternative that hormone increases growth.

The interpretation of main effects in the presence of interactions needs care with parametric ANOVA because interactions may influence the significance of main effects. If the interaction is one of enhancing the additive effect of factors this tends for the randomization tests in Example 7.13, as in parametric ANOVA, to enhance the significance of the main effect, whereas, as the following example shows, if it is an order-based interaction it will diminish or even remove all evidence of a main effect. This applies both to parametric and to permutation ANOVA, and when there is evidence of an interaction it is often more appropriate to look at the effects of one factor separately at each given level of the other factor(s). In that case a permutation test should be applied to the sub-set of the data involved. For instance, in Example 7.13 to test for a beneficial effect of hormone under normal light a two-independent sample test should be performed on the *normal light* data only, ignoring that for extended light. For this small data set the test is not very powerful.

Example 7.14 Consider a data set identical to that in Example 7.13 except that the responses with hormone and extended light (final line) are changed giving

no hormone	*normal light*	1.8	2.1	1.7	2.9	3.6
	extended light	2.4	2.9	3.5	3.9	4.1
hormone	*normal light*	7.3	7.5	8.1	8.4	9.2
	extended light	6.7	6.9	7.4	7.5	7.6

A moment's reflection shows that there is at least a suspicion of an order-based interaction, for while extended light appears slightly beneficial with no hormone it may be slightly deleterious with hormone. The result is that in the main effect of light, estimated by averaging the differences (*mean height with extended light – mean height with normal light*) over the two levels of hormone, is close to zero because of the contrary behaviour at the different hormone levels.

Indeed, StatXact gives $S = 52.6$ (the same value as we obtained in the previous example because the relevant terms in the sum are unchanged), but now the one-tail $p = 0.4784$ so there is clearly no evidence of a main effect of light. Following the procedure outlined in Example 7.13 for a main effect of hormone we again find that $p < 0.00005$.

Because the order-based interaction negates the main effect of light it is sensible to look at the particular effects for light, i.e. to consider separately the effect of light at each hormone level, treating these as separate two-independent sample problems. For no hormone the data are

no hormone	*normal light*	1.8	2.1	1.7	2.9	3.6
	extended light	2.4	2.9	3.5	3.9	4.1

and StatXact gives a one-tail $p = 0.0556$ which conventionally is not quite significant even in a one-tail test.

When the data *with hormone* are tested for an effect of light (but now for one in the opposite direction), the corresponding $p = 0.0357$. It is debatable whether a one-tail test is justifiable in these circumstances, but at least the tests of particular effects indicate, even for so small a data set, slight evidence of opposite effects that are completely masked by interaction when testing for a main effect of light.

In parametric ANOVA for a 2×2 factorial experiment the test for interaction involves two comparisons that are equivalent, namely the difference between the particular effects (both measured in the same direction) of B at each level of A or the difference between the particular effects of A at each level of B. This comparison in terms of expected treatment means is $a_2b_2 - a_2b_1 - a_1b_2 + a_1b_1$. No interaction implies $a_2b_2 - a_2b_1 - a_1b_2 + a_1b_1 = 0$. This concept does not carry over happily to permutation ANOVA because if there are any treatment effects, permutations required for the reference sets will be over units which are already known to show effects whether or not the hypothesis of no interaction holds (because of the presence either of particular or main effects). However, recalling that in parametric ANOVA we define the main effect of A in terms of the expected values of the function $a_2b_2 - a_1b_2 + a_2b_1 - a_1b_1$ of the treatment means, we see that what we do in a permutation ANOVA is to allow for the stratum effect of B by permuting separately within observations corresponding to the first two treatments (i.e. a_2b_2, a_1b_2) and within observations corresponding to the last two treatments (a_2b_1, a_1b_1), in accord with the logic of the meaning of a main effect and the relevant H_0 that specifies that such an effect does not exist. Since the interaction is defined by $a_2b_2 - a_1b_2 - a_2b_1 + a_1b_1$ it seems logical to make a pseudo-stratum split with comparisons of the form $a_2b_2 - a_1b_2$ and $a_1b_1 - a_2b_1$ with randomizations confined to being within each stratum. Now this is a comparison of the particular effect of A at level 2 of B with the **opposite** particular effect (i.e. one with sign reversed) at level 1 of B. This seems appropriate to pick out an interaction in which B has opposing influences on the behaviour of A at the two levels under investigation, i.e. an order-based interaction. In these circumstances the main effect of A is likely to be small. Alternatively, we might make a pseudo-comparison split with comparisons of the form $a_2b_2 - a_2b_1$ and $a_1b_1 - a_1b_2$ which will be sensitive to opposing influences in B at the two levels of A;, that is in situations where there is an interaction of a nature reflected in reduction or elimination of a main effect of B. These situations are particularly relevant to order-based interactions, which, as Cox points out, are not symmetric, so we may have one but not the other. On the other hand interactions that are not order-based, such as those that enhance a main effect as in Example 7.13, could not be expected to show up clearly by the tests proposed above. Indeed there are many unsatisfactory aspects of studies of interactions even in parametric ANOVA especially in more complex experiments where higher-order interactions (interactions between several factors) are involved and as Cox (1984) suggests 'By far the most powerful device for detecting interaction is, however, the critical inspection of . . . tables of means'.

Example 7.15 For the data in Example 7.13 we may calculate p-values for the two interaction tests proposed above by reversing the stratum I sample orders in each of the main effect tests and performing a two stratum analysis on these revised ordered sets. The respective p-values are $p=0.004$ and $p=0.0708$ in a one-tail test. The former suggests some interaction and inspection of the data suggests an interaction that is not order based but enhances the main effect of the light factor.

Example 7.16 Proceeding as in Example 7.15 for testing for interaction in the data for Example 7.14 the relevant p values are $p=0.0047$ and $p=0.2304$. The first p reflects the fact that the response to light is in the opposite direction at each hormone level, and the higher second p-value reflects the fact that at the different light levels the differential response at each hormone level is masked by the much greater effect of hormones relative to that of light.

The interaction tests suggested above are by nature *ad hoc*. Several writers including Good (1994, chapter 4) have suggested a different approach to testing for interactions.   I do not pursue this further, but the interested reader should refer to Good and to Edgington (1995) and to references therein for the detailed arguments and counter-arguments and also to Manly (1997, Chapter 7).

For testing main effects or particular effects the basic ideas illustrated for 2×2 factorial experiments extend to more general factorial experiments. A detailed account is given by Edgington (1995, Chapter 6). Applications, at least ones using exact p-level tests, may be restricted by lack of suitable statistical software although some of the tests may be performed by suitable adaptations of programs in StatXact or TESTIMATE and Edgington gives programs for Monte Carlo estimates of relevant p covering many cases.

Transformation, especially to ranks, are often used and the methods of this section are applicable to ranked data either using software specifically tailored to such data or by using ranks as data in appropriate general permutation tests. Note however, as remarked earlier, that transformations (especially to ranks) may sometimes remove interactions and even more disturbingly, though it would seem in practice only rarely, introduce interactions not evident in the original data.

Because of analogies between the KW test and parametric ANOVA some writers have suggested partitioning the KW statistic along the lines of partitioning of treatment effects into orthogonal components analogous to that in ANOVA. Examples have been given that show that this approach must be used cautiously, especially as it is essentially based on asymptotic results. Writers who have discussed this approach include Bennett (1968), Mack and Skillings (1980), Groggel and Skillings (1986), Shirley (1987), some of whom consider more general partitioning for rank-based tests.

An important aspect of parametric analysis of factorial treatment structures is estimating the magnitude of effects. This has received less attention in the distribution-free context, partly because the meaning of the effects may be more general than just location shift and there may then be difficulty defining what an effect really means other than a tendency for a factor combination to increase yield relative to some other combination and this might be reflected in other distributional properties such as dispersion. Another difficulty, even if there is only location shift, is a lack of software to compute confidence intervals using the principles outlined in section 1.8. In the more complex permutation systems used here application becomes difficult. This is an area where bootstrapping looks to be potentially useful.

I sketch the extension of the ideas from a 2×2 experiment to a 2×3 factorial structure in an analysis of main and particular effects. Further generalizations are given by Edgington (1995).

In all but the simplest experiments main effects or interactions will in general each have more than 1 degree of freedom and may themselves be further partitioned into single degrees of freedom in a nonunique manner, and the choice of how this should be done should be dictated by what is of interest. Although care must be taken when choosing methods for permutation tests there is some freedom from the constraints of orthogonality required to justify more straight-forward permutation ANOVA partitioning of degrees of freedom. Nevertheless at the design stage orthogonality is still a valuable concept, often giving increased efficiency as well as computational simplicity, but a full discussion of this point is beyond the scope of this book.

My notation for a 2×3 structure is to denote the factors by A with 2 levels 1, 2 and by B with 3 levels 1, 2, 3. In parametric ANOVA the main effect of A has 1 degree of freedom and is measured by the difference between the means at the higher and lower levels of A each averaged over the 3 levels of B, i.e. by

$$a_2b_3 + a_2b_2 + a_2b_1 - a_1b_3 - a_1b_2 - a_1b_1$$

whereas the main effect of B has two degrees of freedom and is measured by differences between the averages over both levels of A at each level of B. An overall permutation test for a main effect of B can be formulated but it is sometimes more useful to compare single degree of freedom components of this, especially for a quantitative factor such as increasing levels of fertilizer, or doses of a drug. When there is interaction (i.e. if the effects of A and B are not additive) it may be more interesting to look at particular effects of A and B. For A these are the differences, if any, in responses to A at a given level of B. If the differences are in location these may be written $a_2b_3 - a_1b_3$, $a_2b_2 - a_1b_2$ and $a_2b_3 - a_1b_3$. The particular effect of B at the lower level of A is reflected by differences between the means of a_1b_3, a_1b_2 and a_1b_1 and at the higher level of A by differences between the means of a_2b_3, a_2b_2 and a_2b_1. I describe the relevant permutation tests for some of these possibilities with a small data set to avoid the need to consider computing technicalities requiring special software.

Example 7.17 Eighteen experimental units are allocated at random, three to each treatment in a 2×3 factorial experiment with two levels of factor A and three levels of factor B. The measurement of a characteristic on each unit are as follows:

B level	A level	Measurements		
1	1	2	5	19
	2	7	14	23
2	1	21	26	34
	2	27	31	54
3	1	16	18	21
	2	26	28	32

The data show that the response to A differs at each level of B, but within each B level the response at level 2 of A is greater than that at level 1, suggesting a likely significant main effect of A. The permutation ANOVA for the main effect of the two-level factor A is a generalization of that in Example 7.13. The data are permuted **within** each of the three levels of factor B and a suitable test statistic is the sum of the observations at the lower level of A over all levels of B, i.e. $S = 162$. At each level of B there are $^6C_3 = 20$ permutations giving in all $20 \times 20 \times 20 = 8000$ permutations in the reference set. StatXact gives $p = 0.0137$ for a one-tail test for no effect against the alternative of an increasing effect. If particular effects of A at a specific level of B are of interest the procedure is to apply the standard two independent sample test procedure to the data for that stratum (level of B). Since with only 3 replicates in each factor combination there are only 20 permutations, the test is not useful for such small samples, but is feasible for realistic sample sizes.

To test for main effects of B the data are rearranged into strata with each level of A corresponding to a stratum, i.e.

A level	B level	Measurements		
1	1	2	5	19
	2	21	26	34
	3	16	18	21
2	1	7	14	23
	2	27	31	54
	3	26	28	32

By analogy with the stratification for the two-level factor, in this case the reference set of permutations consists of all possible permutations **within** each of the two strata defined by the levels of A. In each stratum there are $9!/(3!3!3!) = 1680$ permutations and combining these for two stratum gives a reference set of $1680 \times 1680 = 2\,822\,400$ permutations. Programs for a Monte Carlo evaluation of p for this problem exist but I do not know of any program that gives exact p-values. An overall test for differences in response between levels of B is in essence a non-directional or two-sided test and the appropriate statistic is suggested by the parametric ANOVA and for this equal replication situation if we denote the sum of the responses at level i of B summed over both levels of A by B_i, this may be written $T = B_1^2 + B_2^2 + B_3^2$. There are a number of equivalent statistics in use, an important one being the above statistic divided by r, where r is the number of observations in each total B_i, reminiscent of the uncorrected sum of squares for the main effect in parametric ANOVA.

If the levels of B are quantitative differences like increasing levels of fertiliser, increasing temperature or pressure in some chemical process, etc., one is often interested in single degree of freedom components of this main effect. The difference in response at level 3 and level 1 measured here by $B_3 - B_1$ is said to represent a linear component of response and the function $B_3 - 2B_2 + B_1$ represents depatures from linrearity or a possible 'quadratic' component. Since the fomer involves only levels 1 and 3 it requires only relevant permutations over observations at these levels, and we are effectively back in the set-up for a 2×2 experiment with the two levels of A and the levels 1 and 3 as the relevant levels of B. The test procedure is exactly that described for testing a main effect of B in such an experiment. Using the statistic $B_3 - B_1$ is equivalent to using $S = B_1$, since $B_1 + B_3$ is constant over all permutations in the reference set.

The above example gives only a flavour of extension of factorial analyses to general factorial designs. The approach is clearly flexible in the choice of possible tests and associated estimation procedures. There is room for developments in computer software to make exact tests, or Monte Carlo procedures for estimating exact p-values, possible on a routine basis. To date the emphasis has been on asymptotic results, more usually associated with rank transformations of the data. I touch upon asymptotic results for the 2^n design in a more general context in the next section.

7.7 Stratified samples

To test main effects in a 2×2 factorial experiment in section 7.6 I considered a special case of two independent sample stratification. There, given one factor at two levels, stratification was by each of the two levels of the second factor. In the case of a $2 \times r$ structure, for the two level factor stratification was by each of the r levels of the other factor. An essential element was the allocation of units to each factor combination (treatment) at random, subject to the constraint of equal replication. That constraint may be relaxed for some testing and estimation procedures. Stratification when comparing two treatments (which may be regarded as different levels of a single factor) is also possible when the strata are determined by phenomena other than a further factor. In this context that other phenomenon is often called a covariate, each strata corresponding to a different value of that covariate. The covariate may be some measured quantity, an ordering, or simply a qualitative characteristic and it need not, indeed in general will not, be randomized over units. The following example is based on Sprent (1993, Example 6.8) although the analysis below differs in some respects from that given there and a larger data set is used here.

Example 7.18 Tests are made on an up-dated version of a word-processing program designed to make it easier to prepare technical reports containing graphs, mathematical formulae, etc. To see whether the updated version (package A) shows an advantage over the current version (package B) 7 people are chosen at random form 15 and each is asked to prepare a specimen report using package A while the remaining 8 produce the same report using package B. All participants had similar experience in word processing. The numbers of mistakes made by each operator in producing the report were:

| Package A | 2 8 11 24 26 31 44 |
| Package B | 8 12 17 25 28 31 52 63 |

An exact Fisher–Pitman permutation test gives $p = 0.1792$ and using ranks the WMW test gives $p = 0.1905$, neither indicating any advantage for the new package. However, it transpires that some of those doing the testing had been given training in the discipline which was the subject of the report, while others had not. It is likely that those with a detailed knowledge of the subject matter may have tended to make fewer mistakes and this made a case for dividing the testers into strata on this basis, giving

Training	*Package A*	2	8	11		
	Package B	8	12	17		
No training	*Package A*	24	26	31	44	
	Package B	25	28	31	52	63

These might be regarded as four independent samples and the data submitted to a KW or similar analysis. A Fisher–Pitman or KW test both indicate a difference between samples (exact $p = 0.0088$ and exact $p = 0.0013$ respectively) but there is both a theoretical and a practical objection to this approach. The theoretical objection is that if the subjects are not random samples from some larger population, but are simply the only subjects available, then the only randomization is in allocation of subjects to packages; there is not a random allocation to training or no training. This is a fact of life. This means that the random allocation to all four 'treatments' that is required for strict validity of a permutation ANOVA or KW analysis does not hold. This may not imply seriously misleading results in practice except insofar as it limits the validity of any inferences about beneficial effects of prior training in the subject of the report. The practical objection is that the test merely signifies some difference (as is obvious from the data) but what really interests us is whether there is evidence of an overall benefit of using package A, both among those with and among those without previous training in the subject matter. Training in the subject matter of the report is clearly the major source of improved performance, at least for those involved in this experiment.

The appropriate test follows the lines of that for a main effect in a 2×2 factorial experiment. Randomization is performed separately within each stratum (*training* or *no training*). The reference set here clearly consists of ${}^6C_3\times{}^9C_4 = 2520$ permutations. The permutation ANOVA for the given data gives $p = 0.1198$ and if we replace the data by an over-all ranking across both strata (usually referred to as a stratified WMW test we get $p = 0.0929$ (in each case relevant to the one-tail test that package A gives a lower number of errors). Thus we would not conclude from these data that package A is superior. In Exercise 7.5 I ask you to check whether the particular effects of using different packages within each stratum are significant.

The idea of stratification is common in *two treatment* experiments. Other situations where it arises naturally are discussed in sections 12.5.

For larger samples an asymptotic test is available and for k strata in each of which there are two independent samples it is based on the fact that if we use as the statistic the sum S of the sample I data over all strata then the mean and variance for any one strata is computed using (1.4) and (1.5) and the overall mean and variance for S is the sum of those means and variances over all strata. As usual $Z = [S - E(S)]/\sqrt{[\mathrm{Var}(S)]}$ has asymptotically a standard normal distribution. These asymptotic results are clearly applicable also to some of the tests for main effects with factors at two levels described in section 7.6

A scoring modification is often used in stratified two-sample WMW inference. Instead of ranking over all data, the data may be ranked independently in each stratum. Although the results will often be similar to those obtained after ranking over the complete data sets there is evidence that if one is testing for main effects in factorial treatment structures that there are some advantages in ranking separately within strata, especially when asymptotic test procedures based upon ranks are being used. Any difference is likely to be most pronounced if there is a clear data outlier.

Example 7.19   For the following data

Stratum I	*Sample I*	1	3	15	
	Sample II	4	7	11	
Stratum II	*Sample I*	4	6	9	
	Sample II	8	9	10	11

overall rankings are

Stratum I	*Sample I*	1	2	13	
	Sample II	3.5	6	11.5	
Stratum II	*Sample I*	3.5	5	8.5	
	Sample II	7	8.5	10	11.5

while ranking independently within strata gives

Stratum I	*Sample I*	1	2	6	
	Sample II	3	4	5	
Stratum II	*Sample I*	1	2	4.5	
	Sample II	3	4.5	6	7

The respective *p*-values for a one-tail test are $p = 0.1271$ and $p = 0.0586$. The difference reflects the differing weights associated with the value 15 in Sample I of Stratum 1 which might be regarded as an outlier in terms of the general result pattern.

7.8 Tests for specific effects

After a preliminary parametric ANOVA interest often moves to specific treatment comparisons. In factorial experiments these are typically main effects and interactions or components of these when there is more than one associated degree of freedom. In section 7.6 I dealt with the permutation ANOVA equivalent of several parametric tests and most required the specification of a permutation reference set appropriate to the particular analysis, some of these only involving part of the data. This is contrary to the usual parametric approach where all the data are used to obtain an estimate of the error mean square used for specific comparisons. This increases precision because of the additional error degrees of freedom. It, however, assume homogeneity of variance.

In a one-way classification parametric ANOVA to test whether there is a significant mean difference between a prespecified 2 among a total of 6 treatments one would effectively apply a t-test to the mean difference for the selected pair using the error mean square (with its degrees of freedom) obtained in the complete analysis. This is a more powerful test than a t-test using only the sample values for the two treatments of interest to estimate the standard error; however, its validity depends critically on the assumption of homogeneity of variance. The fact that some distribution-free permutation tests avoid the need for this assumption, and indeed may be invoked specifically because such an assumption is implausible, favours a test based on relevant permutations of the data in the two selected samples.

One must also remember that selective significance tests or estimation procedures are only valid for comparisons chosen before the experiment is undertaken. This is because if, for example, there are six treatments then there are 20 possible pairwise comparisons. If we accept results for which $p < 0.05$ as significant, then in the long run even when the null hypothesis is true we are likely to find one significant result even if all tests were independent (which in fact they are not) The classic example of this sort of thing usually quoted in the parametric case is selection of the comparison showing the largest difference between sample means. For even moderate numbers of treatments this is almost certain to give a p-value that (misleadingly) indicates significance.

While exact tests of particular effects should be done using the relevant permutation reference sets and test statistics a number of writers have suggested analogues of parametric procedures that seem reasonable for asymptotic inference. I describe here only one such example, namely analogous procedures for obtaining a least significant difference between two treatments that is valid asymptotically for the KW test using the statistic T defined by (7.2) [or by (7.3) if there are no ties]. Logically the least significant difference test can only be used if the comparison is selected before the experiment is carried out (or at least before inspection of the data) and in addition if the overall KW test indicates significance and any treatment differences are postulated to be location differences.   Further the significance level used in the least significance difference test must be at least as stringent as that used in the KW test. The criteria for accepting a location difference between two preselected samples, or treatments, i and j are:

(i) the overall test indicates significance and (ii) if m_i, m_j are the means of the ranks for samples i and j then the difference is a least significant difference if

$$|m_j - m_i| > t_{N-k,\alpha}\sqrt{[D(N-1-T)(n_i + n_j)/\{n_in_j(N-k)(N-1)\}]}$$

where T is given by (7.2), D is the denominator in (7.2) where ranks are the data, $t_{N-k,\alpha}$ is the t value required for significance at the $100\alpha\%$ significance level with $N-k$ degrees of freedom, other quantities being as defined in the discussion of the KW test in section 7.3.

This idea is often used when one treatment such as a zero dose of a drug or nonapplication of a fertiliser is a 'control' or base reference treatment and one wishes to compare responses in the first case to increasing doses of one drug or to use of a range of different drugs and in the second case to a variety of fertiliser applications differing in quantity or type of fertiliser.

Least significant differences are one of many testing techniques. Parametric techniques called multiple comparisons also have distribution-free analogues.

7.9 Differences in dispersion and other characteristics

Some tests for differences in dispersion for two independent samples introduced in section 6.5 extend in fairly obvious ways to k independent samples, but the use of most of these becomes questionable because of a common breakdown of the assumption that the locations are all the same. Approximate tests are sometimes based upon shifts of location to the combined sample mean or median before application of the dispersion test, but this does not overcome all problems because of the dependence between location and dispersion for many families of distributions. Nevertheless if such shifts are made, the subsequent tests may be used as approximate tests for differences in variance and they should at least have moderate power in picking up differences not covered by tests primarily designed to pick up location differences or stochastic dominance.

Exercises

7.1 Show that for the Fisher–Pitman k-sample permutation test the uncorrected treatment sum of squares may be used as a statistic to test for differences in treatment effects or for a shift in population location means (i.e. show that other terms in the parametric test variance ratio are invariant over the relevant permutation reference set).

7.2 Verify the asymptotic result given in (7.3) for the Kruskal–Wallis test by applying (7.2) to ranks considered as data.

7.3 Verify the computed value of the statistic S for each of the scoring systems for $f(t)$ used in Example 7.9

7.4 In Example 7.13 if a parametric ANOVA were being performed the component sum of squares statistic for testing the main effect of light would be based upon $S_1^2 + S_2^2$ where S_1^2, S_2^2 are the totals for all r observations at normal and extended light levels respectively. Show that this is equivalent to the statistic S used in Example 7.13.

*7.5   For the data in Example 7.18 test separately whether Package A is superior for those with, and separately for those without, previous training in the subject of the report.

*7.6   A professor of English asserts that short stories written by writer A are excellent, those written by writer B are good, those written by writer C are inferior. To test his claim and judgement he is given 20 short stories to read on typescripts that do not identify the author and asked to rank them on merit from 1 to 20 (1 for best, 20 for worst). In fact there are 6 by author A and 7 each by B and C. Rankings given by the professor when checked against authors are:

A	1	2	4	8	11	17	
B	3	5	6	12	16	18	19
C	7	9	10	13	14	15	20

Do these data back his claims of discriminatory ability?

*7.7   The following data are a sub-set of some given by Kempthorne (1952, p. 158) and H and C are two grass mixtures and A and B are two sowing rates per unit area, the rate B being 4 times that of the rate A. The variable recorded is the number of plants after a certain unspecified interval on 3 random samples of 1 square yard in the plot subjected to each treatment. Perform appropriate factorial analyses examining factorial effects using both what you consider to be suitable parametric and nonparametric methods.

HA	94	53	57
HB	303	271	134
CA	25	11	18
CB	90	59	103

*7.8   In a study of working conditions that a company hopes might improve productivity at a routine task assembling machine components 12 workers in each of the age-bands 20–29, 30–39, 40–49 and 50–59 are split into two groups of six and one group carries out the task to a musical background and the other without. The average numbers of items per hour produced by each group are as follows. Do the results support a claim that production is better with a musical background. Is there evidence that production levels decline with age (a) with music and (b) without music?

Age	Music status	Production					
20–29	No music	21	25	31	33	36	41
	Music	23	27	29	38	39	44
30–39	No music	23	24	24	26	32	33
	Music	25	25	28	33	36	39
40–49	No music	19	23	23	27	28	37
	Music	16	35	29	34	34	38
50–59	No music	22	22	23	25	25	26
	Music	18	26	27	28	28	29

*7.9   The following data from Hand et al (1994, Set 95) used in the UK Open University course MDST242 Statistics in Society, Unit C2 were obtained from an experiment in which mustard seedlings were grown in dark or in daylight during the day. After germination stems were cut off some seedlings to allow for the possibility that light affected the vigour of the whole plant through stem and leaves. Data are root lengths (in mm) off all seedlings at a common later time. Does light effect root growth and does the effect depend upon whether stems are cut? There is a case for using both a parametric and a distribution-free analysis because there is some indication of variance heterogeneity or perhaps of outliers.

Grown in light	*Stem cut*	21	39	31	13	52	39	55	50	29	17	
	Stem not cut	27	21	26	12	11	8					
Grown in dark	*Stem cut*	22	16	20	14	32	28	36	41	17	22	
	Stem not cut	21	39	20	24	20						

*7.10 Shirley (1977) developed a nonparametric test for contrasting increasing levels of a dose and applied it to the data below. You may want to consult the reference for details but it is possible to obtain useful results using some of the methods described or referred to in this chapter. The main interest was to decide whether there was evidence of toxicity due to a drug given to mice and if so at what dose level toxicity became apparent. In the experiment there were three increasing dose levels (I, II, III) and a control group receiving zero dose (level 0) and the data are reaction times in (seconds×10) of the mice to stimuli applied to their tails.

Level 0	24	30	20	22	22	22	22	28	30	30
Level I	28	22	38	94	84	30	32	44	32	74
Level II	98	32	58	78	26	22	62	94	78	34
Level III	70	98	94	88	88	34	90	84	24	78

Comment on any special features of the data that influence your choice of methods of analysis.

*7.11 Hand *et al* (1994, Set 304) give the following data from Dolkart, Halperin and Perlman (1971) on amounts of nitrogen-bound bovine serum albumen in mice, commenting that a simple power transformation will solve problems of skewness and variance inequality. Consider alternative distribution-free methods of analysis. The three groups are normal mice, those that are alloxan diabetic untreated and those that are alloxan diabetic and treated with insulin.

Normal	156	282	197	297	116	127	119	29	253	172	349	110	143	64
	26	86	122	455	655	14								
Alloxan	391	49	469	86	174	133	13	499	168	62	127	276	176	146
	108	276	50	73										
Alloxan + insulin	82	106	98	150	243	68	228	131	73	18	20	100	72	
	133	465	40	46	34	44								

8

Designed experiments

8.1 Randomized block experiments

A paired comparison experiment (section 5.6) is a simple randomized block experiment, a common design for comparative experiments. The paired comparison situation with b pairs is a randomized block experiment with 2 treatments in b blocks. To compare k treatments in a randomized block design with b blocks each treatment occurs the same number of times, usually once, in each block giving bk experimental units in all. Units are allocated to treatments **at random within each block**. The aim of blocking is to achieve reasonable homogeneity of units within each block for characteristics other than the applied treatments. For example, for testing the effect of three different feeding regimes on the growth of pigs, the first block might consist of three pigs from one litter, the second of three pigs from another litter, the third of three pigs from a further litter, and so on. Providing the diets are assigned, one to each pig, randomly chosen within each block (litter), valid comparisons of diets may be made **within litters**.

In an industrial context ingredients from four different sources (each regarded as a 'treatment') might be used to produce a plastic moulding where processing involves baking in one of 5 ovens (blocks) for a fixed time. Because of possible temperature or other differences between ovens, batches from each of the four sources might be baked in each of the ovens and comparison of the strength (or some other characteristic of interest) of plastic made between the 4 sources (treatments) within each of the 5 ovens.

In clinical trials and in some psychological studies experimenters are often interested in the response of individuals to each of a number of treatments. Such experiments are sometimes conducted as randomized block experiments regarding each individual as a block, but the use of all treatments on each individual invokes special considerations which I discuss in section 8.5.

In the standard parametric ANOVA the overall test statistic for a treatment effect is the variance ratio

$$V = treatment\ mean\ square/residual\ mean\ square$$

where the residual, or error, mean square is computed after removing a component representing differences between blocks. In broad terms blocking is effective if the differences between blocks is such that the precision of the experiment is improved by subtracting a between-blocks sum of squares from the sum of squares that remains after allowing for treatments.

In a permutation ANOVA, because treatments are allocated to units at random independently within each block, the permutation reference set for testing the hypothesis H_0 of no treatment effect against a hypothesis that there is at least one differential treatment effect is based upon the set of all permutations (i.e. allocations to treatments) of units **within** each block. If k treatments each occur once in each block the number of possible allocations of treatments to units within a block is $k!$ The appropriate reference set of permutations is obtained by combining any of the $k!$ permutations in any one block with any of the $k!$ permutations in each of the remaining $b - 1$ blocks, giving $(k!)^b$ permutations in the complete set. This reference set grows rapidly with increasing k even for moderate b. For example, when $b = 6$ and $k = 2, 3, 4$ the sizes of reference sets are respectively 64, 46 656 and 191 102 976, making exact tests for overall treatment differences virtually impossible even with appropriate software except in small experiments. Monte Carlo or asymptotic results are needed even for not very large experiments.

As in section 7.2 the general test of H_0 against a hypothesis that at least one treatment has an effect is a two-tail test, and we may use the parametric ANOVA test statistic, i.e.. the variance ratio V, but the tabulated F-distribution tables are no longer valid. As in the one-way classification permutation ANOVA in Section 7.2, some of the terms in this ANOVA are invariant under the permutation set used. In particular, the grand total and the block totals are invariant under permutation within blocks, and as in the one-way classification permutation analysis it is easily established that an appropriate statistic is the uncorrected treatment sum of squares, $T = \sum T_i^2/b$ where T_i is the total over all blocks for treatment i and summation is over all k treatments. An alternative but equivalent statistic omits the constant divisor b.

Large values of T indicate significance. However, the same value of T will be obtained for any of the $k!$ possible rearrangements of the treatment labels, thus there are at most only $(k)^{b-1}$ numerically different values of T. It is, in essence, this feature that allowed us to reduce the paired-sample case in section 5.6 to a single-sample problem.

Example 8.1 A trivial example, illustrating principles rather than practice, shows the basics. Three treatments are allocated to three blocks with the following results:

Treatment	Yields		
	Block I	Block II	Block III
A	3	7	22
B	9	14	23
C	17	44	31

The reference set consists of $(3!)^3 = 216$ permutations. In order of treatments, (A, B, C), the six permutations for block I are (3, 9, 17), (3, 17, 9), (9, 3, 17), (9, 17, 3) (17, 3, 9), (17, 9, 3) and these may be combined with any of six permutations of the second block, i.e. (7, 14, 44), (7, 44, 14), (14, 7, 44), (14, 44, 7), (44, 7, 14), (44, 14, 7) and any six for the third block, i.e. (22, 23, 31) (22, 31, 23) (23, 22, 31), (23, 31, 22), (31, 22, 23), (31, 23, 22). The sum of squares of the observed treatment totals is $32^2 + 46^2 + 92^2 = 11\ 604$. The corresponding sum of squares can be worked out for each of the remaining 216 permutations, i.e. that corresponding to (3, 17, 9), (44, 7, 14) and (31, 23, 22) is 10 318. Since we know that there are six outcomes giving the same sum (corresponding to permutations of the treatment labels), we only need to compute 36 potentially different sums. By hand this is an exercise in tedious arithmetic best left to a computer, but the 36 possible values range from 9886 given by 12 possible allocations to treatments, being two sets of six determined by relabelling of the treatments, one of each set being

Treatment	Yields		
	Block I	Block II	Block III
A	3	44	22
B	9	7	31
C	17	14	23

and

Treatment	Yields		
	Block I	Block II	Block III
A	3	44	22
B	9	14	31
C	17	7	23

while the largest $T = 11\ 604$ is given by our original data set, or one of the 5 remaining sets obtained by permuting the treatment labels. The corresponding $p = 1/36 = 0.028$. Common-sense reasoning support these results because, for the given data, treatment A gives the lowest yield in every block, treatment C the highest, and treatment B an intermediate yield, whereas in the permutations giving $T = 9886$ the ordering of the yields differs markedly from block to block.

There are strong indications of variance heterogeneity from plot to plot for the given data and indeed a parametric ANOVA gives a variance ratio for treatments of $F = 4.99$ with a probability of approximately 0.087 of being exceeded if H_0 were true, so that the test would not indicate significance even at the 5% level.

The benefit of blocking is demonstrated if we ignore blocks and perform a one-way classification permutation ANOVA. However, such an analysis would not be valid as the appropriate randomization has not been carried out. All we can do to make a comparison is to assume we would have got this set of results had we not used blocks. In that one-way analysis the permutation reference set consists of $9!/(3!3!3!) = 1680$ permutations and the p-value for a permutation ANOVA is $p = 0.1214$ and for a KW test using ranks it is $p = 0.1679$, giving a clear indication of a not-to-be-unexpected gain in efficiency from blocking.

There are few programs available for exact or Monte Carlo tests for the permutation ANOVA for blocked data, although one for the latter is given by Edgington (1995, Chapter 5), and it is not hard to develop macros for software such as MINITAB or other comprehensive general statistical packages for Monte Carlo testing. An asymptotic approximation valid for reasonably large sample sizes and especially if the treatment differences are ones of location is to use as a test statistic

$$T' = b(k-1)(S_k - C)/(S_T - C) \qquad (8.1)$$

where $S_k = \sum T_i^2/b$, S_T is the sum of squares of all kb data values and C is the square of the data total divided by kb. Thus $S_k - C$ and $S_T - C$ are respectively the usual corrected sums of squares for treatment and total in a parametric ANOVA. This has asymptotically a chi-squared distribution with $k-1$ degrees of freedom, as was the case for the analogous (7.2). The approximation is not satisfactory for small data sets, where it tends to overestimate p and the same is often true even for larger data sets if these have a far-from-homogeneous error structure.

8.2   Tests based on ranks

A widely used procedure for analysing randomized block data is due to Friedman (1937). In essence it is an extension of the sign-test for matched pairs (section 5.6, and see Exercise 8.1). The method removes differences between blocks by ranking observations in ascending order 1, 2, . . ., k within each block, using midranks for ties where necessary. Clearly the test is immediately applicable if the basic data are ranks allocated separately within each block. In this context an equivalent test for consistency of ranking given in section 9.7 was developed independently by M.G. Kendall. I denote the rank for treatment i in block j by r_{ij}, $i = 1, 2, . . ., k$ and $j = 1, 2, . . . , b$. The permutation procedure follows that described in section 8.1 with ranks replacing the original data. The simplest test statistic is $T' = \sum R_i^2$ where R_i is the sum over all blocks of the ranks allocated to treatment i. A more commonly used statistic in practice is $S_k = T'/b = (\sum R_i^2)/b$

and this term appears in the asymptotic test statistic analogous to (8.1) which is

$$T = b(k-1)(S_k - C)/(S_T - C), \qquad (8.2)$$

where now $S_T = \sum_{i,j} r_{ij}^2$ and $C = (\sum_{ij} r_{ij})^2/(bk)$. When there are no ties S_T simplifies to $S_T = bk(k+1)(2k+1)/6$ and C to $C = \frac{1}{4}bk(k+1)^2$ (Exercise 8.2). Asymptotically, under the hypothesis of no treatment differences T has a chi-squared distribution with $k-1$ degrees of freedom. Iman and Davenport (1980) propose a test statistic $T_1 = (b-1)(S_k - C)/(S_T - S_k)$ which under H_0 has for kb large an approximate F-distribution with $k-1$ and $(b-1)(k-1)$ degrees of freedom. If the ranking for treatments is identical in all blocks the denominator of T_1 tends to zero and Iman and Davenport show that this may be interpreted as a result significant at the $100(1/k)^{b-1}$% significance level. This is based on an exact p-value that follows immediately from the permutation theory developed in section 8.1 (Exercise 8.3).

The exact Friedman test is a permutation test applied to ranks within blocks as data. Exact tests are available for small samples both in StatXact and in TESTIMATE. The former also has a program for Monte Carlo estimates of exact p-values and these and many other programs provide asymptotic approximations, but the latter should be used only if the total data set size bk is fairly large.

Example 8.2 Berry (1987) gives the following data for the number of premature ventricular contractions per hour for 12 patients with cardiac arrhythmias when each is treated with one of three drugs A, B, C. Although the actual experimental design was more complex I use the data here to illustrate a randomized block type analysis. Berry gives alternative randomized block parametric ANOVAs for the original data and also after several transformations. I consider here a Friedman analysis. The result is in close agreement with that obtained by Berry for a corresponding parametric ANOVA using his preferred transformation to overcome obvious variance heterogeneity.

Patient	1	2	3	4	5	6	7	8	9	10	11	12
Drug A	170	19	187	10	216	49	7	474	0.4	1.4	27	29
Drug B	7	1.4	205	0.3	0.2	33	37	9	0.6	63	145	0
Drug C	0	6	18	1	22	30	3	5	0	36	26	0

The exact $p = 0.0155$ and the asymptotic chi-squared approximation is 0.0179. This compares with a variance ratio giving $p = 0.078$ for a parametric ANOVA on the untransformed data.

The Friedman scoring system has the attractive property of removing block differences since all block totals are $\frac{1}{2}k(k+1)$. An alternative scoring method with some appeal would be to rank all observations in ascending order and to perform a permutation ANOVA using these ranks. This, however, may be sensitive to outliers. Another alternative is to use normal scores in place of over-all ranks and to perform either a permutation or a parametric ANOVA on such scores. The latter procedure has been proposed also for over-all ranks themselves. Procedures of this kind are discussed by Iman, Hora and Conover (1984).

An alternative that in certain circumstances is more powerful than the Friedman test was suggested by Quade (1979). It also eliminates block differences, but weights the rank scores to give extra weight to blocks where the raw data indicate more marked treatment effects. I use the notation x_{ij} for the original datum for treatment i in block j and r_{ij} for the corresponding Friedman rank, using mid-ranks for ties and ranking separately from 1 to k within each block. The Quade procedure is to compute within each block the range of the x_{ij}. For block j this is $D_j = \max_i x_{ij} - \min_i x_{ij}$. The D_j are then ranked in ascending order from 1 to b, using mid-ranks for ties if needed. Let this rank for block j be q_j. We next compute for each x_{ij} the Quade score $s_{ij} = q_j[r_{ij} - \frac{1}{2}(k+1)]$. These scores may be used in an exact permutation test if a suitable program is available. A useful asymptotic result uses the Iman–Davenport statistic

$$T_q = (b - 1)S_k/(S_T - S_k)$$

which has asymptotically an F-distribution with $k - 1$ and $(b - 1)(k - 1)$ degrees of freedom, where $S_k = \sum_i [(\sum_j s_{ij})^2]/b$ and $S_T = \sum_{i,j} s_{ij}^2$. As for the Friedman test when the corresponding statistic is used, if the denominator is zero the exact p value is $p = (1/k!)^{b-1}$ (Exercise 8.3).

Example 8.3. I obtain the Quade scores for the first four patients only from the data in Example 8.2. The relevant data are

Patient	1	2	3	4
Drug A	170	19	187	10
Drug B	7	1.4	205	0.3
Drug C	0	6	18	1

Here $k = 3$ and $b = 4$. The r_{ij} are

Patient	1	2	3	4
Drug A	3	3	2	3
Drug B	2	1	3	1
Drug C	1	2	1	2

From the original data we find $D_1 = 170 - 0 = 170$, $D_2 = 19 - 1.4 = 17.6$, $D_3 = 205 - 18 = 187$, $D_4 = 10 - 0.3 = 9.7$, whence $q_1 = 3$, $q_2 = 2$, $q_3 = 4$ and $q_4 = 1$, leading to the following s_{ij}:

Patient	1	2	3	4
Drug A	3	2	0	1
Drug B	0	-2	4	-1
Drug C	-3	0	-4	0

The statistic $T_q = 1.675$. For the F-distribution approximation we find $Pr(F \leq 1.675) = 0.2642$. StatXact also gives the exact $p = 0.2870$. For the complete data given in Example 8.2, StatXact gives the exact $p = 0.0243$ for the Quade test and an asymptotic $p = 0.0236$. Even for the reduced set of 4 blocks the asymptotic approximation works well in this case, but this does not imply that the asymptotic result is always satisfactory for small samples. In this example for the complete data, the Quade test gives a slightly higher p-value than the Friedman test.

For two treatments the Quade test is equivalent to the Wilcoxon signed rank test for paired samples (Exercise 8.5), and it may be regarded as an extension of that test. With the clear heterogeneity of variance evident in the data in Example 8.2 one should have reservations about using an extended Wilcoxon-type test.

In section 7.4 I developed the Jonckheere–Terpstra test for ordered treatments. An analogue for randomized block designs was given by Page (1963). If treatments are arranged in ascending order of expected effect under H_1 and Friedman ranks are used and s_i is the sum of these ranks for treatment i the Page statistic is $P = s_1 + 2s_2 + 3s_3 + \ldots + ks_k$. An exact permutation test based on this statistic is included in StatXact and TESTIMATE. Asymptotically P is normally distributed with $E(P) = \frac{1}{4}kb(k + 1)^2$ and $var(P) = bk^2(k^2 - 1)^2/[144(k - 1)]$. Since the treatments are ordered we may perform either a one- or two-tail test. The one-tail test is appropriate for a monotonic trend prespecified under H_1 either as increasing or as decreasing. The two-tail test is for a trend that may be either way. The motivation for the form of the test statistic follows closely reasoning introduced in section 7.5 for using scores in correlation type tests. Effectively the test use row scores 1, 2, 3, ..., k corresponding to the ordered treatments.

Example 8.4   I illustrate some strengths and weaknesses of the tests in this section in a situation where a parametric ANOVA is appropriate using data from an experiment conducted at the North Carolina Agricultural Experiment Station at Rocky Mount, N.C. in 1944. A detailed parametric analysis is given by Cochran and Cox (1950, Chapter 4) and in later editions of that work. The measurements were breaking strengths of cotton measured on an arbitrary scale for crops grown at 5 levels of application of potash, namely 36, 54, 72, 108 and 144 lb. K_2O per acre. There were three replicates of each treatment in a standard randomized block design. Interest attached to whether breaking strength decreased with increasing fertilizer. The breaking strengths were:

Treatment	Replicate		
(Lb. K_2O per ac.)	I	II	III
36	7.62	8.00	7.93
54	8.14	8.15	7.57
72	7.76	7.73	7.74
108	7.17	7.57	7.80
144	7.46	7.68	7.21

A visual scan suggests that strength falls off noticeably but somewhat erratically for fertilizer levels at or above 72 lb.ac^{-1} while remaining fairly steady, or perhaps rising slightly between levels 36 and 54. The conventional parametric ANOVA given by Cochran and Cox shows a significance overall treatment effect at a nominal 5% level (the exact level is about 4.4%). These authors also fit a linear regression of breaking strength on fertilizer level which is significant at a nominal 1% level in a two-tail test (the exact level is about 0.6%).

The Friedman and Quade tests are both tests for over-all unspecified treatment differences in location and as such are competitors to the over-all test of treatment differences in a parametric ANOVA. In this case both fare badly relative to the parametric test failing by a substantial margin ($p = 0.24$ and $p = 0.34$ respectively) to indicate significance. The poor performance of these tests in small samples relative to parametric ANOVA when the latter is justified may be due to the nonlinearity of the rank transformation enhancing differences in observed data that are well within the limits of error under a quite reasonable assumption of homogeneous error. In a small data set such enhancements can have a marked effect on the value of the test statistic relative to the values it may take for other permutations. This is the opposite effect to that of ranking upon an extreme observation or outlier. This problem tends to become less marked with more blocks. For few blocks it is reflected in a comparatively low Pitman efficiency against the normal model. For 3 blocks Lehmann (1975) quotes a Pitman efficiency of 0.72 in this case.

The Page test is a test for monotone trends and as such should pick up linear trends detected by a parametric ANOVA, although one may again fear that in small experiments ranking may enhance differences that are within the range of experimental error. However, in this case the exact and even the asymptotic test give similar results to the parametric test for linear regression. The one-sided $p = 0.0097$ and the two-sided $p = 0.0194$. Cochran and Cox noted, as I did above, that strength only falls off at fertiliser levels above 72 lb.ac^{-1} but they indicate that there is no evidence of a quadratic trend.

8.3 Specified comparisons

Comparison of means on the basis of least significant differences extends immed- iately from that described for the KW test in section 7.8 to the Friedman test and to the Quade test. The appropriate scores are used in each case and the test may be made on treatment means or treatment totals for these scores. The algebraic expressions in terms of totals are simpler and if we denote the total score over all blocks for treatment i by s_i then a least significance difference between treatments i and j in the notation used in the previous section and that in section 7.8 is

$$|s_j - s_i| > t_{(b-1)(k-1),\alpha}\sqrt{[2b(S_T - S_k)/(b-1)(t-1)]}.$$

The provisos about the need to choose comparisons for testing prior to collection of data and only to make the test if overall significance is attained as set out in section 7.8 must be observed.

A fuller discussion of multiple comparison tests for one-way classifications and for randomized blocks in rank-based analyses is given by Shirley (1987).

The rest of this section assumes familiarity with the content of section 7.6 and should be omitted at first reading if that section was.

The permutation analysis of factorial treatment structures in randomized blocks raises many of the problems described in section 7.6. The simplest case is a 2^k experiment. I outline a possible *ad hoc* method of analysis for a 2×2×2 experiment with factors denoted by A, B and C and levels by 1 and 2. I consider specifically the situation where one is interested in effect of the factor A including its interaction with the other factors, although clearly the method may be modified to cover effects of other factors. First one should perform an appropriate overall test of the type previously described, e.g. Fisher–Pitman, Friedman, Quade. If such a test indicates a treatment difference one would wish first to establish whether there are any interactions between the factor A and the other factors. These interactions in conventional factorial notation are denoted by AB, AC and ABC. Using similar arguments to those in section 7.6 if there is no interaction between factor A and the factors B, C, then if we form the differences between responses to A at the higher and lower levels for each fixed combination of the other factors (i.e., $a_2b_2c_2 - a_1b_2c_2$, $a_2b_2c_1 - a_1b_2c_1$, $a_2b_1c_2 - a_1b_1c_2$, $a_2b_1c_1 - a_1b_1c_1$) in each block, each of these should be measuring the same response, i.e. all are measures of the effect of A and their mean is the main effect of A. If they show a significant difference this is an indication of interaction and we should then look at particular effects of A, i.e. at each of the 4 differences individually. The same test procedure as that used for testing for overall treatment differences should be used for this interaction test (e.g. if the Quade test was used for the over-all test this should be used here, treating the four differences $a_2b_2c_2 - a_1b_2c_2$, $a_2b_2c_1 - a_1b_2c_1$, $a_2b_1c_2 - a_1b_1c_2$, $a_2b_1c_1 - a_1b_1c_1$ in each block as data). This is important because, as indicated in the discussion of interactions in section 7.6, any transformation of data may have the effect of removing or of reducing (and occasionally of amplifying) an interaction. If there is no evidence of interaction the remaining question is whether these differences show a significant departure from zero. This may be tested as a matched pair test over the $4×b$ differences in each of the b blocks, i.e. $a_2b_2c_2 - a_1b_2c_2$, $a_2b_2c_1 - a_1b_2c_1$, $a_2b_1c_2 - a_1b_1c_2$, $a_2b_1c_1 - a_1b_1c_1$. Recalling that the Friedman test is a generalization of the sign test, the Quade test is a generalization of the Wilcoxon signed-rank test, one should use here the equivalent test to that used in the overall test and the test for interaction. If this test does not reject the hypothesis of a zero main effect then A has no effect. This may occur even if the overall test indicates a treatment difference, because such a difference may be associated with the main effect of the factors B and C or the interaction BC. Returning now to the situation where there is an indication of an interaction between A and the other factors, it is now no longer appropriate to make an overall test for a main effect of A and instead one should look at the particular effects, i.e. the effect of A at each fixed level of the other factors separately. Since there is only one measure of each effect (e.g. $a_2b_2c_2-a_1b_2c_2$) in each block there are only b replicates for each of the four effects, so that unless b is quite large any test for particular effects may have low power.

8.4 Binary responses

Each of a number of subjects may give a binary response to several stimuli. Formally we regard subjects as blocks and the stimuli as treatments and record 1 for a positive response and 0 for a negative response, where the labels positive and negative are often assigned arbitrarily to the possible outcomes. Typical pairs of binary responses are (win, lose), (alive, dead), (succeed, fail), (male, female).

Example 8.5 Three drugs are given in random order to ten patients all suffering from the same illness, sufficient time being left between administration of each drug to remove any latent effect from the previous drug, and there being confidence in the light of the chemical nature of the drug and past experience that the response to any one drug will, if sufficient time has elapsed, not affect the response to the next drug. For ten patients (blocks) a record is made of the response to each drug, recording 1 for a beneficial effect and 0 for no change. Is there evidence that the drugs differ in their beneficial effects? The results were:

Patient	1	2	3	4	5	6	7	8	9	10
Drug A	1	0	0	0	0	0	1	1	1	0
Drug B	1	1	0	0	1	1	1	1	1	0
Drug C	0	0	0	1	0	1	1	0	0	1

Clearly one approach is to regard this as a heavily tied situation and to convert responses to Friedman ranks within each block (patient) using mid-ranks for ties. Thus in block (patient) 2 the ranks (in order *Drug A, Drug B, Drug C*) would be 1.5, 3, 1.5, and for block 4 they would be 1.5, 1.5, 3, while for block 3 all are ranked 2. For these data the Friedman test gives an exact $p = 0.4856$.

An alternative to the Friedman test would be a permutation ANOVA using the basic data of 1s and 0s if a suitable program is available. The result will be the same as that for the Friedman test! This is because in this particular situation the Friedman statistic (Exercise 8.4) is a linear function of the permutation statistic where we use the sum of squares of the treatment total as an appropriate statistic in each case. This particular test is given a special name and the customary statistic takes a special form which can be deduced from the Friedman statistic used for the asymptotic chi-squared test. The test in this form was proposed by Cochran (1950) and usually bears his name. The formula for Cochran's statistic, usually denoted by Q is

$$Q = \frac{k(k-1)\sum_i S_i^2 - (k-1)N^2}{kN - \sum_j B_j^2}$$

where k is the number of treatments, T_i the total number of 1s for treatment i, B_j the total number of 1s in block j and N the total number of 1s in the complete data. Asymptotically Q

has a chi-squared distribution with $k - 1$ degrees of freedom. If $k = 2$ the Cochran test is equivalent to the McNemar form of the sign test. This is to be expected because since, as already indicated, the Friedman test is a generalization of the sign test and the Cochran test is equivalent to the Friedman test. StatXact provides a special program for the Cochran test, but the result is equivalent to that given by the Friedman test program.

8.5 Using individuals as blocks

In some clinical trials and experiments in psychology involving several treatments the experimenter wishes to apply all treatments to each of a number of individuals. This was the situation in Example 8.5 and you may have noted I included provisos regarding treatment effects. In such experiments a complication is that the response to a particular treatment may be influenced by one or more of the earlier treatments. This is especially true in clinical trials involving the use of several drugs, where the response to one drug may be influenced by residual effects of a drug or drugs previously administered. Similarly in an experiment to compare, say, the use of different tools to carry out a mechanical task the performance of workers with a particular tool, A, may be more satisfactory if it is used before another tool, B, than if it is used after B. This could happen, for example if B created considerable vibration that made the operator's hands 'shaky' when he or she started to use A afterwards. Sometimes residual effects or carry-over effects from a previous treatment can be eliminated by allowing sufficient time to elapse between treatments. If this not possible more sophisticated experimental designs are called for. Similar precautions are needed if there is a learning effect that may depend on the order of treatment. This happens in psychological and educational experiments. A simple example of a design useful in such circumstances, called a cross-over design, is given below.

When a randomized block design is used with subjects as blocks it is important to randomize for each subject the order in which treatments are given to that subject even when sufficient time is allowed between treatment periods to avoid residual effects from earlier treatments. We must do this because, even if there are no residual effects associated with treatments, other temporal effects will often influence responses to treatments. For example, if treatments involve a manual task tiredness or a learning effect may well affect responses in later periods, irrespective of which treatments have been associated with earlier periods. If there is reasonable confidence that there are no residual effects of earlier treatments associated with the order of application and treatments are allocated to units at random a parametric or permutation ANOVA for a randomized block design is often appropriate, the choice of the particular analysis depending upon relevant assumptions about the nature of measured responses.

When there may be residual effects from a previous treatment a **cross-over** design (sometimes also called a change-over design) is often appropriate. I consider here only the simplest of such designs that involves the comparison of two treatments in what is called a two-period cross-over design. The designs may be extended to comparisons of several treatments over more than two periods and are widely used in pharmaceutical studies. A comprehensive account of their use together with descriptions of both parametric and nonparametric analyses is given by Senn (1992).

An early account of a parametric analysis of a cross-over design in a clinical context is given by Grizzle (1965). Koch (1972) gives an interesting example of a nonparametric analysis of one such design, some features of which are considered in Example 8.6. He illustrates many facets of cross-over trials in a situation where two treatments A, B are used on 10 (more generally an even number N) patients. Five of these (generally $\frac{1}{2}N$), selected at random, are given treatment A first, followed by treatment B. The treatment order is reversed for the remaining 5 (generally $\frac{1}{2}N$) patients. This experimental design enables us to test hypotheses about direct effects of treatments, about residual effects of treatments (i.e. whether one ordering of treatments gives different overall responses to the other) and also what are called period effects. This last test is of interest if there are no residual treatment effects, but a genuine difference of responses in each period (period I or period II) that is independent of the actual treatment applied in each period.

Example 8.6   Koch (1972) analysed data from an experiment at the Dental Research Center, University of North Carolina. The children were randomly assigned to two groups of 5. The two treatments administer to each child were

A: Drink 100 ml. grapefruit juice followed by an elixer of Pentobarbital.
B: Drink 100 ml. water followed by an elixer of Pentobarbital.

The first group of 5 received A in period I and B in period II. This sequence is described below as AB. The second group received the treatments in reverse order, described as BA. Data are the amounts of drug in a 10 ml blood sample taken 15 minutes after administration expressed in units of $\mu g.ml^{-1}$. The data were:

Child	1	2	3	4	5	6	7	8	9	10
Sequence	AB	AB	AB	AB	AB	BA	BA	BA	BA	BA
Period I	1.75	0.30	0.35	0.20	0.30	7.20	7.10	0.75	2.15	3.35
Period II	0.55	1.05	0.63	1.55	8.20	0.35	1.55	0.25	0.35	1.50

The large values for treatment B, relative to the rest of the data, for subjects 5, 6 and 7 suggest heterogeneity and therefore a nonparametric analysis is appropriate. Koch proposed analyses based on WMW tests. He first considered the data for each child summed over Period I and Period II. If there is a differential carry-over effect (i.e. if the joint response summed over the two periods are different for the sequences AB and BA) we would expect an appropriate test to indicate a significant difference between the sums for child 1 to 5 (the first group or AB sequence and the sums for child 6 to 10 (the second group or BA sequence).

The sums are easily obtained, e.g. that for child 3 is $0.35 + 0.63 = 0.98$, and the sums for each child in the order given above are:

Sequence AB	2.30	1.35	0.98	1.75	8.50
Sequence BA	7.55	8.65	1.00	2.50	4.85

The nonnormal appearance of these data suggests a WMW test or a normal scores test, remembering that both are reasonably robust. Using StatXact for both gives two-tail test p-values respectively $p = 0.3095$ and $p = 0.2778$. Thus the data give no evidence of a differential residual effect. This does not mean that there is no residual effect from a previous treatment, but only if there are any, that they apply equally to either sequence.

The next step, and when there are no differential residual effects, the most important step, is to test for a direct effect of treatments. This is based on the differences, period I – period II for all individuals. If there is a differential effect between treatments we should expect the differences for the first five children to exhibit a different location pattern to those same differences for the remaining children because one is measuring the effect *treatment A – treatment B*, and the other *treatment B – treatment A*. If a null hypothesis of no differential response to treatments holds both differences should have the same distributions

The relevant differences are:

Sequence AB	1.20	−0.75	−0.28	−1.35	−7.90
Sequence BA	6.85	5.55	0.50	1.80	1.85

Applying the WMW test or the normal scores test using StatXact gives a two-tail $p = 0.0159$ for either test, implying significance at an exact 1.59% level.

A third test proposed by Koch which is only appropriate if the hypothesis of a differential residual effect is rejected is a test for a period effect, namely that the responses *treatment A – treatment B* may be different for each of the sequence orders. Clearly for the sequence AB the *treatment A – treatment B* difference for each individual is the difference given above, while for the sequence BA it is the difference above after reversing the sign. Again we may apply a WMW, or a normal scores test. In this case StatXact gives $p = 0.3095$ and $p = 0.3175$ for the relevant two-tail tests, so there is no indication of a period effect.

TESTIMATE provides a specialized program for exact tests with cross-over designs computing rather different statistics that are often relevant specifically to pharmacological applications and drug testing. The reader should consult the program manual for details. Most practical cross-over designs are more complex than that in the above example.

Exercises

8.1 Show that the Friedman test is a generalization of the matched pair sign-test. (Hint: If in each pair in a matched-pair experiment the responses are ranked, 1, 2 show that the number of plus signs in the differences is equal to the number of Sample I units ranked 2 and that the difference between the rank sums for each treatment equals the difference between the numbers of plus and minus signs and confirm that these provide alternative statistics for the Friedman test in this case).

8.2 Confirm that the statistic (8.2) for the Friedman test is consistent with (8.1) when data are ranks.

8.3 For the Iman–Davenport statistics on p.214 and 215 establish the p-value quoted there for the situation where the denominator of the statistic is zero is the exact test p-value in that case..

8.4 In Example 8.5 show that the Friedman statistic is a linear function of the permutation test statistic..

8.5 Show that for two samples the Quade test reduces to the paired-sample Wilcoxon signed rank test.

*8.6 At the beginning of a class 12 names are read out in random order to 10 students. Four are names of prominent sporting personalities (Group A), four of national and international politicians (Group B) and four of local dignitaries (Group C). At the end of the class students are asked to recall as many of the names as possible. The results are:

Student	1	2	3	4	5	6	7	8	9	10
Group A	3	1	2	4	3	1	3	3	2	4
Group B	2	1	3	3	2	0	2	2	2	3
Group C	0	0	1	2	2	0	4	1	0	2

Use a Friedman test to assess evidence of a difference between recall rates for groups. In particular is the rate for group C significantly lower than that for groups A and B. Carry out a parametric ANOVA on the data. Do the conclusions agree with the Friedman test? If not, why not?

*8.7 Four share tipsters each predict on 10 randomly selected days whether the London FTSE Index of share prices, commonly called the Footsie, will rise or fall on the following day. A correct prediction is scored as 1 and an incorrect prediction as 0. Do the results below indicate differences between the tipsters' abilities to predict accurately?

Day	1	2	3	4	5	6	7	8	9	10
Tipster 1	1	0	0	1	1	1	1	0	1	1
Tipster 2	1	1	1	1	0	1	1	0	0	0
Tipster 3	1	1	0	1	1	1	1	1	0	1
Tipster 4	1	1	0	0	0	1	1	1	0	1

*8.8 Cohen (1983) gave data for the numbers of births in Israel for each day in 1975. I give below his data for numbers of births on each day of the 10th, 20th, 30th and 40th weeks of the year. Use a Quade test to decide whether the data indicate (i) a difference in birthrate between days of the week that shows consistency over the four selected weeks and (ii) any differences between rates in the 10th, 20th, 30th and 40th weeks.

Day Week	Mon	Tue	Wed	Thu	Fri	Sat	Sun
10	108	106	100	85	85	92	96
20	82	99	89	125	74	85	100
30	96	101	108	103	108	96	110
40	124	106	111	115	99	96	111

*8.9 Snee (1985) gives data on average liver weights of chicks given three levels of growth promoter (none, low, high). Blocks correspond to different bird houses. As the dose levels are ordered, use a Page test to see if there is evidence of an effect of promoter.

Block	1	2	3	4	5	6	7	8
No promoter	3.93	3.78	3.88	3.93	3.84	3.75	3.98	3.84
Low dose	3.99	3.96	3.96	4.03	4.10	4.02	4.06	3.92
High dose	4.08	3.94	4.02	4.06	3.94	4.09	4.17	4.12

*8.10 Berry (1987) gave data for the resistance (in kilo-ohms) when five different types of electrodes were applied to arms of 16 subjects. The data below are for sub-group of 6 subjects from the 16. Is there evidence that all electrodes behave similarly on subjects (clearly there are large differences between subjects)? Subject VI had excessively hairy arms and it was thought that might explain the two extreme readings (outliers?) he gave. Would omitting the results for this subject alter your conclusions? Would you consider it valid to omit this subject from an analysis?

Electrode type	A	B	C	D	E
Subject I	660	600	600	75	310
Subject II	72	140	240	33	54
Subject III	135	300	450	430	70
Subject IV	15	45	75	88	80
Subject V	200	290	320	280	135
Subject VI	66	1000	1050	280	220

*8.11 A cross-over design is used to test reading skills. Numbers of errors are recorded for each of 16 children when reading a passage from Shakespeare (A) and from Dickens (B). Half the children read A first (Per. I) and half read B first. The numbers of mistakes are recorded as follows:

Sequence	AB	AB	AB	AB	AB	AB	AB	AB	BA	BA	BA	BA	BA	BA	BA	BA
Errors Per. I	12	0	8	12	17	4	0	9	1	14	2	5	3	11	0	4
Errors Per. II	4	1	7	2	5	1	1	2	4	13	1	5	6	11	5	6

Do these data indicate differences in numbers of mistakes between A and B and is there evidence that numbers of mistakes depend upon the reading order?

*8.12 Bartlett (1936) gave the following data for counts of numbers of surviving leatherjackets on plots of 1 sq. ft for two control areas and areas subjected to each of four toxic emulsions in six randomized blocks. Counts were made on two plots on each experimental unit. The data are also given by Hand et al (1994, set 137) who suggest a parametric ANOVA after transformation. Consider an appropriate analysis using ranks and in particular the use of relevant specific comparisons between controls and toxic treatments.

				Treatment			
Block	Control 1	Control 2	Emul.1	Emul.2	Emul.3	Emul.4
I	33 59	30 36	8 11	12 17	6 10	17 8
II	36 24	23 23	15 20	6 4	4 7	3 2
III	19 27	42 39	10 7	12 10	4 12	6 3
IV	71 49	39 20	17 26	5 8	5 5	1 1
V	22 27	42 22	14 11	12 12	2 6	2 5
VI	84 50	23 37	22 30	16 4	17 11	6 5

*8.13 Grizzle, Starmer and Koch (1969) give the following data on frequencies of possible response patterns favourable (F) or unfavourable (U) for 46 patients to each of 3 drugs. Is it reasonable to conclude that the drugs give similar responses?

Drug				Response patterns			
1	F	F	F	F	U	U	U	U
2	F	F	U	U	F	F	U	U
3	F	U	F	U	F	U	F	U
Frequency	6	16	2	4	2	4	6	6

9

Correlation and concordance

9.1 Measures of bivariate association

The Pearson product-moment correlation coefficient, often referred to simply as 'the correlation coefficient', is the best-known measure of bivariate association. Classical inferences about correlation usually assume that n paired observations (x_i, y_i) are a sample from a bivariate normal distribution and make tests about, or estimates of, the population correlation coefficient, ρ, but correlation is widely recognised also as a more general indicator of association. Under normality assumptions $\rho = 0$ implies independence, but in general $\rho = 0$ only indicates lack of linear or of certain other patterned associations. With normality, $\rho = \pm 1$ implies a degenerate bivariate normal distribution confined to a straight line and in a more general context indicates an exact linear relationship between variables X and Y.

Given n paired observations (x_i, y_i) the Pearson coefficient, r, is computed as

$$r = s_{xy}/(s_{xx}s_{yy})^{\frac{1}{2}}$$

where

$$s_{xx} = \sum x_i^2 - (\sum x_i)^2/n, \quad s_{yy} = \sum y_i^2 - (\sum y_i)^2/n, \quad s_{xy} = \sum x_i y_i - (\sum x_i \sum y_i)/n.$$

Values of r close to zero imply that independence, i.e. that $\rho = 0$, may be a reasonable assumption for a bivariate normal distribution. More generally, near zero values of r imply a lack of linear association although some nonlinear relationship may exist. For example, the value of r calculated for y, the time of sunrise in minutes after 3 am in New York, or any other fixed location, recorded on each xth day after 1 January in any year where $x = 0, 30, 60, 90, 120, 150, 180, 210, 240, 270, 300, 330, 360$ is close to zero, but such data do contain information about how the time of sunrise changes during a year! Values of r near $+1$ or -1 indicate strong correlation, i.e. a high degree of linear or even of some kinds of monotonic or near-monotonic dependence between X and Y. Pitman (1937b) devised permutation tests for $H_0: \rho = 0$ against the one-tail alternative $H_1: \rho > 0$ (or $H_1: \rho < 0$) or against the two-tail alternative $H_1': \rho \neq 0$, where no assumption is made about the form of the population. The test considers all pairs (x, y) formed by permutation of the x_i and the y_i and is carried out on similar lines to other permutation tests for raw data.

Example 9.1   The numbers of supermarkets and independent butchers in four towns of comparable size are:

Town	A	B	C	D
Number supermarkets (x)	3	4	6	8
Number butchers (y)	6	3	5	2

Test the hypothesis

H_0: *There is no correlation between supermarket numbers and independent butcher numbers*

against

H_1: *The number of butcher shops decreases as the number of supermarkets increases,*

or, stated more formally, test $H_0: \rho = 0$ against $H_1: \rho < 0$.

   This unrealistically small example illustrates basic ideas while avoiding tedious computation. Clearly we get all possible pairings (permutations) of the x and y if we write down the observed x in ascending order (as above) and associate with these all possible orderings of the y. We then ask if the observations represent an ordering we would happily associate with H_0, or if it is one we would more readily associate with H_1? We must be more precise about what we mean by *happily* and *more readily associate*. Standard permutation theory tells us that with 4 observations there are $4! = 24$ different pairings of x, y because we have 4 choices for the y value associated with x_1, then 3 choices from the remaining y that we associate with x_2 and so on. In general for n paired observations there are $n!$ permutations.

   The critical region for testing H_0 against H_1 is formed by those permutations that give negative values of r of largest magnitude, i.e. values closest to -1. There are at most 24 different values of r among the correlation coefficients for all permutations. A critical region that includes only the largest negative value of r has size $1/24 = 0.0416$. If the observations correspond to that permutation the result would be significant at the 4.16% level. If it were the negative r of second greatest magnitude the appropriate critical region would include 2 from the 24 possible permutations giving $p = 1/12 = 0.0833$. In this example visual inspection of the data shows we do not reject H_0 at the 4.16% level in a one-tail test because it is well-known that the negative value of greatest magnitude corresponds to the permutation in which, if the x values are in **ascending** order, then the corresponding y are in **descending** order, i.e. 6, 5, 3, 2, not the order in the data.

   We may use as a test statistic either r or more simply $S = \sum_i x_i y_i$ because $s_{xx}, s_{yy}, \sum x_i, \sum y_i$ are all invariant under permutation of the x_i or the y_i. Table 9.1 gives the permutations of the y_i when the x_i are in ascending order and corresponding values of S associated with some of the least possible values of r.

   Good computing facilities are needed for realistic samples. With 10 paired observations, as small a sample as any likely to be met in practice, there are $10! = 3\,628\,800$ permutations, so in a one-tail test at a significance level not exceeding 5%, we need to know if the observed data are among the 181 440 permutations

Table 9.1 Permutations of y for x_i in ascending order that correspond to least values of the correlation coefficient r

x	3	4	6	8	$S = \sum x_i y_i$
y	6	5	3	2	72
y	5	6	3	2	73
y	6	5	2	3	74
y	5	6	2	3	75
y (obs.)	6	3	5	2	76

that comprise the appropriate tail, i.e. those with the smallest or those with the largest values of $S = \sum x_i y_i$.

This test may be put in a contingency table format. For n paired observations (x_i, y_i) if there are no tied values of x or of y we form a table with n rows and columns, i.e. an $n \times n$ table. Each row category corresponds to an x value and each column category to a y value. It is convenient, and for some applications important, to arrange rows and columns in ascending order of x, y. When there are no ties in the x or in the y all cell entries in the table are either 0 or 1. For the observed data we place 1 in each cell corresponding to an observed (x, y) pair and zeros everywhere else. This implies there is one and only one entry of 1 in each row and in each column, all other entries being 0. The row and column totals are all 1. Table 9.2 illustrates these points.

The test statistic, $S = \sum x_i y_i$ used above can be written $S = \sum u_i v_j n_{ij}$, a similar form to that given for the contingency table formulation of the Fisher–Pitman test in section 1.6. The row scores are now $u_i = x_i$, $v_j = y_j$ and n_{ij} is the count (here either 0 or 1) in cell (i, j) and the summation is over all cells. The permutations are over all possible tables with exactly one 1 in each row and one 1 in each column (so that row and column totals remain all unity).

When the x, y are in ascending order a moment's reflection shows that the permutation giving the largest positive r has all the 1s on the diagonal that runs

Table 9.2 Contingency table format for the data in Example 9.1

y	2	3	5	6	Row total
x					
3	0	0	0	1	1
4	0	1	0	0	1
6	0	0	1	0	1
8	1	0	0	0	1
Column total	1	1	1	1	4

Table 9.3 Contingency table format for the data in Example 9.2

y	2	3	5	6	Row total
x					
3	0	0	0	1	1
4	1	1	0	0	2
6	0	0	1	0	1
8	1	0	0	0	1
Column total	2	1	1	1	5

from top left to bottom right and the permutation giving the largest negative r has all 1s on the diagonal from top right to bottom left. A dispersed scatter of 1s over the table corresponds to near zero values of the coefficient r. The table has analogies with, but because of the uniform spacing, is not completely analogous to, a rotation clockwise through 90° of a conventional scatter diagram. Modifications are needed for data ties.

Example 9.2 Consider the problem in Example 9.1 after adding the information that in one further town there are 4 supermarkets and 2 butcher shops, giving the revised paired data set

Supermarkets (x)	3	4	4	6	8
Butchers (y)	6	2	3	5	2

There are tied values at $x = 4$ and at $y = 2$. The permutation test can again be based on a contingency table format. We need one row for each distinct x and one column for each distinct y. We score 1 in any cell corresponding to an observation and 0 in all others giving Table 9.3.

Row and column totals now reflect tied values in the relevant row x or column y. In this table all cell entries are unity, but after permutation to form all possible tables with the given row and column totals, cell entries may be 2 for some permutations. This occurs if we have a tied pair with both x and y identical. Clearly one possible permutation is

x	3	4	4	6	8
y	3	2	2	6	5

giving the contingency table format

y	2	3	5	6	Row total
x					
3	0	1	0	0	1
4	2	0	0	0	2
6	0	0	0	1	1
8	0	0	1	0	1
Column total	2	1	1	1	5

The permutation test is based on selecting the required number of extreme values of S, as defined above, among all possible permutations conditional upon the given row and column totals. The data in this example are not continuous, but necessarily consist of integers. When continuous data are used to provide row and column scores, it helps to think of ties as a limiting case of two observed values converging to equality. When this happens corresponding adjacent rows or columns of the $n \times n$ table are merged to form a single row or column. The process extends to three or more ties, where three or more adjacent rows or columns are merged.

The two-sample Fisher–Pitman test for location difference is computationally equivalent to a special case of the Pearson coefficient with ties. It uses a dummy variable x taking the values 0, 1 to denote whether an observation is in sample I or in sample II. Taking the data values as the observed y, we have a Pearson coefficient model with heavy tying of the x values (either 0 or 1) in a contingency table format like that in Table 1.1. With 0, 1 assigned as row scores u_i and the observed values in the combined samples as column scores v_j, the critical region is formed by the largest values of $S_2 = \sum u_i v_j n_{ij}$ because our choice of row scores makes S_2 the sum of the second sample in that table. Interchanging the row scores gives a test based on the first sample sum, S_1. This is the situation described in Section 1.6 using a contingency table format.

The contingency table format for the Pearson coefficient and for many generalizations or particular cases of it, using the statistic $S = \sum u_i v_j n_{ij}$ is, as already indicated, an example of a linear-by-linear association model in the context of log-linear models discussed in section 12.3 and more specifically in section 12.4.

The permutation tests above may be applied to any appropriate paired variables (x_i, y_i), $i = 1, 2, \ldots, n$ that are not necessarily continuous variables but may be derived paired variables such as the corresponding ranks of the x_i and ranks of the y_i. Denoting these paired ranks by (r_i, s_i) and assuming at this stage the absence of any ties the analogue of the Pearson coefficient using ranks is a well-known rank correlation coefficient first proposed by Spearman (1904). There are several other distribution-free or nonparametric measures of correlation and most have links with other exact test procedures.

To allow consistent comparisons, correlation coefficients are usually defined so that they take values only in the closed interval $(-1, 1)$. Lack of association implies a value near zero, although as indicated on p. 225, the converse need not be true; however any association is then usually nonlinear. For the Pearson coefficient values near 1 or -1 imply a near straight-line relationship between x and y, this being exactly a straight line when $r = 1$ or $r = -1$. In the former case the line has a positive slope, i.e. y increases as x increases, while in the latter case it has a negative slope, i.e. y decreases as x increases. If we replace continuous paired data by ranks and compute the Spearman coefficient, which I denote by r_S, then clearly, since this simply changes the 'scores' (x_i, y_i) to ranks (r_i, s_i) it follows that r_S also takes values in the interval $(-1, 1)$, but now, although values of 1 or -1 imply a linear relationship between ranks, these values no longer imply a linear relationship between the original (x_i, y_i). Linearity of ranks implies **monotonicity in the x, y relationship**, i.e. that y_i increases as x_i increases if $r_S = 1$ or that y_i decreases as x_i increases if $r_S = -1$. Values close to ± 1 imply near monotonicity or a slightly less specific association than that implied by monotonicity. The

Spearman coefficient is a distribution-free statistic widely used in the social sciences partly because data in these disciplines often consist basically of ranks and also because it is often robust in situations where there are measurement data and these contain, or may contain, a few rogue observations. Data of the latter kind are also common in the life sciences where responses to, say, increased doses of a drug may be progressively more favourable up to a certain level but doses above this level cause an unfavourable response due to a toxic reaction.

Example 9.3   Carroll and Rupert (1988, pp. 46–7) give 115 readings for concentration of an enzyme esterase (x) and number of bindings (y), but because the data were proprietary they were unable to give further information such as the unit of measurement for concentration or how accurately those concentrations were measured. They were interested in the data in another context, but a subset of 16 of the paired data values is given in Table 9.4. The data are shown on a scatter diagram in Figure 9.1

The first observation (11.0, 423) represented by the point labelled A on the scatter diagram is clearly an outlier. Outliers sometimes, but not always, have a strong influence on the value of the Pearson coefficient and may blur an otherwise strong indication of a relationship that suggests nonzero correlation (i.e. a degree of association). Such outliers suggest nonnormality in the data. In this example the data, apart from that first pair (and to a lesser extent perhaps the last pair), suggest a positive association and if there were theoretical grounds for believing that any association would be of that nature a one-tail test would be appropriate. Otherwise a two-tail test should be used. StatXact has a permutation program for the Pearson product-moment coefficient to test the hypothesis $H_0\!:\!\rho = 0$ against one- or two-sided alternatives specifying non-zero ρ. Using that program with the exact permutation test Monte Carlo option gave $r = 0.3998$, and estimated exact p-values under H_0 together with 99% confidence intervals for these estimates were:

One-tail	$p = 0.0605$ (0.0577, 0.0632)
Two-tail	$p = 0.1243$ (0.1205, 0.1281).

Asymptotic values based on a bivariate normality assumption were $p = 0.0625$ and $p = 0.1249$ in close agreement, illustrating once again that exact permutation tests on measurement data often give results similar to normal theory tests even when the validity of the latter is in doubt, implying that the approach may often not be robust. We would not reject H_0 at a conventional significance level, e.g. 5%. In Table 9.5 the ranks r_i, s_i are given in the same order as in Table 9.4 (where the x were already ordered).

Again, the first observation is a clear outlier. However, it would have had the same rank pair (1, 16) for any y_1 greater than 342, so one might hope an analysis based on the corresponding rank statistic Spearman's r_S might show some robustness in that the outlier effect might not be quite so marked and consequently lead to smaller p values. The differences are however slight and StatXact gives the following Monte Carlo estimates of the p-values:

One-tail	$p = 0.0522$ (0.0496, 0.0547)
Two-tail	$p = 0.1044$ (0.1009, 0.1079).

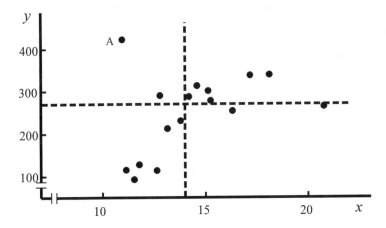

Figure 9.1 A scatter diagram for the data in Table 9.4. The broken lines represent axes when the origin is transferred to the *x*, *y* sample medians.

Table 9.4 Data from Carroll and Rupert (1988) on esteraze concentration (*x*) and binding (*y*)

x	11.0	11.1	11.6	11.8	12.6	12.8	13.1	13.8	14.1	14.6	15.1	15.2	16.4	17.1	18.1	20.8
y	423	116	97	131	115	291	215	226	288	317	301	278	256	340	342	260

Table 9.5 Ranked data derived from Table 9.4

rank x (r)	1	2	3	4	5	6	7	8	9	10	11	12	13	14	15	16
rank y (s)	16	3	1	4	2	11	5	6	10	13	12	9	7	14	15	8

Asymptotic results based on an approximation I give in section 9.3 are in close agreement. There is a slight indication of significance at or slightly above the 5% level if a one-tail test is thought appropriate.

Spearman's rank correlation coefficient is often referred to as Spearman's ρ (rho), but I denote its population value by ρ_s to distinguish it from Pearson's ρ.

9.2 Two important rank correlation coefficients

There are several distribution-free or nonparametric measures of correlation, the best known being the Spearman coefficient discussed briefly above and the Kendall coefficient. This section gives some of their properties and uses.

Spearman rank correlation coefficient. Because this is effectively the Pearson coefficient for ranks the usual formulae or a computer program for the Pearson coefficient may be used for Spearman's r_S, using paired ranks as data. Most statistical packages provide a program specifically for calculating r_S, making the transformation to ranks automatically. If there are no ties a simplified formula is obtained by algebraic manipulation of the Pearson formula using properties of the sum and sum of squares of ranks, namely:

$$r_S = 1 - \frac{6T}{n(n^2 - 1)} \qquad (9.1)$$

where $T = \sum (r_i - s_i)^2$, i.e. T is the sum of squares of the differences between ranks for each pair. With perfect matching of ranks clearly $T = 0$ and $r_S = 1$. If there is a complete reversal of ranks, i.e. if for all i, $s_i = n + 1 - r_i$ tedious elementary algebra establishes that $r_S = -1$. If there is no correlation between ranks it can be shown that $E(T) = n(n^2 - 1)/6$ so that r_S then has expectation zero. If the observations are a random sample from a bivariate distribution with X, Y independent, or at least uncorrelated, one expects near zero rank correlation. Because n is constant under the reference set of data permutations T itself may be used as a test statistic. However, inspecting numerical values of T gives no immediate picture of the degree of association.

For ties one uses mid-ranks. If exact p-values are needed these may be obtained from programs such as one in StatXact. Without such a facility if there are only a few ties critical values at the nominal 5% or 1% level valid for the no-tie case and available in many published tables may not be too misleading, but with many ties a program giving at least Monte-Carlo estimates of p-values is desirable. Thomas (1989) showed that if (9.1) is used to compute r_S with ties represented by midranks it gives a value of r_S that is always either greater than or equal to the correct value given by the Pearson formula applied to those mid-ranks.

Kendall rank correlation coefficient. Computation for this coefficient is reminiscent of that used for the WMW or Jonckheere–Terpstra tests. Indeed, both those test statistics are special cases of the Kendall coefficient with approp-riate allowance for ties. As in those tests ranks are not needed to compute the statistic, but they may be used if they are available. We can calculate the statistic directly if we know only the ordering of the data pairs with respect to each observed characteristic and measured data values are not required. Computation is simplified if we assume the x_i or their ranks are arranged in ascending order with the corresponding y-data directly below as in Tables 9.4 or 9.5. However, this ordering is not essential. Note the trend, except for the first observed point, in both those tables for the y values or their ranks to increase as the x_i increase.

The Kendall correlation coefficient, often referred to as Kendall's tau, is based on the principle that if there is association between the ranks of the x and the y then if we arrange the x or their ranks in ascending order then the y or their ranks should show an increasing trend if there is a positive association and a decreasing trend if there is negative association. Kendall (1938) therefore proposed that if the observations are arranged in order of increasing x that we score each paired y-rank difference $s_j - s_i$ for $i = 1, 2, \ldots, n-1$ and $j > i$ as $+1$ if the difference is positive and as -1 if the difference is negative. Temporarily, I assume there are no ties in x or y values. Clearly, with the x in ascending order we may equally well use the signs of the differences $y_j - y_i$, since these are equivalent to the signs of the corresponding rank differences. Kendall called a positive difference a **concordance** and a negative difference a **discordance**.   If the x_i are not in ascending order one must determine $\text{sgn}[(y_j - y_i)/(x_j - x_i)]$ for $i = 1, 2, \ldots n-1$, for all $j > i$ or the corresponding signs for rank difference quotients; positive values correspond to concordances and negative values to discordances. Denoting the number of concordances by n_c and the number of discordances by n_d respectively, Kendall's tau, which I denote by t_K, although historically it has usually been denoted by the Greek letter τ (tau), is

$$t_K = \frac{n_c - n_d}{\frac{1}{2}n(n-1)} \tag{9.2}$$

I use t_K to distinguish between the sample estimate t_K and the population value τ.

If we are comparing, say, preferred ranks assigned by two judges to a finite set of objects such as eight different varieties of raspberries then the eight varieties are usually the entire populations of interest so our calculated coefficient is a measure of the agreement between these two judges in relation to that population. On the other hand for continuous data like those in Table 9.4 the calculated t_K using (9.2) estimates some underlying τ that is a measure of order association between variables X, Y in some larger population.

If the x_i are in increasing order, since there are $\frac{1}{2}n(n-1)$ pairs of y differences $y_j - y_i$ and if all are concordances then $n_c = \frac{1}{2}n(n-1)$ and $n_d = 0$, whence $t_K = 1$. Similarly, if all are discordances then $t_K = -1$. If the ranks of x and y are independent we expect a fair mix of concordances and discordances and t_K should be near zero. With no ties $n_c + n_d = \frac{1}{2}n(n-1)$ and so one could base an exact permutation test on n_c only since n_d is expressible in terms of n and n_c.

A permutation test of $H_0:\tau = 0$ against a one- or two-sided alternative of a nonzero τ is based on evaluating t_K or an equivalent statistic (Exercise 9.1) for all $n!$ permutations of the y values (or the ranks 1 to n). For a small data set one might compute t_K corresponding to all permutations.

Example 9.4 Two people rank four varieties of apple in order of flavour preference:

Variety	A	B	C	D
First Taster	1	2	3	4
Second Taster	3	1	2	4

Is there statistical evidence of at least some agreement about flavour preference? For strict validity if we want to make inferences applying to many varieties of apples one must assume that, or know that, the 4 chosen varieties were a random selection from that larger population of apple varieties. If this were so the permutation test becomes a valid test of $H_0: \tau = 0$ against the alternative $H_1: \tau > 0$ when we are interested in at least some measure of positive agreement between these two tasters (ignoring the possibility of complete disagreement) as our alternative. There are $4! = 24$ possible orderings of the ranks 1 to 4. Clearly for the one-tail test formulated above, negative, zero or sufficiently small positive values of t_K will support H_0 while larger positive t_K will suggest rejection of H_0. The 24 possible permutations together with the corresponding values of t are:

Permutation				t
1	2	3	4	1
1	2	4	3	2/3
1	3	2	4	2/3
1	3	4	2	1/3
1	4	2	3	1/3
1	4	3	2	0
2	1	3	4	2/3
2	1	4	3	1/3
2	3	1	4	1/3
2	3	4	1	0
2	4	1	3	0
2	4	3	1	-1/3
*3	1	2	4	1/3
3	1	4	2	0
3	2	1	4	0
3	2	4	1	-1/3
3	4	1	2	-1/3
3	4	2	1	-2/3
4	1	2	3	0
4	1	3	2	-1/3
4	2	1	3	-1/3
4	2	3	1	-2/3
4	3	1	2	-2/3
4	3	2	1	-1

The asterisk marks the permutation corresponding to the data. From these permutations it is easy to check that the probabilities associated with each possible t_K are:

t_K	1	2/3	1/3	0	-1/3	-2/3	-1
Probability	1/24	1/8	5/24	1/4	5/24	1/8	1/24

Not surprisingly t_K is distributed symmetrically about zero since reversing the order of any permutation simply changes its sign. For the given data $t_K = 1/3$ and the corresponding one-tail p-value is $1/24 + 1/8 + 5/24 = 9/24$, so there is no question of rejecting H_0. Indeed with 4 paired ranks we would reject H_0 at a conventionally acceptable level only if $t_K = 1$.

Computing t_K is straightforward. For the given data the first y rank of 3 gives one concordance (being followed by 4) and 2 discordances (being followed by 1 and 2). Similarly the next rank, 1, gives 2 concordances and the rank 2 gives 1 concordance, giving $n_c = 4$ and $n_d = 2$, when, since $n = 4$, (9.2) gives $t_K = (4 - 2)/6 = 1/3$.

For small samples the permutation distribution of t_K is markedly discrete. StatXact provides exact tests or Monte Carlo estimates of p for small or moderate sample sizes together with an asymptotic approximation given below for large samples. Most general statistical packages have programs to compute t_K and some also giving an exact value of, or asymptotic estimate of, p.

Example 9.5 In both Tables 9.4 and 9.5 the data are in ascending order of x. Since there are no ties among the x or y it is easy to count the numbers of concordances among the y (Table 9.4) or the corresponding ranks (Table 9.5). For example, in Table 9.4, $y_1 = 423$ is followed by 15 lower values implying 15 discordances, $y_2 = 116$ is followed by two lower values (discordances) and 12 higher values (concordances) and proceeding this way we finally note that $y_{15} = 342$ is followed by one lower value (discordance). It is easily verified that $n_c = 82$, $n_d = 38$, whence $t_K = 0.3667$. StatXact confirms this value and although the sample size is too large for an exact test, a Monte Carlo approximation based on 10 000 samples gave a one-tail $p = 0.0269$ [with 99% confidence interval (0.0227, 0.0311)], a smaller p-value than that for either the Pearson or Spearman coefficients.

Many writers have compared relative merits of the Spearman and Kendall coefficients and illustrated connections and differences between them. I give one approach that helps to clarify these connections and differences although it provides little more than broad pointers as to which test is likely to be more powerful in a particular situation. This approach shows that the Spearman coefficient uses more information, but as in other situations where there are outliers or other data peculiarities ignoring, or reducing, the influence of information at variance with a dominant trend may increase robustness.

An alternative way to compute Spearman's r_S highlights its relationship to Kendall's t_K. Assume all observations are arranged in ascending order of x-ranks and that there are no ties in either the x or the y. If s_i is the y rank for the observation with x rank i, let $s_{ij} = s_j - s_i$, $i = 1, 2, 3, \ldots, n-1$, for all $j > i$. Clearly the number of concordances, n_c, in Kendall's t_K equals the number of positive s_{ij}, and the number of discordances, n_d equals the number of negative s_{ij}.

Tedious algebra (Exercise 9.2) shows that if n_{cs} is the sum of the values of the positive s_{ij} (i.e. the sum of the values that are concordant) and n_{ds} is the magnitude of the sum of the values of the negative s_{ij} (i.e. the sum of the values that are discordant) then $n_{cs} + n_{ds} = n(n^2-1)/6$ and (9.1) reduces to

$$r_S = \frac{n_{cs} - n_{ds}}{n(n^2 - 1)/6} \tag{9.3}$$

This expression has a formal similarity to (9.2). Kendall's coefficient involves only signs of the s_{ij}, but Spearman's coefficient depends on signs and magnitudes of the s_{ij}. It is convenient to write the s_{ij} as an upper triangular matrix

$$
\begin{array}{cccccccc}
s_2 - s_1 & s_3 - s_1 & s_4 - s_1 & \cdot & \cdot & \cdot & \cdot & s_n - s_1 \\
& s_3 - s_2 & s_4 - s_2 & \cdot & \cdot & \cdot & \cdot & s_n - s_2 \\
& & s_4 - s_3 & \cdot & \cdot & \cdot & \cdot & s_n - s_3 \\
& & & \cdot & \cdot & \cdot & \cdot & \cdot \\
& & & & \cdot & \cdot & \cdot & \\
& & & & & \cdot & \cdot & \\
& & & & & s_{n-1} - s_{n-2} & s_n - s_{n-2} \\
& & & & & & s_n - s_{n-1}
\end{array}
$$

Example 9.6 The triangular matrix of s_{ij} is easily found for the ranks in Table 9.5. The first term $s_{21} = s_2 - s_1 = 2 - 16 = -14$; the next $s_{31} = 16 - 1 = -15$; the final term $s_{16,15} = 8 - 15 = -7$. The complete matrix is:

−14	−15	−12	−13	−5	−11	−10	−6	−3	−4	−7	−9	−2	−1	−8
	−2	1	−1	8	2	3	7	10	9	6	4	11	12	5
		3	2	10	4	5	9	12	11	8	8	13	14	17
			−1	7	1	2	6	9	8	5	3	10	11	4
				9	3	4	8	11	10	7	5	12	13	6
					−6	−5	−1	2	1	−2	−4	3	4	−3
						1	5	8	7	4	2	9	10	3
							4	7	6	3	1	8	9	2
								3	2	−1	−3	4	5	−2
									−1	−4	−6	1	2	−5
										−3	−5	2	3	−4
											−2	5	6	−1
												7	8	1
													1	−6
														−7

In Exercise 9.3 I ask you to use this data matrix to verify that (9.3) gives the same values of r_S as (9.1). It is easily verified by counting the numbers of positive and negative signs in the matrix that the number of concordances is $n_c = 82$ and the number of discordances is $n_d = 38$, as obtained in Example 9.5.

Inspecting the matrix in Example 9.6 indicates why the outlying observation (the first) has more influence on r_S than upon t_K. There is a preponderance of positive s_{ij} (indicated by the excess of concordances over discordances) and the first row of the matrix has some of the s_{ij} of greatest magnitude and these are all negative. While the Spearman coefficient takes these magnitudes into account the Kendall coefficient gives each concordance and discordance equal weight.

In the case of ties leading to mid-ranks I pointed out that the Spearman coefficient using (9.1) tends to overestimate the magnitude of r_s compared to that given by the correct Pearson form applied to ranks. Thomas (1989) also showed that with ties (9.3) tends to underestimate the correct value.

For Kendall's tau with ties (9.2) will no longer takes the value +1 or −1 even when all mid-ranked pairs lie on a straight line and published tables of critical values for the no-ties case are misleading with more than a small proportion of ties, but the calculations may be modified to allow for ties. Ties among the x or the y or both are scored zero in Kendall's coefficient because they are neither concordances nor discordances.

Example 9.7 An example with heavy tying illustrates the above points. Two judges each rank 12 Cabernet-Sauvignon wines in order of preference. Each agree on which wine they rank 1 and upon two that they rank as equally least favourable, but neither can discriminate between the remaining 9 wines. Thus the allocated mid-ranks are:

Taster A (x)	1	6	6	6	6	6	6	6	6	6	11.5	11.5
Taster B (y)	1	6	6	6	6	6	6	6	6	6	11.5	11.5

Clearly, if we plot these rank pairs they lie on a straight line of slope 1, so it is intuitively reasonable to expect a meaningful rank correlation coefficient to take the value 1.

Since, the x-ranks are ordered, n_c is equal to the number of concordances in the y values that correspond to distinct x ranks. Clearly the y-rank $s = 1$ gives rise to 11 concordances and each of the mid-rank values of 6 gives two concordances, giving a further $2 \times 9 = 18$ concordances, thus $n_c = 11 + 18 = 29$ and there are no discordances. Since $n = 12$, (9.2) gives $t_K = 29/66$ rather than the anticipated $t_K = 1$. This is because $n_c + n_d$ no longer equals $\frac{1}{2}n(n - 1)$.

It can be shown (see, e.g. Kendall and Gibbons, 1990, Chapter 3) that we obtain a coefficient that takes the value +1 or −1 for a linear relationship involving tied ranks if we replace the denominator in (9.2) by $\sqrt{[(D - U)(D - V)]}$ where

$$D = \tfrac{1}{2}n(n - 1), \ U = \tfrac{1}{2}\textstyle\sum u(u - 1), \ V = \tfrac{1}{2}\textstyle\sum v(v - 1)$$

and u, v are the numbers of consecutive ranks in a tie within the x- and within the y-rankings respectively, and the summations in U, V are over all sets of tied ranks. If there are no tied ranks $U = V = 0$ and the modified denominator reduces to that in (9.2). This modified statistic, denoted by t_b, is called **Kendall's tau-b**.

Example 9.7 (continued).   There is one set of 9 tied ranks and also one set of 2 tied ranks among each of x and y ranks giving $U = V = \frac{1}{2}(9 \times 8 + 2 \times 1) = 37$. Since $D = 66$ when $n = 12$, we replace the denominator in (9.2) by $\sqrt{(66 - 37)(66 - 37)} = 29$. This is equal to the numerator $n_c - n_d = 29 - 0 = 29$, whence $t_b = 1$.

238 Correlation and concordance

With heavy tying like that in Example 9.7 it helps to display the results in a contingency table. It is then easy to compute n_c and n_d. I discuss this in a wider context in section 12.3, but here I use the method for the data in Example 9.7.

Example 9.8. Since the wines are placed in only three categories which we may conveniently describe as ordered categories I (best), II (intermediate) and III (worst), the results may be set out in a 3×3 contingency table:

		TasterA preference category		
		I	*II*	*III*
TasterB	*I*	1	0	0
peference category	*II*	0	9	0
	III	0	0	2

We find the number of concordances by working down the table by rows, considering all columns except the last and multiplying the number in each cell by the number of entries in cells below and to the right of that cell. This process is carried out for all rows except the last. Thus the number in cell (1,1) is 1 and there are $9 + 2 + 0 + 0 = 11$ entries to the right and below giving 1×11 concordances. The next nonzero row entry is the second diagonal entry 9 in cell (2,2) and there is only one entry of 2 below and to the right giving a further $9 \times 2 = 18$ concordances, making a total of 29 concordances, as found in Example 9.7. Here there are no discordances, but in general the number of these is obtained by working in like manner starting from the top right and for each cell, multiplying that cell entry by a count of the number of entries below and to the left.

StatXact includes a program for the Kendall coefficient which in the case of tied ranks computes both t_K given by (9.2) and t_b. For only a few ties the difference between t_K and t_b is often small.

Example 9.9 Table 9.6 gives the year of death and age at death for 13 males of one clan buried in the Badenscallie burial ground in Wester Ross, Scotland. Is there evidence that life expectancy for this clan is increasing with the passage of time (i.e. do members of the clan tend to live longer now than was the case 50, 100 or more years ago?)

The problem is too large for the exact correlation options in StatXact but the Monte Carlo options gives estimates of the relevant coefficients and one-sided p-values. A 99%confidence interval for the Monte Carlo approximation is given in brackets after each p-value.

Pearson coefficient	$r = 0.5351$	$p = 0.0274\ (0.0232, 0.0316)$
Kendall coefficient	$t_K = 0.3846$, $t_b = 0.3896$	$p = 0.0365\ (0.0317, 0.0413)$
Spearman coefficient	$r_S = 0.5069$	$p = 0.0390\ (0.0340, 0.0440)$

All coefficients lead to p-values that indicate rejection of H_0:*life expectancy is not increasing* in favour of a one-sided alternative H_1:*life expectancy is increasing* at a significance level less than 5%. For Kendall's coefficient the numerical values of t_K and t_b differ, but the permutation distribution is the same for each leading to the same p-value. Although the numerical values of r, t_K, t_b and r_S may differ appreciably, each sometimes lead to p-values of similar magnitude.

Table 9.6 Year of death and age at death of 13 males belonging to one Scottish clan

Year	1827	1884	1895	1908	1914	1918	1924
Age	13	83	34	1	11	16	68

Year	1928	1936	1941	1964	1965	1977
Age	13	77	74	87	65	83

The procedure outlined in Example 9.8 may be used to demonstrate the equivalence already mentioned of the Kendall coefficient to the Jonckheere–Terpstra test. Here the $x = 0, 1, 2, \ldots$ may be regarded as sample labels (remember the samples are ordered) and the n_i observations in Sample i are the number of ties at that x-value. If there are k samples there will be k rows and a column corresponding to each distinct observed value in the complete data set. This provides a contingency table for the counting procedure outlined in Example 9.8.

Example 9.10 There is no practical advantage in formulating the Jonckheere–Terpstra test this way but it shows the equivalence. Three secretaries with no training in using a new word-processing program, 3 with one week's training and 2 with two week's training all prepare the same document with the program. Do the following results indicate that numbers of mistakes decrease with training?

Number of weeks training	Number of mistakes by each secretary
2 weeks	3, 7
1 week	4, 9, 11
None	10, 16, 19

You should check that the extended Mann–Whitney type count in Example 7.10 gives $U = 19$. The formulation proposed above leads to the 3×8 contingency table

Training weeks	Number of mistakes							
	3	4	7	9	10	11	16	19
2	1	0	1	0	0	0	0	0
1	0	1	0	1	0	1	0	0
0	0	0	0	0	1	0	1	1

The entry in the first row and column, cell (1,1), is 1 and sum of all entries below and to the right is 6, similarly there are 5 entries below and to the right of the entry 1 in cell (1,3), and respectively 3, 3 and 2 entries below and to the right of the entry 1 in cells (2,2), (2,4) and (2,6) indicating by the rule given before this example that the number of concordances is $U = 6 + 5 + 3 + 3 + 2 = 19$

With ties the table is scanned in the same way, but any entry immediately below that in a cell being considered is scored as ½. For example if there were an entry of 3 in cell (2,4) and an entry of 5 below it in, say, cell (3, 4) this would be scored as $3 \times ½(5) = 7.5$. A moment's reflection shows this is equivalent to the usual procedure in a Mann–Whitney type count with a scoring system 1, ½, 0. Modification is needed (Exercise 9.4) for a 1, 0, –1 scoring system.

In the special case where the number of samples, k, is two, this reduces to the Mann–Whitney formulation of the WMW test, showing this to be equivalent to computing Kendall's tau for a heavily tied situation in X, i.e. all X values tied at one of two possible values, one corresponding to each sample. I pointed out in section 9.1 that the Fisher–Pitman two-sample permutation test may be treated as a special case of the Pearson correlation coefficient test, and a moment's reflection shows that the Wilcoxon form of the WMW test is a special case of the Spearman coefficient test because the y data are replaced by ranks. We know the Wilcoxon test and Mann–Whitney test are equivalent. Thus, in this special situation the Spearman coefficient and the Kendall coefficient lead to equivalent tests. This comes about because, for two samples, when testing for a location difference only or for dominance of one population over the other the only alternative to no difference is an ordered alternative of *greater than* or *less than* for the second sample relative to the first, making it automatically a special case of the Jonckheere–Terpstra test. In section 7.3 I also indicated that the Kruskal–Wallis test was a generalization of the Wilcoxon test. It, however, is not equivalent to either the Kendall or Spearman tests for it does not involve an ordering of samples in a contingency table format like that proposed above for the Jonckheeere–Terpstra test. Columns may be arranged in ascending orders of observations, but rows are not ordered in terms of increasing treatment effects.

9.3 Some asymptotic results

The mean of both r and r_S under the hypothesis H_0 of zero population correlation are relatively easily shown to be zero and each has variance $1/(n–1)$. These results are well known for the Pearson coefficient in the bivariate normal case and Maritz (1995, sections 7.2.1 and 7.2.2) shows that they carry over to the permutation test situation. A more detailed discussion of the distribution of r_S under H_0 is given in Gibbons and Chakraborti (1992, section 12.3) and in Kendall and Gibbons (1990, section 4.14). Convergence to normality tends to be slow and it is often suggested that asymptotic results should not be used for $n < 30$ or even for $n < 35$. For larger n it is usual to assume that $z = r\sqrt{(n-1)}$ and $z = r_S\sqrt{(n-1)}$ will under H_0 have approximately a standard normal distribution. A better approximation that is often reasonable for $n > 10$ suggested in Kendall and Gibbons and elsewhere and applicable to r or r_S takes the form

$$t_{n-2} = \frac{r\sqrt{(n-2)}}{\sqrt{(1-r^2)}}$$

and this has approximately a t-distribution with $n - 2$ degrees of freedom. This approximation is used in StatXact for asymptotic estimation of p both for the Pearson and Spearman coefficients.

For Kendall's tau the distribution of t_K under H_0 is not easily obtained. A full discussion is given in Gibbons and Chakraborti (1992, section 12.2) and in Kendall and Gibbons (1990, section 4.8) who show under H_0 that $E(t_K) = 0$ and $\text{Var}(t_K) = 2(2n + 5)/[9n(n - 1)]$, and for reasonably large n (about $n > 15$) that

$$z = 3t_K\sqrt{\{[n(n - 1)]/[2(2n + 5)]\}}$$

has approximately a standard normal distribution. This approximation is used for asymptotic results in StatXact and in many general statistical programs, not all of which give an exact p-value. Modifications needed with ties are discussed by Gibbons and Chakraborti.

There are two difficulties in obtaining confidence intervals for the population coefficients ρ, ρ_S and τ. Use of the asymptotic results given above may not only give limits outside the interval $(-1, 1)$ but the distributions of the of the sample statistics are no longer simple for nonzero population coefficients. In particular the variances are no longer the same as those given under a zero correlation hypothesis. The problem is discussed further by Gibbons and Kendall (1990, Chapters 4 and 5) where some useful approximations are given. In section 9.5 I consider bootstrap confidence intervals.

9.4 Some other measures of correlation

Measures of correlation discussed so far in this chapter were developed either for continuous data or for data on individual units that could be ordered even if exact continuous observations were not available or not relevant. I show in Chapter 12 that by developing some of the ideas associated with ties that many of the methods introduced in this chapter extend to categorical data given in ordered contingency tables; indeed simple examples of this situation were illustrated in Examples 9.8 and 9.10.

In this section I consider some other measures of association. The first was proposed by Blomqvist (1950, 1951) and shows obvious analogies with the median test in section 4.6. Suppose the medians of the marginal distributions of X, Y are respectively θ_X, θ_Y. In a sample one expects about half the observed x to be below θ_X and about half the y to be below θ_Y. If X and Y are independent we expect a good mix of high, medium and low values of Y to be associated with the various values of X. If we knew θ_X and θ_Y we could shift the origin in a scatter diagram of the observed (x_i, y_i) to (θ_X, θ_Y). Then under $H_0:X, Y$ are independent we would expect about one quarter of all points to be in each of the four quadrants determined by these new axes. Usually, as was the case for the median-test in sections 4.6 and 7.4, we do not know θ_X, θ_Y so we replace them by their estimates, the sample medians, M_X, M_Y. In Figure 9.1 the dotted lines represent new axes with origin at the sample medians for the data in Table 9.4. From that table it is easy to see that $M_X = \frac{1}{2}(13.8 + 14.1) = 13.95$ and $M_Y = \frac{1}{2}(260 + 278)$ $= 269$. In Figure 9.1 using the new axes and working anticlockwise from the first, or top right, quadrant the numbers of points in the quadrants are 6, 2, 6, 2. The concentration of points in the first and third quadrants suggests positive dependence between X and Y.

A formal test of H_0 against a one-sided alternative of either *positive* or *negative association* or a two-sided alternative of *some association* follows closely the procedure in section 4.6. If n is even and no values coincide with the sample medians, one counts the number of points a, b, c, d in each of the four quadrants after transferring the origin to the median point (M_X, M_Y). Because of the sample median properties we get a 2×2 contingency table with marginal totals shown in Table 9.7. Modifications discussed after Example 9.11 are needed if some sample values equal the relevant median (giving a point or points either at the new origin or on one of the new axes).

The row and column totals in Table 9.7 follow from the properties of the sample medians and the restriction to even n with no x, y equal to their sample medians. In this situation if X, Y are independent, then clearly a, b, c, d each has expectation $\frac{1}{4}n$. Marked departures from this value indicate association between X and Y. The relevant test procedure is based upon Fisher's exact test. I illustrate this in Example 9.11 but first I give two possible correlation measures that have the desirable properties that they take values within the interval $(-1, 1)$ with values near ± 1 indicating near monotonic dependence while independence or lack of association results in near-zero values (though the converse may not be true). The two coefficients are

$$r_D = (a + d - b - c)/n \tag{9.4}$$

and

$$r_M = 4(ad - bc)/n^2 \tag{9.5}$$

Table 9.7 A contingency table for Blomqvist's median test for correlation

	Above M_X	Below M_X	Row total
Above M_Y	a	b	$\frac{1}{2}n$
Below M_y	c	d	$\frac{1}{2}n$
Column total	$\frac{1}{2}n$	$\frac{1}{2}n$	n

It is easy to verify that in the circumstances of Table 9.7 (no values equal to the medians) that $a = d$, $b = c$ and since $a + b = \frac{1}{2}n$, that $r_D = r_M$ and that each takes the same set of $\frac{1}{2}n+1$ possible values with the same associated probabilities (given by Fisher's exact test) under the null hypothesis of zero correlation or no association. While (9.4) has arithmetic simplicity, (9.5) generalizes more readily to other contexts, including the case n odd, so is usually preferred.

Example 9.11   Inspection of Figure 9.1 indicates that for the Table 9.7 format $a = d = 6$ and $b = c = 2$, whence it is easily verified since $n = 16$, that $r_D = r_M = 0.5$. Any program for Fisher's exact test gives $p = 0.066$ for a one-tail test, close to the value for the Pearson coefficient permutation test. The test has relatively low power for small samples because little of the available information is used and there are marked discontinuities between possible p-values.

If n is odd, or more generally, if there are one or more x or y values equal to the respective sample medians, if the corresponding points are ignored the marginal totals in tables like Table 9.7 will no longer all be $\frac{1}{2}n$. The Fisher test should then be used with the observed totals and if a correlation coefficient is appropriate a modification of r_M in (9.5) should be used, namely

$$r_M = (ad - bc)\sqrt{[(a + b)(c + d)(a + d)(b + d)]} \qquad (9.6)$$

where a, b, c, d are the observed cell values. This reduces to (9.5) when all marginal totals are $\frac{1}{2}n$. Clearly the test is unaltered if ranks are used in place of the data, e.g. in Example 9.11 the test could be based on the ranks in Table 9.5, the 'rank' medians now each being 8.5. As for any Fisher exact test for a 2×2 table, because marginal totals are fixed, we need only know the number of points in one quadrant since the others are then fixed to give correct marginal totals.

A less well-known test for rank correlation was proposed by Gideon and Hollister (1987). It is relatively easy to calculate and is often more powerful than the Blomqvist test, and in certain outlier situations also more powerful than the Kendall or Spearman tests. A description with an example is given in Sprent (1993, section 7.1.6).

9.5 The bootstrap and correlation

The correlation coefficient for a bivariate normal distribution is estimated from a sample of n paired observations (x_i, y_i) by the Pearson product-moment correlation coefficient, r, (section 9.1), which is a plug-in estimator of the population coefficient, ρ. Except in the case $\rho = 0$, the distribution of r is complicated, and r is a biased estimator of ρ even under an assumption of population bivariate normality. The correlation coefficient is often of interest when assumptions of bivariate normality strain credulity beyond breaking point.

Bootstrapping provides estimates for se(r) but in view of the marked non-normality in the distribution of r even when sampling from a bivariate **normal** distribution, rule-of-thumb guides like the 'two-standard error' approximation in hypothesis testing about, or establishing confidence intervals for, ρ are inappropriate, not least because the latter often lead to limits with $|\rho| > 1$. I discuss a transformation that overcome some of these difficulties after an example of the bootstrap applied to a small correlation problem.

Example 9.12   Hand *et al.* (1994, p.69) give data on average outside temperature and gas consumption for one house in the UK before and after insulation. What was of interest was how temperature and gas consumption were related and whether this relationship changed after insulation, but I use only part of the data here. Table 9.8 gives a random sample of 7 from 26 observations before insulation. Although regression is the appropriate approach to answer the questions of interest I use the data to illustrate application of the bootstrap to estimating the population correlation coefficient where that population may be the 26 observations in the one house before insulation, or perhaps an extended hypothetical population of data for houses of that type subject to similar climatic conditions. Of course, given the full data set of 26 observations one would normally compute the coefficient based on all those data and use that for comparisons with similar coefficients computed for other houses if this was of interest. An appreciable variation in the correlation coefficient from house to house night occur, reflecting use of different building materials or methods of construction and different levels of insulation.

The Pearson coefficient for the data in Table 9.8 is $r = -0.9383$, close to $\rho = -0.9715$ for the data set of 26 readings given by Hand *et al* for that house.

For the data pairs in Table 9.8 I computed r^* for 50 bootstrap samples, i.e. samples of 7 sets of paired values obtained by sampling with replacement. One bootstrap sample gave the points (3.6, 5.6), (8.0, 4.0), (3.6, 5.6), (6.9, 3.7), (0.4, 6.4), (3.6, 5.6), (8.0, 4.0), for which $r^* = -0.965$. Setting $B = 50$ and $s(X^{*b}) = r^{*b}$ in (2.3) the 50 bootstrap samples gave se(r) = 0.040. For all 26 data the 'true' population coefficient value, -0.9715, lies within one standard error of the sample $r = -0.9383$, but it reflects the non-normality characteristic of the sampling distribution of the correlation coefficient in that a range of two standard errors includes values of the coefficient greater than 1.

Table 9.8 Outside temperature and gas consumption for a sub-set of data for one house after Hand *et al* (1994).

Average outside temperature($^\circ$C), x	*Gas consumption (1000 cubic feet), y*
0.4	6.4
3.6	5.6
4.2	5.8
6.3	4.6
6.9	3.7
7.6	3.5
8.0	4.0

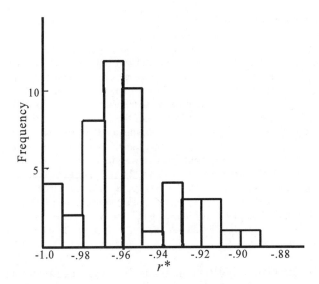

Figure 9.2 Histogram of 50 bootstrap sample values of $r*$ based on the data in Table 9.8. One outlier with r* between –0.72 and –0.73 is not shown on this histogram.

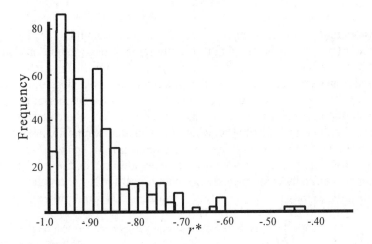

Figure 9.3 Histogram of 500 bootstrap sample values of $r*$ based on the data in Table 9.8. A value r* = 0.2459 is not shown.

Figure 9.2 shows a histogram of $r*$ for my 50 sample values which reflects the nonnormal distribution of $r*$.

This nonnormality is not a small-sample freak, but nevertheless for small samples the bootstrap distribution of a statistic $s(x*)$ sometimes exhibits bizarre behaviour because sampling with replacement leads to a few samples placing a heavy emphasis on one or two data points. This became noticeable when I took 500 bootstrap samples from the data in Table 9.8. Figure 9.3 is a histogram of the values of $r*$ for those 500 samples. Here $E(r*) = -0.8924$ and $se_{500}(r*) = 0.1197$. One bootstrap $r* = 0.2459$ is not shown on the histogram. It is clearly an outlier, well out of line with any other value represented on the histogram . In Exercise 9.9 I ask you to show that at least one possible bootstrap sample gives rise to an even greater value of $r*$. Also included in the 500 values of $r*$ twice was the value $r* = -1$ exactly, indicating perfect correlation. A moment's reflection shows that values of $r* = \pm1$ will occur if the replacement sampling results in only 2 of the 7 points appearing in a sample of 7 or if 3 or more collinear points only are selected. These are events of small, but not negligible probability for sample sizes less than about 10. Extreme or wildly outlying values are less likely when bootstrapping larger samples, but for the correlation coefficient the inherent nonnormality persists as I indicate in Example 9.13.

When an assumption of bivariate normality is made a commonly used device to overcome problems arising from the non-normality of the sampling distribution of r is to apply the transformation

$$z = \tfrac{1}{2}\ln \left(\frac{1 + r}{1 - r} \right) \tag{9.7}$$

The distribution of z is approximately normal with mean $\tfrac{1}{2}\ln[(1 + \rho)/(1 - \rho)]$ and variance $1/(n - 3)$. It is tempting to use the same transformation on bootstrap samples converting each $r*$ to a corresponding $z*$ using (9.7). However, this approach should be used with reserve, especially for small n, since the procedure of sampling **with replacement** means that there will be a number of tied values within bootstrap samples and this makes such samples inherently atypical of samples from a normal distribution where ties should theoretically not occur and in practice should not dominate. Also, difficulties arise if some $r* = \pm1$, when $z = \pm\infty$. Also the transformation may not be entirely satisfactory if the original data depart markedly from bivariate normality even for large samples, although small departures may not be too serious. I discuss this further in Example 9.14.

Example 9.13   To see how the bootstrap performs for correlation with larger samples I computed $r*$ for $B = 500$ for the complete data set of 26 readings for the house insulation data given by Hand *et al.*, from which the sample of 7 in Table 9.8 was obtained. A plot of the 26 data points is given in Figure 9.4. If the data were distributed bivariate normally one would expect a somewhat elliptic scatter about a straight line. There is however a suggestion of some nonnormality in that if a straight line were fitted to the data for values of x greater than 4 the observations for smaller values of x would tend to lie below such a line, or putting it another

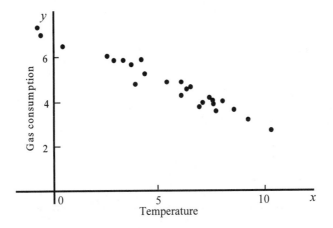

Figure 9.4 Scatter plot of house insulation data in Example 9.13.

way, some curve such as a quadratic might fit the data a little better. This in itself indicates slight departure from bivariate normality, but this would not be expected to seriously affect the usual 'normal theory' properties of the correlation coefficient.

Figure 9.5 is a histogram of 500 bootstrap r^* for these data. No r^* fell outside the interval $(-0.9903, -0.908)$ and $E(r^*) = 0.9686$ and $se(r^*) = 0.0146$. For the original data $r = -0.971$. Clearly the few unstable values obtained with small samples arose because they involved a small proportion of the data points, and this situation has a lower probability of occurring with larger samples. Also, as is reasonable, $se(r^*)$ is lower for the larger sample of 26, reflecting the higher precision attainable with a larger sample. Further, good estimates of $se(r^*)$ are often obtainable for values of B as low as 20. For these dates the estimates I obtained for $B = 20$ and $B = 50$ were respectively 0.0134 and 0.0135, each reasonably close to the value 0.0146 when $B = 500$. For small samples the instability noted in Example 9.12 makes estimates of standard error erratic even for values of B greater than 100.

Example 9.14. For the problem introduced in Example 9.12 I computed z^* in (9.7) for each of the 50 bootstrap samples and computed $se(z)$ by appropriate substitutions in (2.3), i.e. setting $s(x^{*b}) = z^{*b}$ and $B = 50$. This gave the estimated standard error $se(z) = 0.487$ compared to the theoretical value 0.5 given by putting $n = 7$ in the expression $1/\sqrt{(n-3)}$. Figure 9.6 is a histogram of the z values for my 50 bootstrap values of z^*. This shows less evidence of non-normality than the histogram for r^* in Figure 9.3. even though there is some doubt about an assumption of bivariate normality of the data and also a possibility of a few outlandish values of r^* arising from certain random samples that involve only a few data points. Indeed, as indicated in Example 9.12, when I took 500 samples in the case $n = 7$ there was one bizarre reading $r^* = 0.2459$. This gives $z^* = 0.2510$, far out of line with any values shown in the histogram in Figure 9.6. Also, the values $r = -1$ occurring in the sample of 500 gives $z^* = -\infty$. Thus $se(z^*)$ is undefined (infinite) unless the sample giving $r^* = -1$ is omitted. These difficulties are unlikely to occur if the original samples we are bootstrapping are larger. Indeed

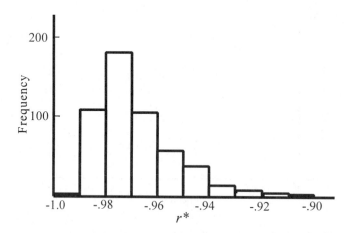

Figure 9.5 Histogram of $r*$ for $B=500$ bootstrap samples for house insulation data in Example 9.13

for the 500 bootstrap samples from the full date set of $n = 26$ paired observations considered in Example 9.13 transformation of $r*$ to $z*$ works well and the histogram, shown in Figure 9.7 is reasonably symmetric and normal looking. For these data $E(z*) = -2.1164$ and $se(z*) = 0.2182$. The latter is close to the theoretical $se(z) = 1/\sqrt{(n-3)} = 1/\sqrt{23} = 0.208$

Despite the reasonable results for bootstrap samples using (2.3) for the larger sample in Example 9.13 potential difficulties with an occasional rogue observation or a possible $r* = \pm 1$ must be borne in mind. Also, unless one has reasonable confidence one is sampling from a near bivariate normal distribution there must be reservation about using this transformation as a routine procedure. If one is confident of near normality, an alternative approach would be to use a parametric bootstrap. This approach is sketched by Efron and Tibshirani (1993, section 6.5). With this approach the z-transformation has much to commend it.

I indicated in section 2.6 that the rule-of-thumb approach for quoting approximate 95% confidence limit for a parameter θ given an unbiased estimator T, say, as $T \pm 1.96[se(T)]$ often does not work well in the bootstrap analogue with $se(T)$ replaced by the bootstrap estimator $se(T*)$ for several reasons. One is the somewhat erratic behaviour of bootstrap samples for small n but even if n is reasonably large our troubles are not over, firstly because the distribution of our sample estimator, or more importantly the bootstrap estimator, may be far from normal and far from symmetric (e.g. for the correlation coefficient). Bias in the bootstrap estimator is also common and may be hard to correct. In practice, as indicated in section 2.6, more sophisticated methods are required to estimate

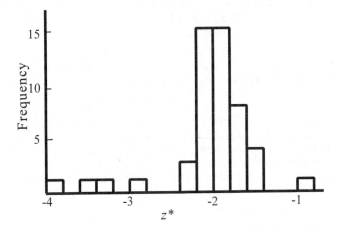

Figure 9.6 Histogram for 50 bootstrap values of z^* for the data in Table 9.8.

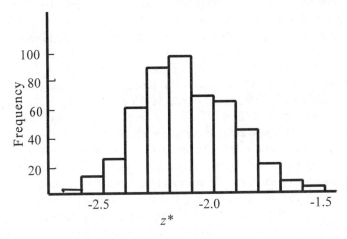

Figure 9.7 Histogram of 500 bootstrap values for z^* using the full house insulation data set where $n=26$.

confidence intervals, the simplest of these being the quantile-based intervals. All this sounds depressing, but it is important to remember that we may obtain bootstrap standard errors with minimal assumptions, indeed no assumptions about the population that generates our random sample. This makes it important to look at the characteristics of bootstrap samples and not only at numerical estimates.

That is why in this section I have looked at histograms of the bootstrap $s(x^*)$ and compared results in some situations with theoretical results even when these rely heavily on distributional assumptions.

Example 9.15   For the data in Example 9.12 I generated 1000 bootstrap samples computing r^* for each. I used these to form 95% quantile based confidence interval for r, obtaining an interval $(-0.993, -0.874)$. A further 1000 bootstrap samples gave an interval $(-0.995, -0.879)$ in good agreement despite the small sample size $n = 7$. Here there is no advantage in transforming to z and then back-transforming as this would lead to the same interval. However, for more sophisticated intervals such as the BC$_a$ or ABC methods mentioned in section 2.6 there may be advantages in applying these methods using z^* determined by (9.7) and back-transforming, as different corrections will apply for intervals based on r^* or z^*. I do not pursue that approach here, but details of these method are given by Efron and Tibshirani, Chapter 14.

9.6 Partial correlation

In both parametric and nonparametric contexts partial correlation is concerned with whether an association between two variables X and Y reflects only that each is correlated with a third variable Z or whether an association between X and Y would still exist if Z were fixed. For example, a group of pupils may be tested in mathematics, in a foreign language and also be given a general intelligence test. If their scores in mathematics and in the foreign language are highly correlated we may ask whether this simply reflects general intelligence or whether this correlation would still show if all had a fixed common level of general intelligence. To assess whether an association between X and Y exists irrespective of Z the usual procedure is to calculate a partial correlation coefficient which relates to a fixed Z. The classical procedure for doing this, originally developed in the multivariate normal case, is to calculate the correlation coefficient between the conditional distributions of X and Y with Z held fixed. The usual notation in this context is to denote the ordinary pairwise Pearson correlation coefficients between X and Y, between X and Z and between Y and Z by r_{XY}, r_{XZ} and r_{YZ} respectively and the partial coefficient between X and Y conditional on fixed Z by $r_{XY.Z}$. It is well-known, see, e.g. Stuart and Ord (1991, Chapter 27) that

$$r_{XY.Z} = \frac{r_{XY} - r_{XZ}r_{YZ}}{\sqrt{[(1 - r_{XZ}^2)(1 - r_{YZ}^2)]}} \tag{9.8}$$

The concept extends to partial correlations conditional on fixed values of several variables $Z_1, Z_2, Z_3 \ldots$. Details are given by Stuart and Ord and in many general statistical texts that cover this topic.

Partial correlations are often relevant in situations other than that of multivariate normality although in many cases formal tests and interpretation may then be more difficult. In particular, if data consists of ranks a partial Spearman's rho may be computed by (9.8). Distribution-free partial correlations may also be based on Kendall's tau. I describe these briefly, but they are dealt with more fully by Kendall and Gibbons (1990, Chapter 8) and by Siegel and Castellan (1988, section 9.5).

Example 9.16 Suppose three variables X, Y, Z for 7 subjects have the following rankings:

Subject	a	b	c	d	e	f	g
X	2	1	3	4	7	5	6
Y	4	2	3	1	6	7	5
Z	1	2	3	4	5	6	7

The table is arranged in order of increasing Z ranks; this loses no generality and is convenient when we consider a partial Kendall coefficient corresponding to fixed Z which I denote by $t_{XY.Z}$. Denoting the Kendall coefficients between X and Y by t_{XY}, etc. it is easy to show (Exercise 9.5) that for these data $t_{XY} = (15 - 6)/21 = 3/7$, $t_{XZ} = (18 - 3)/21 = 5/7$ and $t_{YZ} = (14 - 7)/21 = 1/3$.

Since the Z are in ascending rank order a concurrence score based on the usual count of consistent orderings as used for Kendall tau in Section 9.2 gives $^7C_2 = 21$ concordances among the Z. Table 9.9 records for each relevant pair whether there is concordance or discordance in X and in Y when the subjects are arranged in the order above. A concordance is indicated by a + and a discordance by a –. All Z are necessarily concordant.

In Table 9.9 we see that in some cases X and Y are both concordant (both +) when Z is concordant, in other cases X and Y are both discordant (both –) when Z is concordant, in other cases X is concordant and Y discordant or *vice versa*. The cell entries in the 2×2 contingency Table 9.10 give the numbers in each of the X, Y categories that are (+,+),(+,–),(–,+),(–,–).

Table 9.9 Concordances and discordances among pairs of ranks for 7 subjects.

Pair	X	Y	Z	Pair	X	Y	Z
a,b	–	–	+	c,d	+	–	+
a,c	+	–	+	c,e	+	+	+
a,d	+	–	+	c,f	+	+	+
a,e	+	+	+	c,g	+	+	+
a,f	+	+	+	d,e	+	+	+
a,g	+	+	+	d,f	+	+	+
b,c	+	+	+	d,g	+	+	+
b,d	+	–	+	e,f	–	+	+
b,e	+	+	+	e,g	–	–	+
b,f	+	+	+	f,g	+	–	+
b,g	+	+	+				

Table 9.10 Numbers of pairs in each (X,Y) concordance pattern relative to concordant Z.

		Y rankings		
		Pairs +	Pairs −	Total
X rankings	Pairs +	13	5	18
	Pairs −	1	2	3
	Total	14	7	21

Table 9.11 Numbers of pairs in each (X,Y) concordance pattern relative to concordant Z for n items.

		Y rankings		
		Pairs +	Pairs −	Total
X rankings	Pairs +	a	b	$a+b$
	Pairs −	c	d	$c+d$
	Total	$a+c$	$b+d$	$a+b+c+d={}^nC_2$

Inspecting Table 9.10 bearing in mind its structure makes it clear that the row totals are the numbers of concordances and discordances relative to calculating t_{XZ} and the column totals the analogous numbers relevant to calculating t_{YZ}. The grand total of 21 equals in each case the sum of the number of concordances and discordances, i.e. the denominator in (9.2). Further consideration shows that the main diagonal sum $13 + 2 = 15$ is the number of concordances relevant to computing t_{XY} and that the other diagonal sum $5 + 1 = 6$ is the corresponding number of discordances. These results may be generalized to rankings of n items for criteria X, Y, Z leading to Table 9.11, where the general entries a, b, c, d are not of course the same as the a, b, c, d in Table 9.9.

Clearly using the arguments above, and for convenience writing $N = {}^nC_2$, we have

$$t_{XZ} = [(a + b) - (c + d)]/N,$$
$$t_{YZ} = [(a + c) - (b + d)]/N, \qquad (9.9)$$
$$t_{XY} = [(a + d) - (b + c)]/N.$$

The Kendall partial tau is defined as

$$t_{XY.Z} = \frac{ad - bc}{\sqrt{[(a + b)(c + d)(a + c)(b + d)]}} \qquad (9.10)$$

Note the formal similarity between (9.9) and (9.4) and between (9.10) and (9.6).

Example 9.16 (continued) Substituting appropriate values from Table 9.10 in (9.10) gives

$$t_{XY.Z} = (13 \times 2 - 5 \times 1)/\{\sqrt{[(18 \times 3 \times 14 \times 7)]} = 0.2886.$$

This is less than t_{XY} and for reasons discussed below this result suggests that any rank association between X and Y is largely a reflection of their joint association with Z.

The formal similarity between (9.10) and (9.6) indicates that (9.10) is an appropriate measure of association. Values of $t_{XY.Z}$ near zero imply a lack of association other than any arising because both of X and Y are associated with Z. Values near ± 1 imply a strong association between X and Y irrespective of any association each may have with Z.

More formally from (9.10), $t_{XY.Z} = \pm 1$, if and only if

$$(ad - bc)^2 = (a + b)(c + d)(a + c)(b + d).$$

giving

$$4abcd + a^2(bc + cd + bd) + b^2(ac + ad + cd) + $$
$$c^2(ab + ad + bd) + d^2(ac + ab + bc) = 0.$$

Since all of a, b, c, d are necessarily non-negative integers a moment's reflection confirms that the necessary and sufficient condition for this to hold is that at least two of a, b, c and d must be zero. If two in the same row or column of Table 9.11 are zero this implies that either X or Y are in complete agreement, or in complete disagreement, with Z. Cases of interest are $a = 0$ and $d = 0$ or $b = 0$ and $c = 0$, i.e. zeros on the diagonals. The first implies X and Y disagree completely in their concordances with Z and then $t_{XY.Z} = -1$. The latter implies X and Y agree completely in their concordances with Z and then $t_{XY.Z} = 1$. Clearly when $ad = bc$ then $t_{XY.Z} = 0$, implying that any association between X and Y is a consequence only of a common association with Z which is, of course, removed if Z is fixed.

The above procedure for computing $t_{XY.Z}$ is tedious for moderate to large n. From (9.9) remembering that $a + b + c + d = N$ it follows that

$$1 - t_{XZ}^2 = \{[a + b + c + d]^2 - [(a + b) - (c + d)]^2\}/N^2 = 4(a + b)(c + d)/N^2. \quad (9.11)$$

Similarly

$$1 - t_{YZ}^2 = 4(a + c)(b + d)/N^2. \quad (9.12)$$

Also from (9.9) straightforward algebra (Exercise 9.6) gives

$$t_{XY} - t_{XZ} t_{YZ} = 4(ad - bc)/N^2. \quad (9.13)$$

and from (9.11) – (9.13) it follows that

$$t_{XY.Z} = \frac{t_{XY} - t_{XZ} t_{YZ}}{\sqrt{[(1 - t_{XZ}^2)(1 - t_{YZ}^2)]}} \quad (9.14)$$

which is analogous to (9.8). It is easily verified (Exercise 9.7) that substituting t_{XY}, t_{XZ}, t_{YZ} in (9.14) gives $t_{XY.Z} = 0.2886$ for the data in Example 9.16.

For an exact permutation test of H_0:*X and Y are uncorrelated for fixed Z* against an alternative of some association, the appropriate reference set is based on all orderings resulting from the $(n!)^2$ permutations of the n observed x and the n observed y keeping the order of the Z fixed. For each permutation the corresponding $t_{XY.Z}$ or an equivalent statistic is computed and large values (of appropriate sign for a one-tail test) indicate significance and rejection of H_0. Tables exist for rejection at conventional significance levels (see, e.g. Kendall and Gibbons, 1990, Appendix Table 11). Monte Carlo simulations of the permutation test or bootstrapping are ways to get approximate p-values in particular cases and appropriate macros are easily written for Minitab or other standard statistical software. The asymptotic test for large samples is identical to that for Kendall's τ in section 9.3; i.e. for large n, (say $n > 15$), $z = 3t_{XY.Z}\sqrt{\{[n(n-1)]/[2(2n+5)]\}}$ has approximately a standard normal distribution.

9.7 Concordance

If more than two variables are ranked for each experimental unit we often want to know if there is consistency or agreement in rankings between the variables as opposed to complete independence. If there are 6 candidates for a job vacancy and each of 3 members of an interviewing panel rank the applicants from 1 to 6 in order of suitability for a vacancy a possible outcome is given in Table 9.12.

This table suggests reasonable agreement between panel members on applicant suitability with a preference for applicant II whilst applicant IV is also well regarded. With no agreement between panel members we should have expected a mix of low, medium and high ranks for each candidate. The situation is analogous to that for the Friedman test in section 8.2 and indeed the exact permutation test (or its asymptotic approximation) given there may be used for testing H_0:*panel members rank applicants independently* against H_1:*there is a degree of consistency in rankings*. This implies that Friedman test programs may be used for testing the above hypotheses, but for interpretive purposes there are advantages in looking at problems of agreement (concordances) between rankings in a way that extends the idea of bivariate correlation to a related multivariate concept called **concordance**. A widely used measure of concordance is Kendall's **coefficient of concordance** fully discussed by Kendall and Gibbons (1990). It incorporates some, but not all, features of a conventional correlation coefficient and interestingly is in some respects more closely related to Spearman's rank correlation coefficient than to Kendall's correlation coefficient.

Table 9.12   Rankings for 6 job applicants by 3 members of an interviewing panel.

Applicant	Panel member A	Panel member B	Panel member C	Rank totals
I	3	2	4	9
II	1	1	2	4
III	6	4	6	16
IV	2	3	1	6
V	5	5	3	13
VI	4	6	5	15

The motivation is easily explained by referring to Table 9.12. The final column of rank totals allocated to each applicant range from 4 to 16 showing wide variation about the mean total of 10.5. With no agreement between panellists the expected value of the rank sum for each candidate would still have been 10.5 and we would expect the individual rank sums all to be close to this mean. If there were complete agreement the rank sums for applicants would be some permutation of 3, 6, 9, 12, 15, 18. By analogy with the desirable property for a correlation coefficient we seek a measure of concordance that takes the value 1 in these circumstances. I denote by S the sum of squares of deviations of the applicant rank totals from the mean of 10.5; i.e. for the Table 9.12 data

$$S = (9 - 10.5)^2 + (4 - 10.5)^2 + (16 - 10.5)^2 + (6 - 10.5)^2 +$$
$$(13 - 10.5)^2 + (15 - 10.5)^2 = 121.5$$

or more generally if n applicants are ranked by m panel members and s_i with mean $\bar{s}$ is the sum of ranks for the ith applicant

$$S = \sum(s_i - \bar{s})^2 = \sum s_i^2 - (\sum s_i)^2/n. \tag{9.15}$$

Clearly the minimum value of S is zero when all sums equal the mean (indicating complete disagreement between panel members) and the maximum occurs when all are in complete agreement. Kendall's coefficient of concordance is

$$W = \frac{S}{\text{Max}(S)} \tag{9.16}$$

and obviously $0 \leq W \leq 1$ with $W = 1$ corresponding to complete agreement in rankings. It is easy to verify that with no ties $\text{Max}(S) = m^2 n(n + 1)(n - 1)/12$.

Agreement as measured by W is particularly interesting for ranking of performance of n competitors by each of m judges. Similar ideas are relevant in tasting tests where m tasters may rank each of n related products (e.g. varieties of raspberries or brands of tomato soup) in order of preference. For simplicity of framing arguments I refer in the rest of this section where appropriate to those being ranked as competitors and to those performing the rankings as judges.

Unlike a correlation coefficient, W cannot take negative values, but this need cause no alarm, for disagreement is no longer symmetric with agreement. For example, with four judges assessing 5 competitors the ranks given might be

Item	Judge A	B	C	D	Rank total
I	1	5	1	5	12
II	2	4	2	4	12
III	3	3	3	3	12
IV	4	2	4	2	12
V	5	1	5	1	12

or they could be

Item	Judge A	B	C	D	Rank total
I	1	3	3	5	12
II	3	4	2	3	12
III	5	1	5	1	12
IV	4	5	1	2	12
V	2	2	4	4	12

For both allocations clearly $S = 0$, and hence $W = 0$. In the first case there is agreement between judges A and C and between judges B and D, but the former pair are in complete disagreement with the latter, leading to overall disagreement. In the second case there is a diffuse pattern of rankings. A different analysis is needed to distinguish between these two types of disagreement.

Example 9.17 For the data in Table 9.12 I have already established that $S = 121.5$ and also noted that for complete agreement, corresponding to max(S), the sums of ranks for each applicant (not necessarily in this order) would be 3, 6, 9, 12, 15, 18 giving max(S)=157.5, whence from (9.17), $W = 121.5/157.5 = 0.7714$.

StatXact computes W and gives exact, Monte Carlo and asymptotic tests computing p-values relevant to significance in relation to a null hypothesis of disagreement. For these data the program confirms $W = 0.7714$ and gives an exact test $p = 0.0117$. The asymptotic p-value based on modification of the chi-squared approximation used for the Friedman test is $p = 0.0414$, in poor agreement with the exact value. As pointed out at the start of this section identical p-values may be obtained using a Friedman test program.

The relationship between W and the Friedman test statistic T given by (8.2) is (Exercise 9.8):

$$W = \frac{T}{m(n-1)} \qquad (9.17)$$

An intuitively reasonable alternative measure of agreement between ranking of competitors by several judges is the mean of all pairwise rank correlation coefficients. It can be shown that the mean of all $\frac{1}{2}m(m+1)$ such Spearman coefficients is $\overline{r}_S = (mW - 1)/(m - 1)$. If $\overline{r}_S = 1$ it follows that $W = 1$ and when $W = 0$ that $\overline{r}_S = -1/(m-1)$. This is the least possible value for the mean.

Modifications are needed for tied ranks. These are similar to those for ties for the Friedman test given in section 8.2. To get a coefficient of concordance that takes the value 1 for complete agreement modification is needed since $\max(S)$ no longer equals that for the no tie case. There is no difficulty if we compute T for the tied Friedman case and then derive W from (9.17). Kendall and Gibbons (1990, Chapter 6) and Siegel and Castellan (1988. section 9.6.4) give a treatment of ties based directly on modification of W and this leads to a W identical with that obtained using the method suggested above.

Interpretation of high values of W is not always straightforward. Some potential problems are discussed in section 11.6.

Exercises

9.1 Obtain the exact permutation distribution for 5 paired observations that form the reference set for tests using Kendall's t_K and confirm that either the number of (i) concordances, or (ii) discordances are equivalent statistics in the no-tie situation. What would be the number of permutations in the reference set for 8 paired observations?

9.2 Verify that the formula for Spearman's r_S given in (9.3) is equivalent to that given by (9.1) or the usual Pearson coefficient formula applied to ranks.

9.3 For the data matrix in Example 9.6 verify that using (9.3) gives the same value of the Spearman coefficient as (9.1).

9.4 For data like that in Example 9.10 what modification would be needed for the Jonckheere–Terpstra test if the alternative WMW type scoring system (−1, 0, 1) is used? How would you score ties in this system?

9.5 Verify the values of all Kendall correlation coefficients quoted in Example 9.16.

9.6 Verify the result given in (9.14).

9.7 Confirm that substitution of the appropriate values in (9.14) for the data in Example 9.15 gives the correct value of $t_{XY,Z}$.

9.8 Verify the relationship between the coefficient of concordance W and the Friedman statistic T given in (9.17).

9.9 Show that at least one possible bootstrap sample from the data in Table 9.8 gives a value of r^* greater than 0.2459.

*9.10 Bardsley and Chambers (1984) gave numbers of beef cattle and sheep on 19 large farms in a region. Explore the evidence for correlation between sheep and cattle numbers evaluating any coefficients you consider appropriate and using any bootstrap methods that you think will be informative about the nature of the correlations.

Cattle	41	0	42	15	47	0	0	0	56	57
Sheep	4716	4605	4951	2745	6592	8934	9165	5917	2618	1105

Cattle	707	368	231	104	132	200	172	146	0
Sheep	150	2005	3222	7150	8658	6304	1800	5270	1537

*9.11   Select a random sample of 8 data pairs from the data in Exercise 9.10 and use appropriate bootstrap analyses to determine how well the Pearson and Spearman correlation coefficients for your samples perform as estimators for the corresponding coefficients for the complete data set in Exercise 9.10.

*9.12   A china manufacturer is investigating likely customer reaction to 7 designs of dinner sets. His main markets are Britain and America. To assess likely preferences in each market he carries out a survey of 100 housewives in each asking each participant to rank the designs 1 (most favoured) to 7 (least favoured). For each country the 100 rank scores are totalled and the design with the lowest total ranked 1, the next lowest 2 and so on. The resulting overall rankings for each country are:-

Design	A	B	C	D	E	F	G
British	1	2	3	4	5	6	7
America	3	4	1	5	2	7	6

The manufacturer later decides to carry out a similar survey in the Canadian and American markets with these results:

Design	A	B	C	D	E	F	G
Canadian	5	3	2	4	1	6	7
Australian	3	1	4	2	7	6	5

Explore the level of agreement about preferred designs first between the British and American markets and then between the Canadian and Australian markets. Examine also the nature of any agreement across all four markets and explain what useful information for the manufacturer stems from your analyses.

*9.13   Nanji and French (1985) and Hand *et al* (1994, Set 122) give the following data on cirrhosis of the liver, alcohol and pork consumption for Canadian provinces:

Province	Cirrhosis mortality (per 100,000)	Alcohol consumption (l per head/year)	Pork consumption (kg per head/year)
Prince Edward Is.	6.5	11.00	5.8
Newfoundland	10.2	10.68	6.8
Nova Scotia	10.6	10.32	3.6
Saskatchewan	13.4	10.14	4.3
New Brunswick	14.5	9.23	4.4
Alberta	16.4	13.05	5.7
Manitoba	16.6	10.68	6.9
Ontario	18.2	11.50	7.2
Quebec	19.0	10.46	14.9
British Columbia	27.5	12.82	8.4

There is a well known relationship between alcohol consumption and cirrhosis and the main interest in the study was whether there was also a relationship between pork consumption and cirrhosis. Inspection of the data suggests one or more possible outliers. A possible analytic tool might be appropriate partial correlations. Bootstrapping may also be useful to study the potential effects of outliers, etc. The authors suggested leaving Newfoundland and Prince Edward Island out of the analysis on the grounds that more seafood was consumed in these provinces. Do you consider this is justified? If it is done would it alter your conclusions?

*9.14 The data below are given in Newman, Freeman and Holzinger (1937), Snedecor and Cochran (1967, Chapter 10) and Hand *et al* (1994, Set 372) and represent numbers of finger ridges per individual for 12 sets of identical twins, Tw1 and Tw2. A point of interest is the nature of the correlation between identical twins. As the ordering within pairs is arbitrary there is no *a priori* reason for labelling Tw1 as x and Tw2 as y. Snedecor and Cochran suggest a way to overcome this arbitrary feature. Explore the nature of the correlation and obtain confidence intervals for any correlation coefficients you think appropriate.

Set	1	2	3	4	5	6	7	8	9	10	11	12
Tw1	71	79	105	115	76	83	114	57	114	94	75	76
Tw2	71	82	99	114	70	82	113	44	113	91	83	72

*9.15 Part of a data set used in a different context by Feigl and Zelen (1965) gives time to death in weeks (TD), and the white blood count (WBC) for a group of leukaemia patients. Explore the use of correlation measures to indicate a relationship between TD and WBC.

Patient	1	2	3	4	5	6	7	8	9	10
TD	56	65	17	7	16	22	3	4	2	3
WBC	4400	3000	4000	1500	9000	5300	10000	19000	27000	28000

Patient	11	12	13	14	15	16
TD	8	4	3	30	4	43
WBC	31000	26000	21000	79000	100000	100000

*9.16 The following data on mortgage arrears is given in *Social Trends 21* (1991) published by the UK Central Statistical Office. Examine whether any correlation between number of mortgages in arrears by 6–12 months and numbers of mortgage in arrears by over 12 months at the end of a given year is more than a reflection of how each changes with time. Consider any implications for the validity of your analysis arising from the fact that some mortgages outstanding for 6–12 months in a given year will be included in those outstanding over 12 months in the following year. The data for number of mortgages is in thousands.

Year	1982	1983	1984	1985	1986	1987	1988	1989	1990
Arrears 6–12 mth	23.8	25.6	41.9	49.6	45.2	48.2	37.2	58.0	76.3
Arrears over 1 yr	4.8	6.5	8.3	11.4	11.3	13.0	8.9	12.0	18.8

10

Bivariate regression

10.1 Bivariate straight line regression

Correlation assesses qualitative aspects of a relationship between paired data such as the existence and general nature of an association; e.g., whether it is linear, monotonic, linear in ranks, etc. Regression is concerned with quantitative aspects. What, in some defined sense, is the best straight line fit to some data? What are the best estimates of slope and intercept for this line? If data trends are not adequately described by a straight line can they be described by a polynomial of degree p? If so, what are the estimated coefficients? Correlation and regression approaches overlap; e.g. in straight line regression a test of zero slope is equivalent to a test of zero correlation and a value of ± 1 for the Pearson product-moment correlation coefficient for a sample of n observations (x_i, y_i) implies all observed points lie on a straight line. In the multivariate case there are relationships between partial correlations and regression. These and other relationships let us apply correlation ideas to many problems of hypothesis testing and estimation described in this and the next chapter.

I follow the approach of Maritz (1995, Chapter 5), who gives a more formal exposition of the theory. Computer programs specifically for exact permutation-based tests and estimation procedures in regression are not widely available, but if appropriate derived variables are used programs for correlation like some in StatXact are useful in regression analysis.

Least squares is the classical method for fitting lines, curves or planes to observational data. Its application to linear regression is one aspect of its use in a range of problems based on linear models including the parametric analysis of variance for designed experiments.

A classic model in bivariate regression specifies that corresponding to each of a series of n fixed numbers x_i, $i=1, 2, \ldots, n$ there exist independent symmetrically distributed random variables Y_i with mean $E(Y_i|x_i) = \alpha + \beta x_i$ and $\text{Var}(Y_i) = \sigma^2$. The problem is to estimate α, β and sometimes σ^2 and under these conditions least squares provides an optimal solution given n paired observations (x_i, y_i). Under the additional assumption that the Y_i are normally distributed standard methods based on normal distribution theory for hypothesis tests and for deriving confidence intervals for estimates are well-known.   The symmetry and the variance homogeneity assumptions often break down. There are tests for validity of assumptions and robust methods of estimation when breakdowns seem likely. Many of these generalize to multivariate regression.

In this chapter I develop some exact permutation test and estimation procedures allied to measures of correlation met in Chapter 9.

Dropping the normality assumption for the Y_i complicates the making of joint inferences about α and β and so I consider first estimation of β only.   The problem is then equivalent to making inferences only about **differences** in location between the distributions of the Y_i corresponding to given values of i. For any independent pair Y_i, Y_j with distributions $F_i(Y_i \mid x_i)$, $F_j(Y_j \mid x_j)$ that differ only in location, when we are given x_i, x_j (fixed) then for a linear model the **difference** in location do not depend on α, e.g. $E[(Y_j - Y_i)\mid x_j, x_i] = \beta(x_j - x_i)$. For symmetric distributions, where the mean (if it exists) and median are equal, the location difference for either measure is clearly $\beta(x_j - x_i)$.   Throughout this section and section 10.2–10.5 I assume that the distributions of the Y_i differ only in an appropriate measure of location. For generality it is convenient to think of this measure as the median, remembering that for symmetric distributions the mean (if it exists) and median coincide.

The location difference assumption implies that for all i

$$D_i(\beta) = Y_i - \beta x_i$$

are identically and independently distributed with location parameter α. We can always adjust the x_i by subtracting an appropriate constant so that $\sum x_i = 0$ and clearly this does not affect β (although it will generally change α) so unless otherwise indicated I assume this has been done, or I adjust data so that it is so. This simplifies many arguments. To test the hypothesis $H_0: \beta = \beta_0$ against an appropriate one- or two-sided alternative implying some other value of β, I write $d_i = y_i - \beta_0 x_i$; then clearly one appropriate test of H_0 is one for zero correlation, or lack of association, between the d_i and the x_i based on the Pearson coefficient. I refer to the d_i as residuals although this is an unorthodox use of the term which more conventionally refers to the $e_i = y_i - \alpha_0 - \beta_0 x$ where α_0 is some hypothesized value of α or where α_0, β_0 are estimates of the unknown α, β. I established in Section 9.1 that a suitable statistic for this test of zero correlation was a statistic which in the current notation is

$$T(\beta_0) = \sum x_i d_i = \sum x_i (y_i - \beta_0 x_i). \tag{10.1}$$

Following arguments similar to those used in Section 1.8 leading to (1.3) we may estimate β by the value of b satisfying the equation

$$T(b) = \sum x_i (y_i - bx_i) = 0. \qquad (10.2)$$

with solution $\beta = (\sum x_i y_i)/(\sum x_i^2)$. Since we have set $\sum x_i = 0$ this is the least squares solution optimal under conditions indicated at the start of this section. If we wish to test $H_0 : \beta = 0$, (10.1) reduces to $T(0) = \sum x_i y_i$, the usual test statistic for zero correlation between x and y given in section 9.1, where I called the statistic S.

The regression problem viewed as one of looking for location differences in samples indexed by fixed x_i is a generalization of the Fisher–Pitman two-sample test. This should give no surprise because I pointed out in section 9.1 that that test was a special case of the Pearson product-moment correlation test. In the present notation the regression test reduces to a Fisher–Pitman two-sample test if there are only two distinct x_i values, i.e. $x_1 = 1$ repeated m times and $x_2 = 0$ repeated n times. Here the mean of the x_i is not zero, but the general argument is not altered if a necessary adjustment is made so that it is, although we are then led naturally to slightly different, but equivalent test statistics. A Fisher–Pitman test for any specific location difference may then be based on the statistic $T(b)$. This is the equivalent to a test statistic using a linear function of the first sample sum and is easily seen to be a valid alternative statistic to that sample sum.

More generally, as in the Fisher–Pitman test and the Pearson coefficient test, the 'raw' data permutation test in regression using the statistic (10.1) tends to give similar results to the normal theory inferences whether or not assumptions for validity hold and in that sense it is often not robust. Transformation such as a rank transformation will sometimes produce more robust inference procedures. Nevertheless, its fundamental nature makes the raw data test worth further study. I call this approach **exact least squares** estimation because it leads to the usual least squares estimator.

10.2 Permutation test least squares

I illustrate the principles of exact inference for a small sample to highlight without computational complexity what happens in practice and to show where implementation difficulties may arise without suitable computer software. Computing exact confidence intervals for β is difficult, but hypothesis testing and point estimation are straightforward.

I have shown that solving (10.2) provides the usual least squares point estimate of β and that a test for acceptability of any hypothesized value β_0 for β is essentially a test of whether the Pearson correlation coefficient test for correlation between x_i and $d_i = y_i - \beta_0 x_i$ indicates acceptability of zero correlation. The statistic (10.1) is the appropriate test statistic.

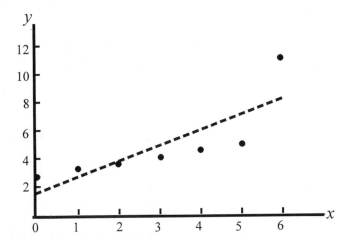

Figure 10.1 Scatter diagram and classic least squares line of best fit for the data in Example 10.1.

Example 10.1 In Sprent (1993, section 8.1.1) I used the following data to illustrate non-robustness of classic least squares estimators.

x	0	1	2	3	4	5	6
y	2.5	3.1	3.4	4.0	4.6	5.1	11.1

Figure 10.1 is a scatter diagram for these data. Classic least squares gives a best-fit equation

$$y = 1.508 + 1.107x$$

This line is also shown in Figure 10.1 and is clearly an unsatisfactory fit. This is an obvious consequence of the point (6, 11.1) being at variance with the straight line trend suggested by the remaining points. This could be because that point indicates departure from a straight-line relationship (something that might be established if additional data were available for values of $x > 5$) or there could be an error in the recorded y value for the last point (perhaps someone forgot to divide by 2 before recording it). Yet another possibility is that x is recorded wrongly, a possibility I consider in Example 11.4.

I use these data, which clearly throw doubt either on the validity of the homogeneity of error assumption, or else on the validity of a straight line fit, in several examples in this chapter because it indicates robustness or otherwise of alternative methods. If the point (6, 11.1) is omitted the classic least squares fit to the remaining points is

$$y = 2.491 + 0.517x$$

and fits these remaining data well.

To use the approach suggested in the previous section for inferences about slope based on the Pearson correlation coefficient I subtract 3 from all x-values to obtain adjusted x_i with mean zero, and henceforth work with the data

x	−3	−2	−1	0	1	2	3
y	2.5	3.1	3.4	4.0	4.6	5.1	11.1

Solving (10.2) for b gives $\hat{\beta} = 1.1071$, the classic least squares point estimate. We may want to test whether the hypothesis $H_0: \beta = 1$ is acceptable. This implies the d_i in (10.1) are given by $d_i = y_i - x_i$, e.g., $d_1 = 2.5 - (-3) = 5.5$, etc., whence

x	−3	−2	−1	0	1	2	3
d	5.5	5.1	4.4	4	3.6	3.1	8.1

The Pearson correlation coefficient program in StatXact used for testing for association between x and d gives an exact two-tail $p = 0.7925$ strongly supporting the hypothesis $\beta = 1$.

A more interesting problem is to find confidence intervals for β and in particular to consider how those given by permutation methods compare with those for classic least squares. In view of the clear breakdown in the homogeneity of variance assumption, at least for the point (6, 11.1), one would like to know whether exact confidence intervals from the permutation model show robustness.

Programs for determining exact least squares permutation confidence intervals are not readily available at present although the principle for obtaining them is described by Maritz. Computation is complicated even for a small problem like that in Example 10.1. With 7 observations the reference set of permutations involves 7!=5040 permutations. If each generates a different value of $T(b)$ following arguments similar to those developed in section 1.8 for a confidence interval for a location difference in the Fisher–Pitman test to get as close as possible to, say, a 95% confidence interval we must determine the maximum r so that $r/5040 \le 0.025$, i.e. $r = 126$ and reject all possible values of β except those that lead to a value of $T(b)$ between 126th largest and 126th smallest value given by the 5040 possible permutations for that value of b. For any specified value b of β_0, Maritz derives a formula for the number, $N(b)$, of statistics of the form (10.1) computed for all permutations other than that observed, that are less than that for the given data. The task is to determine values of b such that $N(b) = r$ or $N(b) = n! - r$. This is not easy to do and is further complicated by the fact that $N(b)$ does not always change by a unit step at some computed values of b given by the formula. Maritz (1995, section 5.2) gives details.

In the absence of computer software for the exact computations made possible by the theory there are some asymptotic results which, although unlikely to be reliable for small samples, provide a starting point for iterative methods that are easily carried out with exact test programs for the Pearson coefficient.

In (10.1) if we apply the usual sampling theory results regarding d_i as observed value of a random variable D we estimate E(D) by

$$\bar{d} = \Sigma(d_i/n)$$

and var(D) by

$$s_d^2 = \sum[(d_i - \overline{d})^2/(n - 1)]$$

Since the x_i are fixed and have mean zero it follows that

$$E(T) = \sum x_i \overline{d} = 0$$

and

$$\text{Var}(T) = s_d^2 \sum x_i^2 \tag{10.3}$$

Providing $\text{Var}(T)$ remains finite as $n \to \infty$ the distribution of $Z = T/\sqrt{\text{Var}(T)}$ tends to standard normal for large n and this provides a basis for hypothesis tests for large samples. Asymptotic confidence intervals for β cannot be obtained directly from this result, but Maritz (section 5.2.1) shows that we may get them by applying a result due to Fieller (1940, 1944) concerning the ratio of normal variables. If z_q is the $1 - \frac{1}{2}\alpha$ quantile of the standard normal distribution then Maritz indicates that approximate $100(1 - \alpha)\%$ confidence limits are given by

$$b = \hat{\beta} \pm z_q \frac{\sqrt{[s^2(n - 2)]}}{\sqrt{[(n - 1 - z_q^2)(\sum x_i^2)]}} \tag{10.4}$$

where s^2 is the usual residual mean square in a regression analysis, i.e.

$$(n - 2)s^2 = \sum(y_i^2) - (\sum y_i)^2/n - [\sum(x_iy_i)]^2/\sum(x_i^2) \tag{10.5}$$

Confidence limits given by (10.4) are not the same as those given by classic least squares assuming normality, which are based upon the t-distribution.

Example 10.1 (continued). For the data in this example (10.5) gives $s^2 = 3.2546$. For a 95% confidence interval we set $z_q = 1.96$. Also $n = 7$ and we established above that $\hat{\beta} = 1.1071$. Substituting these values in (10.4) gives the limits

$$b = 1.1071 \pm 1.0170. \tag{10.6}$$

i.e the confidence interval is $(0.0901, 2.1241)$
Classic least squares confidence limits take the form

$$b = \hat{\beta} \pm t_q s/\sqrt{(\sum x_i^2)}$$

where t_q is the appropriate quantile of the t-distribution with $n - 2$ degrees of freedom. For these data it is easily shown that this gives 95% confidence limits

$$b = 1.1071 \pm 0.8767 \tag{10.7}$$

implying a confidence interval $(0.2304, 1.9838)$.
For this example one should not have much faith in either of these intervals. The first set may be unreliable for so small a sample even had there been no outlier and the latter set are unlikely to be reliable in the presence of an outlier. However one might test their reliability by

inserting the confidence limit values of b into (10.2) and using the StatXact or some other exact program for the Pearson coefficient, test the hypothesis that these are acceptable values of β. If the limits are correct 95% limits then each should give an exact one-tail p-value of $p = 0.025$ in such a test. Carrying out that test with StatXact gave the following one-tail p-values for the limits:

b	One-tail p
0.0901	0.0002
0.2304	0.0002
1.9838	0.0248
2.1391	0.0167

Clearly the lower-tail p-values are less than the required exact values, but the classic upper limit is close to the desired $p = 0.025$. To get a more appropriate lower limit one may use a trial and error process using other values of b. I tried $b = 0.5$ and this gave $p = 0.0381$ and $b = 0.4$ gave $p = 0.001$. I tried in succession $b = 0.48, 0.49, 0.495$ giving respectively $p = 0.0111$, 0.0179, 0.0220. From these results we deduce that exact permutation theory gives a 95.3% confidence interval (0.495, 1.984).

Here the exact interval has a higher lower limit than that given either by classic least squares or asymptotic permutation theory. This reflects upward pressure by the point (6, 11.1) on the slope estimate. Least squares is unsatisfactory because of the heterogeneity of variance imposed upon the location differences specified as the **only** differences in the regression model. Figure 10.1 shows the reason the fitted line is not useful is because the point (6, 11.1) is influential. The Pitman efficiency of permutation test least squares is 1 if the classic least squares normality assumptions hold, but it falls off rapidly for long tailed 'error' distributions. The difficulties with the permutation method here are not unlike those arising where the Fisher–Pitman test is used with inappropriate data. Then better solutions are often provided by rank based transformations.

10.3 Regression associated with Spearman's rho

The first rank-based statistic I consider replaces each d_i in (10.1) by rank(d_i). This leads to major simplification if the x_i are equally spaced. In many real life problems this is the position, but I show in section 10.5 that we still get some simplification even if they are not. In this section I assume the x_i are equally spaced and transformed if necessary to have mean zero. With these assumptions a moment's reflection shows that the statistic

$$T_S(\beta_0) = \sum x_i \operatorname{rank}(d_i) = \sum x_i \operatorname{rank}(y_i - \beta_0 x_i) \qquad (10.8)$$

is a linear function of Spearman's rho and hence that (10.8) is an appropriate statistic for tests of zero correlation between rank(x_i) and rank(d_i). The restriction to equally spaced x_i is needed to ensure that the x_i are linear, not just monotonic, functions of the ranks r_i of the x_i. The test is no longer conditional on either the observed x_i or y_i because for any β_0 the **same** sets of ranks (or linear tranformation of ranks) from 1 to n form the permutation reference sets.

Given a program for exact tests for no association based on Spearman's rho this may be used to test the hypothesis $H_0: \beta = \beta_0$ against a one- or two-sided alternative. Point estimation of β is less straightforward than it was for least squares because the function $T_S(b)$ associated with (10.8) is a step function in b and changes value only as b passes through values which change the rank order of the d_i and thus $T_S(b)$ may sometimes not take the value zero. Steps occur only when b equals one of the $b_{ij} = (y_j - y_i)/(x_j - x_i)$, $i = 1, 2, \ldots, n-1, j > i$, so $T_S(b)$ has the same value for all b between successive b_{ij}. However, if $x_i > 0$ then as b increases the residuals $d_i = y_i - bx_i$ decrease, hence their ranks do not increase and so $T_S(b)$ is a nonincreasing step function of b. A similar argument shows this is also true when $x_i < 0$. Thus $T_S(b)$ is a nonincreasing step function of b for fixed (x_i, y_i), $i = 1, 2, \ldots, n$. Unless three or more observations are collinear or have the same pairwise slopes, if all (x_i, y_i) are distinct these changes occur at all $\frac{1}{2}n(n-1)$ pairwise slope estimators b_{ij}. If there is collinearity involving three or more observations some of the b_{ij} will be equal. Even for equally spaced x_i the steps in $T_S(b)$ are not all of equal height.

Example 10.2. The data in Example 10.1 with mean x_i set to zero become:

x	-3	-2	-1	0	1	2	3
y	2.5	3.1	3.4	4.0	4.6	5.1	11.1

The b_{ij} may be recorded in a triangular matrix like that for the s_{ij} in section 9.2. A typical term is $b_{24} = (y_4 - y_2)/(x_4 - x_2) = (4.0 - 3.1)/[0.0 - (-2)] = 0.450$. Some statistical software packages have facilities for computing the b_{ij}, e.g. in MINITAB the menu command *pairwise slopes* does this. The complete matrix is

0.600	0.450	0.500	0.525	0.520	1.433
	0.300	0.450	0.500	0.500	1.600
		0.600	0.600	0.567	1.925
			0.600	0.550	2.367
				0.500	3.250
					6.000

Repeated values among the b_{ij} imply that some points are collinear or have parallel joins. It is interesting to look at the values of $T_S(b)$ over the range of b values from $-\infty$ to ∞, although this is not essential for inference. Table 10.1 is easily formed using (10.8). Since $T_S(b)$ remains constant for all b between successive b_{ij}, we only need to evaluate (10.8) for any one b value in each of the intervals between consecutive ordered b_{ij}. When $b = b_{ij}$ there are ties in the ranks of the d_i and using mid-ranks for these gives a value of $T_S(b_{ij})$ which is the mean of the values

Table 10.1 Intervals for b values corresponding to each possible $T_S(b)$ arising from the data in Example 10.2. The values of $T_S(b)$ when $b = b_{ij}$ are the means of its values in adjacent b intervals.

b-interval	mid-b value	$T_S(b)$
$-\infty$ to 0.300–		28
0.300+ to 0.450–	0.375	27
0.450+ to 0.500–	0.475	23
0.500+ to 0.520–	0.510	12
0.520+ to 0.525–	0.5125	7
0.525+ to 0.550–	0.5375	3
0.550+ to 0.567–	0.5583	1
0.567+ to 0.600–	0.5835	-2
0.600+ to 1.433–	1.0165	-7
1.433+ to 1.600–	1.5165	-13
1.600+ to 1.925–	1.7625	-18
1.925+ to 2.367–	2.146	-22
2.367+ to 3.250–	2.8085	-25
3.250+ to 6.000–	4.625	-27
6.000+ to ∞		-28

of T_S for b immediately above and below that b_{ij}. Also, in Table 10.1 I give (Exercise 10.1) for each fixed $T_S(b)$ the mid-point of the interval of b values (mid-b value) giving rise to that value of $T_S(b)$.

Maritz suggests that a reasonable point estimate of β is obtained by regarding the step function $T_S(b)$ as an approximation to a continuous function and using linear interpolation between the mid-b values corresponding to the nearest values of $T_S(b)$ to zero. Thus in Table 10.1 we interpolate between $T = 1$, $b = 0.5583$ and $T = -2$, $b = 0.5835$ to obtain a point estimate of b corresponding to $T = 0$, i.e.

$$b = 0.5583 + (0.5835 - 0.5583)/3 = 0.567.$$

It is fortuitous that this is identical with the median of the b_{ij}, though it is common experience that this estimate will be close to that median, so in practice if only a point estimate of β is required one need calculate only a few entries in Table 10.1 for values of b in intervals close to med (b_{ij}) and then use linear interpolation.

Exact tests of hypotheses about β may be performed with any exact test program for the Spearman coefficient such as that in StatXact. For example, to test $H_0: \beta = 1$ we use the values of x_i and d_i computed in Example 10.1, but here we perform a Spearman test for non-association rather than the Pearson coefficient test used in that example. StatXact gives a Spearman coefficient of -0.25 with a one-tail $p = 0.2974$. I pointed out that $T_S(b)$ is a linear function of Spearman's rho. It is easily verified that the extreme values $T = \pm 28$ in Table 10.1 correspond to $r_S = \pm 1$ and indeed that division of all entries in the last column of that table by 28 converts these values to the corresponding r_S. From that table we see that for $b = 1$, $T = -7$, whence $r_S = -7/28 = -0.25$, the value given by StatXact.

Exact confidence intervals are easily obtained for small to medium sample sizes using the StatXact or an equivalent program. The program provides an option to give the complete distribution of the statistic r_S for not-too-large samples. To find an exact confidence interval close to a desired level we use this distribution to find values of r_S, or equivalently of $T_S(b)$ with the requisite associated p-values.

Example 10.2 (continued) For $n = 7$ the exact distribution of r_S indicates that the one tail p-value associated with $r_S \geq 0.7857$ is $p = 0.0241$. This is equivalent for these data to $T_S(b) = 28r_S = 22$. So for an exact two-sided confidence interval at level $(1 - 2\times0.0241)100 = 95.18\%$ we should choose as limits the most conservative b values in Table 10.1 that give $T = \pm22$. From that Table we see that a value of b slightly greater than 1.925 ensures $T = -22$. Our statistic for these data never takes the value $T = 22$, so the best we can do is set $T = 23$, for which the exact one-tail associated $p = 0.0175$. Inspecting Table 10.1 indicates that the shortest appropriate interval is (0.499, 1.926) and from the argument above the exact confidence level is $[1 - (0.0175 + 0.0241)]100 = 95.84\%$. This interval is slightly shorter than those obtained in Example 10.1 but more interestingly the point estimate $\hat{\beta} = 0.567$ is considerably smaller than that given by least squares and is closer to the lower end of the confidence interval. It is also close to the least-squares estimate 0.517 obtained in Section 10.1 when the rogue point (6,11.1) is omitted, indicating robustness resulting from use of ranks.

Clearly if confidence intervals are required at or near conventional levels only the first few and the last few entries (corresponding to large values of $|T_S(b)|$) need be computed in Table 10.1.

For large n exact procedures for confidence intervals may not be readily available but asymptotic theory is usually adequate. The only modification to the asymptotic result in section 10.2 arises because the d_i are replaced by ranks so that the variance s_d^2 simplifies to $(n^2 - 1)/12$ whence (10.3) is replaced by

$$\text{Var}(T_S) = \{[n(n + 1)]/12\}\sum x_i^2.$$

Clearly $E(T_S) = 0$, so providing $\sum x_i^2$ remains finite as $n \to \infty$, $Z = T_S/\sqrt{\text{Var}(T_S)}$ has for large n approximately a standard normal distribution. The approximation is often reasonable even for moderate n. In Example 10.2 we found T = 22 when $n = 7$ corresponded to an exact one-tail $p = 0.0241$. For that example $\sum x_i^2 = 28$ so the normal approximation is $Z = 22/\sqrt{[(56/12)\times28]} = 1.924$. $\Pr(Z > 1.924) = 0.0271$; an approximation that is certainly reasonable for so small a value of n.

The inference procedures in this section are generalizations of the WMW test, so not surprisingly they have the same Pitman efficiencies.

10.4 Regression associated with Kendall's tau

The statistic considered in section 10.3 is only appropriate for a permutation test based on Spearman's rho when the x_i are equally spaced. That restriction is not needed for a test based on Kendall's tau, which depends only on the order of observations and not on magnitudes of differences between the data values or between their ranks. Further, there is no need to assume that the x_i have mean zero. We simplify the description by assuming that the data are so arranged that $x_1 < x_2 < \ldots < x_n$, without loss of generality where the x_i are all different. I discuss briefly the case where some x_i are equal in section 10.7. The statistic used was proposed by Theil (1950) and the procedure is widely known at **Theil's** method. Sen (1968) highlighted the close association with Kendall's tau so I refer to it as the **Theil–Kendall** method. The statistic used for inferences about β is

$$T_T(b) = \sum \text{sgn}[d_{ij}(b)]$$

where

$$d_{ij}(b) = (y_j - \alpha - bx_j) - (y_i - \alpha - bx_i) = (y_j - bx_j) - (y_i - bx_i) =$$

$$d_j - d_i = (y_j - y_i) - b(x_j - x_i)$$

and summation is over $i = 1, 2, \ldots, n - 1, j > i$.

Since the x_i are all different and in ascending order clearly $T_T(b)$ is the numerator in (9.2), i.e. the number of concordances minus the number of discordances arising in computing Kendall's tau for the pairs (x_i, d_i).

Further, with x_i in ascending order, $T_T(b)$ is unaltered if we replace $d_{ij}(b)$ by $b_{ij} - b$ where $b_{ij} = (y_j - y_i)/(x_j - x_i)$ because $x_j > x_i$. Clearly $\text{med}(b_{ij})$ is an appropriate point estimator of β, for then $T_T(b) = 0$. It is easy to see that $T_T(b)$ only changes in value as b passes through a value of b_{ij} and since the statistic is the numerator of Kendall's t_K statistic, then if all b_{ij} are distinct as b increases from $-\infty$ to ∞, T_T is a step function decreasing by steps of 2 from $\tfrac{1}{2}n(n-1)$ to $-\tfrac{1}{2}n(n-1)$. Clearly division of T_T by $\tfrac{1}{2}n(n-1)$ gives Kendall's tau statistic, t_K. If the b_{ij} are not all distinct, some steps will be multiples of 2. Specifically, if a b_{ij} occurs r times the step will be of height $2r$. The value of T_T at any b_{ij} is the mean of the values of T_T for b immediately above and below that b_{ij}.

Example 10.3 I return to the data in Example 10.2 for which the b_{ij} were:

0.600	0.450	0.500	0.525	0.520	1.433
	0.300	0.450	0.500	0.500	1.600
		0.600	0.600	0.567	1.925
			0.600	0.550	2.367
				0.500	3.250
					6.000

Inspection gives the point estimator $b = \text{med}(b_{ij}) = 0.567$. To test $H_0: \beta = 1$ against a one- or two-sided alternative we subtract $b = 1$ from each b_{ij}, and clearly then $T_T(1) = 6 - 15 = -9$. Since $n = 7$ this implies Kendall's $t_K = -9/21 = -0.4286$. StatXact gives the exact one-sided

Table 10.2 Intervals for b values corresponding to each $T_T(b)$ arising from data in Example 10.2. The values of $T_T(b)$ when $b = b_{ij}$ are the means of its values in adjacent b intervals.

b-interval	$T_S(b)$
$-\infty$ to 0.300–	21
0.300+ to 0.450–	19
0.450+ to 0.500–	15
0.500+ to 0.520–	7
0.520+ to 0.525–	5
0.525+ to 0.550–	3
0.550+ to 0.567–	1
0.567+ to 0.600–	−1
0.600+ to 1.433–	−9
1.433+ to 1.600–	−11
1.600+ to 1.925–	−13
1.925+ to 2.367–	−15
2.367+ to 3.250–	−17
3.250+ to 6.000–	−19
6.000+ to ∞	−21

$p = 0.1194$, so one would not reject H_0 at any conventional level.

Using the argument above and noting that there are respectively 2, 4 and 4 tied values of b_{ij} at 0.450, 0.500 and 0.600 we easily establish the values for T_T corresponding to increasing values of b. These are given in Table 10.2.

To obtain a confidence interval for β we need to determine a value of $|T_T|$ with appropriate small one-tail p. The exact distribution when $n = 7$ given by StatXact shows $p = 0.0151$ for $T_T = 15$ and $p = 0.0345$ for $T_T = 13$. Alternatively, for $4 \le n \le 10$ Kendall and Gibbons (1990, Appendix Table 1) give one- tail p-values for T_T. From Table 10.2 we deduce that the shortest $(1 - 2 \times 0.0151)100 = 96.985\%$ interval based on $|T_R| = 15$ is (0.5000–, 1.925+) in close agreement with the exact 95.84% interval found in section 10.3 and based on Spearman's rho.

It can be shown (see, e.g. Maritz, 1995, section 5.2.5) that $E(T_T) = 0$ and that $Var(T_T) = n(n - 1)(2n + 5)/18$. For large n the distribution of $Z = T_T/[\sqrt{Var(T_T)}]$ approaches standard normal.

Example 10.4 For the data in Example 10.3 since $n = 7$ we easily establish $Var(T_T) = 44.3333$ and assuming near normality and setting $Z = \pm 1.96$ indicates that approximate 95% confidence limits would be associated with $T_R = \pm 1.96 \times \sqrt{44.3333} = \pm 13.05$. The exact theory associates $|T_R| = 13$ with an exact 93% interval indicating the approximation here is not unrealistic even for small n.

10.5 Some alternative approaches

Rank-based inference when the x are unequally spaced. Unless the x_i are equally spaced the statistic T_S introduced in section 10.3 will not in general be a linear function of a Spearman rho statistic. Without the equal spacing restriction it becomes a linear function of the Pearson correlation coefficient statistic between the x_i and rank(d_i), which I call T_P. The situation now differs from that in section 10.2 in that for **given** x_i the distribution is invariant under permutation of the d_i since only the ranks of the latter are involved. This avoids the complications that occurred in section 10.2 when determining confidence intervals. If we have a facility to determine the complete distribution of T_P for **given** x_i with $\sum x_i = 0$, the procedure for obtaining confidence intervals or testing hypotheses follows closely that in section 10.3.

Example 10.5 Consider the small data set

x	−10	0	2	3	5
y	1	7	8	9	11

Determine an approximate 95% confidence limit for β based on the statistic

$$T_P(b) = \sum x_i \, \text{rank}(y_i - bx_i).$$

A moment's reflection shows that the value of $T_P(b)$ changes only when b passes through a value of b_{ij} defined in sections 10.3 and 10.4. For the given data these values are easily computed (Exercise 10.2) as

0.600	0.583	0.615	0.600
	0.500	0.667	0.600
		1.000	0.667
			0.500

StatXact gives the exact distribution of the Pearson coefficient between given x_i and rank(d_i) where for these data the ranks are a permutation of 1, 2, 3, 4, 5. It is easily verified that if we denote the Pearson coefficient by r_P then, for a sample of n, $T_P = r_P\sqrt{[(\sum x_i^2)n(n^2 - 1)/12]}$. For this example a few of the values of r_P and T_P in the top tail of the distribution and the corresponding probabilities are:

r_P	*probability*	T_P
0.8883	1/120	33
0.8614	1/120	32
0.8345	2/120	31
0.7807	3/120	29
0.7537	2/120	28

whence $\text{Pr}(T_P \geq 31) = 1/30 = 0.0333$. Clearly T_P is a step function with unequal steps and has a maximum value of 33 for this particular set of x_i. The steps and associated probabilities are symmetric about zero and T_P decreases as b increases. Because of the discontinuities possible two-sided confidence levels near to conventional ones are 98.67% or 93.33%. It is easily verified that for $b < 0.500$ (the minimum b_{ij}), $T_P = 33$, and to determine 93.33% confidence limits for β we must find appropriate values of b for which $|T_P| \geq 31$. We need only compute

T_P for one arbitrary chosen value between relevant successively increasing b_{ij} (Exercise 10.3) to confirm the following:

b-interval	$T_P(b)$
$-\infty$ to 0.500–	33
0.500+ to 0.583–	29
. . . . . .	
0.615+ to 0.667–	–26
0.667+ to 1.000–	–32
1.000+ to ∞	–33

Any intermediate values are not required for present purposes. If computed at any b_{ij}, T_P has a value that is the mean of that for b immediately above or below. Thus for $b = 0.500$, $T_P = 31$. At points corresponding to b_{ij} there are ties in the rank(d_i) and the corresponding probability for the exact permutation distribution with ties may not be the same as that above which was computed for the no-tie situation. Thus we can only say $\Pr(\beta \leq 0.500-) \leq 0.0333$. Similarly we may state $\Pr(\beta \geq 0.667+) \leq 0.0333$. Thus the required interval is $(0.500-, 0.667+)$. In fact this is a conservative interval for the given level based on symmetric tail contributions, because, in fact, due to the discontinuous steps in T_P as computed for these data we can make the stronger assertion that $\Pr(\beta \geq 0.667+) = 1/60 = 0.01667$, implying that the interval is indeed a nonsymmetric at least 95% confidence interval.

A point estimate of β is obtained by linear interpolation as in section 10.3, requiring evaluation of the statistic for values close to the median b_{ij}.

A sign-test analogue. I now consider what turns out to be a special heavily tied case of the Spearman-based estimation procedure of section 10.3. This requires equally spaced x_i although it is apparent that the procedure could be extended for unequal spacing to a special case of the procedure just considered. In section 10.3 I indicated that the Spearman-based test could be regarded as an extension of the WMW test and the test that follows may be regarded as an extension of the sign-test in the manner in which it arose as the Cox–Stuart test for trend. We replace the rank(d_i) used in T_S by $\operatorname{sgn}[d_i - \operatorname{med}(d_i)]$ giving a statistic

$$T_D = \sum \{ x_i \operatorname{sgn}[d_i - \operatorname{med}(d_i)] \},$$

Effectively this means that for n odd we have a linear function of the Spearman statistic with ranks replaced by $\tfrac{1}{2}(n-1)$ values of $+1$, the same number of -1 and one zero, whereas for n even there are equal numbers $(\tfrac{1}{2}n)$ of $+1$ and of -1 (equivalent to a linear transformation of tied mid-ranks). The null distribution of T_D is easily obtained using either the Pearson or Spearman correlation programs in StatXact using x_i and appropriate sgn values of 1, 0, –1. The b values at which the value of T_D changes turn out to be a subset of the b_{ij}.

Table 10.3 Probability associated with each possible value, t, of T_D based on the sign statistic.

t	$\Pr(T_D = t)$	t	$\Pr(T_D = t)$
−12	1/140	1	12/140
−11	2/140	2	8/140
−10	2/140	3	10/140
−9	4/140	4	7/140
−8	3/140	5	8/140
−7	4/140	6	6/140
−6	6/140	7	4/140
−5	8/140	8	3/140
−4	7/140	9	4/140
−3	10/140	10	2/140
−2	8/140	11	2/140
−1	12/140	12	1/140
0	6/140		

Example 10.6 Consider again the data in Example 10.1 and later examples, i.e.

x	−3	−2	−1	0	1	2	3
y	2.5	3.1	3.4	4.0	4.6	5.1	11.1

and compute a confidence limit for slope based on the sign statistic discussed above. Clearly for any b the signs of d_i − med(d_i) are 3 minus, 3 plus and 1 zero, e.g. if $b = 2.0$ the $d_i = y_i - 2x_i$ are in order 8.5, 7.1, 5.4, 4.0, 2.6, 1.1, 5.1 and clearly med(d_i) = 5.1 and the relevant signs in order are +, +, +, −, −, −, 0. The exact distribution of T_D for this case derived from StatXact tabulation of the complete distribution of Spearman's r_s for these tied data is given. in Table 10.3.

It is clear from Table 10.3 that we need only know b values for which $|T_D| \geq 10$ to establish a two-sided 92.86% two-sided confidence interval since $\Pr|T_D| \geq 10 = 10/140 = 0.0714$. Since any changes in b can only occur at a subset of b_{ij} we need only detect a few changes at low and high values of b_{ij} by calculating the values of T_D at arbitrary values of b lying between the b_{ij}. When $b = 0$, $d_i = y_i$ and the relevant sgn in order are −1, −1, −1, 0, 1, 1, 1 giving $T_D = 12$. Trying values between successive increasing b_{ij} indicates a change to the ordering of the sgn values first occurs when b passes through $b_{ij} = 0.45$, for when $b = 0.48$ the relevant sgn ordering is easily verified (see Exercise 10.4) to be −1, 0, −1, −1, 1, 1, 1 giving $T_D = 10$ and a further change occurs at $b_{ij} = 0.5$ where for b slightly greater than 0.5 it is easily verified that $T_D = 6$. Similarly for high values of b we find $T_D = -12$ for values of $b > 2.367$ and $T_D = -9$ for values of b immediately below 2.367. From these results we conclude (Exercise 10.5) that an appropriate 92.86% confidence interval is (0.500−, 2.367+). If we drop the restriction that we should have equal tail probabilities since $T_D = 12$ at the upper limit the true confidence level is 95.71%. This interval is longer than broadly comparable ones for the Spearman or Theil–Kendall methods and more like those given by classical or permutation least squares.

The abbreviated Theil–Kendall method. A modification of the Theil approach leads directly to a Cox-Stuart trend test. It is often referred to as the abbreviated

Theil method and is based on a preselected subset of the b_{ij}. It uses rather limited information and is not recommended if one has computing facilities for methods such as those already described in this chapter. If n is even the only b_{ij} that are included in the abbreviated method are $b_{i, i+\frac{1}{2}n}$ with $i = 1, 2, \ldots \frac{1}{2}n$ and if n is odd only the $b_{i, i+\frac{1}{2}(n+1)}$ are considered with $i=1, 2, \ldots, \frac{1}{2}(n-1)$. These estimators all involve different data pairs and hence are independent, so test and estimation procedures based on those for the sign test may be used. The point estimator of β is the median of the b_{ij} that are calculated. Tests or confidence intervals at conventional levels are only possible if $n > 12$. An example is given by Sprent (1993, section 8.1.3) who also shows how to determine confidence intervals in section 8.1.4.

Transformation of ranks. As alternatives to procedures involving correlation between the x_i and rank(d_i) one may use transformation of ranks to, for instance, normal or van der Waerden scores. For given x_i we may then use a procedure similar to that in Example 10.5 except that ranks are replaced by the appropriate rank transformation. Such transformations are discussed by Adichie (1967) but do not seem to have been widely used in practice and one might expect that their performance would show little improvement over using ranks of the d_i.

Weighted median estimates. Methods using ranks or the Theil–Kendall method both depend heavily on the b_{ij}. Clearly the use of median estimators based on these pairwise slope estimators give robustness to point estimators in the presence of rogue observations or outliers. In Example 10.3, for instance, the observation (6,11.1) which is obviously out of line with the other observations influences only the b_{ij} in the final column of the matrix of values and has little influence on the med(b_{ij}) as a point estimator. There is however an influence on confidence limits with a tendency to raise the upper limit.

However, one may feel, especially if there are no rogue observations, that more weight ought to be given to b_{ij} associated with larger values of $x_j - x_i$. This is precisely what classical least squares does, for it can be shown that that estimator is a weighted mean of the b_{ij} with weights w_{ij} proportional to $(x_j - x_i)^2$. Jaeckel (1972) recommended taking as a point estimator of β the median of weighted b_{ij} with weights $w_{ij} = (x_j - x_i)/\sum_{i<j}(x_j - x_i)$. The procedure has some optimum properties when there are no outliers but these and other weighting schemes that have been proposed by various workers may, when there are outliers, lead to less robust estimators than, for instance, the Theil–Kendall method. Hussain and Sprent (1983) found in simulation studies that the Theil–Kendall method performed almost as well as least squares when appropriate assumptions held and showed a marked improvement in efficiency for long-tailed error distributions, whereas in the latter circumstances weighted medians performed no better than, and often not as well as, Theil–Kendall.

In the methods proposed in Sections 10.2 to 10.5 the test statistic was inspired by the least squares requirement of zero correlation between x_i and the residuals, extending this to cases where the d_i are replaced by a function such as ranks. Is there a case for replacing the x_i by some such function? Maritz (1995) points out that Brown and Mood (1951) used $\text{sgn}[(x_i-\text{med}(x_i)]$ for a test closely related to the Blomqvist test for correlation described in Section 9.4, but that using the x_i themselves generally leads to tests with optimal Pitman efficiencies.

*10.6 Joint estimation of slope and intercept

Estimating slope is a primary aim in regression analyses but we sometimes want also to estimate α or more generally to estimate $E(Y|x) = \alpha + \beta x$ for some specified x; the intercept α is the case $x = 0$. In classic least squares with the usual homogeneity and normality assumptions estimates of α and β are in general correlated. In the special case when $\sum x_i = 0$ and the estimator of α reduces to $a = \sum y/n$ it is uncorrelated with that for β, and then a normality assumption implies independence, so in that case we can make inferences about α without having to worry about what is now a nuisance parameter β.

Unfortunately when we move to distribution-free procedures this simplification does not hold because, in general, even when $\sum x_i = 0$, the statistics used to estimate α, β are correlated.

In section 10.2 I pointed out that inferences about β only were relevant to location differences between populations indexed by the x_i. However, inferences about α or about α and β jointly are inferences about absolute measures of location, not just differences. Indeed, inferences about α or about α and β jointly correspond to generalizations of one-sample location tests such as the Wilcoxon signed rank test that tell us about actual location parameters of the distributions of the Y_i. In this section I assume the appropriate location parameters to be the median of the corresponding distributions or assume that the distributions are symmetrical and identical apart from location differences.

The correlation between most of the statistics used leads naturally to exact confidence **regions** for joint estimation of the parameters, rather than confidence **intervals** for each, and the nature of these regions is indicated by a simple numerical example in Maritz (1995, section 5.3.1), who points out some of the difficulties in defining critical regions and exact confidence regions. You should refer to Maritz for details, but as he point out we can make exact inferences about β free from the nuisance parameter α as I did in sections 10.2 to 10.5. However,

*This section gives a slightly more advanced treatment of a specialist topic and may be omitted at first reading.

it is generally not possible to make exact inference about α free from the nuisance parameter β, or one can only do so with considerable loss of efficiency. It is, however, possible to get approximate confidence intervals for α. Because of such complexities the treatment here is less detailed than that for estimation of β alone, but it gives the flavour of some approaches that have been suggested.

Although it is generally not robust, I commence with the exact least squares procedure. I assume the data, after adjustment if needed, are such that $\sum x_i = 0$. The usual statistics associated with least squares estimation are:

$$T_1(a,b) = \sum(y_i - a - bx_i)$$
$$T_2(a,b) = \sum x_i (y_i - a - bx_i).$$

The least squares point estimators are obtained by equating each to zero giving

$$\hat{\alpha} = \sum y_i /n, \quad \hat{\beta} = \sum x_i y_i / \sum x_i^2.$$

I discussed confidence limits for β in section 10.2. To obtain confidence intervals for α, Maritz proposes an approximate asymptotic method, where similar arguments to those leading to (10.4), give limits

$$a = \hat{\alpha} \pm z_q \frac{\sqrt{[s^2(n-2)]}}{\sqrt{\{n[n - z_q^2]\}}} \qquad (10.9)$$

where, as in (10.4) z_q is an appropriate standard normal deviate. Here s^2 is the residual mean square calculated on the basis of $\hat{\beta}$. The normality assumption is only likely to be reasonable for large n, and unlike when estimating β alone there is no straightforward way of improving the estimate by recourse to an exact test, for exact tests apply to joint confidence regions for both parameters. The procedure is unlikely to be useful in practice in view of the nonrobustness of least squares when the classic assumptions fail. If these assumptions hold this approach has no advantage over the usual normal theory approach.

The statistics used in least squares can be written in a modified form which highlights the link to the signed rank estimation procedure that I consider next. The modified forms of $T_1(a,b)$, $T_2(a,b)$ are

$$T_1(a,b) = \sum \text{sgn}(y_i - a - bx_i) |y_i - a - bx_i|$$
$$T_2(a,b) = \sum x_i \text{sgn}(y_i - a - bx_i) |y_i - a - bx_i|.$$

These forms suggest signed-rank statistics where the magnitude of the deviations term $d(a,b) = (y_i - a - bx_i)$ is replaced by ranks and signs allocated as above, i.e.

$$W_1(a,b) = \sum \text{sgn}(y_i - a - bx_i) \text{ rank } |y_i - a - bx_i|$$
$$W_2(a,b) = \sum x_i \text{sgn}(y_i - a - bx_i) \text{ rank } |y_i - a - bx_i|.$$

These and related statistics were discussed by Adichie (1967) and lead to inferences conditional upon the x_i since the covariance matrix of W_1, W_2 depends upon the x_i, which appear directly in W_2.

Point estimates of α and β are obtained by solving iteratively the equations $W_1(a,b) = 0$ and $W_2(a,b) = 0$. For equally spaced x_i a useful starting point is the estimate of β obtained in the closely related rank method in section 10.3. If the x_i are not equally spaced the estimation method used in Example 10.5 provides a preliminary estimate. Denoting this estimate by β^*, then under the assumption that the Y_i are identically symmetrically distributed apart from location differences it follows from the form of $W_1(a,\beta^*)$ and because the $d_i = y_i - \beta^* x_i$ are identically distributed with mean a, that an appropriate estimator of a given β^*, by analogy with the signed rank procedure given in section 5.3, is the median of the Walsh averages of the d_i . Denoting this estimator by α^* we may determine the values of $W_1(\alpha^*,\beta^*)$ and $W_2(\alpha^*,\beta^*)$. If these are both zero we have the required estimators. This is unlikely in practice. In general W_1 and W_2 are step functions taking values between $-\frac{1}{2}n(n-1)$ and $\frac{1}{2}n(n-1)$ for W_1 and over a range that depends upon the values of the x_i for W_2. If the values of $W_1(\alpha^*,\beta^*)$ and $W_2(\alpha^*,\beta^*)$ are both small in magnitude (about 2 or less in the case of W_1) these estimators will be near optimum. If not, iteration is needed taking values of b_{ij} close to that giving optimum β^* and recomputing α^* with this new estimate of β. Maritz (1995, section 5.3.3) gives an alternative graphical method.

Example 10.7 Revisiting the data in Example 10.1 in the modified form

x	−3	−2	−1	0	1	2	3
y	2.5	3.1	3.4	4.0	4.6	5.1	11.1

obtain estimates of α and β based on the signed rank statistics W_1, W_2.

In Example 10.2 we estimated $\beta^* = 0.567$, from which we easily compute $d_i = y_i - \beta^* x_i$; e.g. $d_3 = 3.4 - 0.567 \times (-1) = 3.967$. The complete set of d_i are 4.201, 4.234, 3.967, 4, 4.033, 3.967, 9.399. The matrix of Walsh averages may be obtained using statistical software that produces them or in this trivial example they may be obtained in triangular matrix format like that in Example 5.6 and the median of these is $\alpha^* = 4.10$. This Hodges-Lehmann estimator may be obtained directly using the d_i as data in the appropriate program in StatXact. To check the fit to the estimating equations we compute $e_i = y_i - 4.10 - 0.567 x_i$ and rank these e_i (Exercise 10.6) with appropriate signs. There are ties among these ranks and using mid-ranks the signed rank(e_i) are 2.5, 5, −5, −2.5, −1, −5, 7.

Since $W_1(\alpha^*,\beta^*)$ is the sum of these signed ranks $W_1(\alpha^*,\beta^*) = 14.5 - 13.5 = 1$. To evaluate $W_2(\alpha^*,\beta^*)$ we multiply each signed rank by the corresponding x_i and add, i.e., $W_2(\alpha^*,\beta^*) = (-3) \times 2.5 + (-2) \times 5 + (-1) \times (-5) + 0 + 1 \times (-1) + 2 \times (-5) + 3 \times 7 = -2.5$. Little improvement can be obtained in these estimators by taking different potential values of α^* or β^*.

The data together with the fitted line $y = 4.10 + 0.567x$ are shown in Figure 10.2. It lies close to the data points other than the outlier (6, 11.1) and the procedure is in that sense robust.

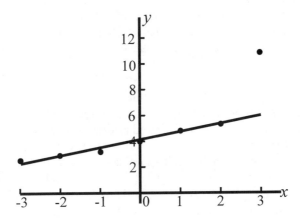

Figure 10.2 The data and line of best fit determined in Example 10.7.

As indicated at the start of this section a joint confidence region for α, β may be obtained, but this is often less useful than approximate confidence intervals for each separately. For β an interval obtained in the way described in section 10.3 will usually be appropriate.

If we knew β it would be appropriate to base confidence intervals for α on the signed rank procedure of section 5.3 applied to the $d_i = y_i - \beta x_i$. We do not know β and Maritz suggest that a sensible estimate may be obtained by replacing β by the estimator β^* obtained above. Estimated confidence intervals can be obtained in the usual way using Walsh averages as illustrated in Example 5.6

Example 10.8 For the data in Example 10.7 using the estimate $\beta^*=0.567$ obtained in that example determine an estimated nominal 95% confidence interval for α.

The relevant d_i were obtained in Example 10.7 and inserting these in the StatXact program for a nominal 95% confidence interval for the median of the distribution represented by the sample values d_i gives the interval (3.97, 6.82). Because the interval is only approximate it is not worth working out an exact confidence level.

The above approach to estimating a confidence interval for α is only appropriate if $\sum x_i = 0$, for α is then a measure of the y-ordinate corresponding to the centered x value. In this sense our estimator is analogous to the estimator $\bar{y}$ in ordinary least squares. For the more important problem of estimating a confidence interval for y corresponding to an arbitrary x estimated by $y^* = \alpha^* + \beta^* x$ a simple approximate interval is not obvious since the error of estimation in both α^* and β^* must be taken into account and clearly that of β^* may be a major contributing factor when $|x|$ is large. However, y^* may still provide a reasonable point estimator of $E(Y|x)$.

In the Theil–Kendall method discussed in section 10.4 I pointed out that the restriction $\sum x_i = 0$ is not essential. Inferences about α based on something like the Theil procedure are available in this situation. Maritz (1979) pointed out that if we write $e_i = y_i - a - bx_i$ then the statistic

$$V(a) = \sum_{j>i} \mathrm{sgn}(x_j e_i - x_i e_j) = \sum_{j>i} \mathrm{sgn}[x_j y_i - x_i y_j - a(x_j - x_i)] \qquad (10.10)$$

does not involve β. This suggests that the methods of section 10.4 might extend to pairwise estimates of α based on all $a_{ij} = (x_j y_i - x_i y_j)/(x_j - x_i)$. However, Maritz shows that unlike the statistic T_T introduced in section 10.4 the statistic $V(a)$ depends on the distribution of the Y_i and hence we cannot resort directly to the inference methods based on Kendall's tau since the probabilities of concordances and discordances may no longer be equal under a particular null hypothesis.

However, Maritz shows that methods similar to that of the abbreviated Theil method which I referred to in section 10.5 may be used if we choose subsets of the a_{ij} which are independent. This is so if no i or j occur more than once in the chosen subsets of $\frac{1}{2}n$ or $\frac{1}{2}(n-1)$ of the chosen a_{ij} depending upon whether n is even or odd. The optimum subset of a_{ij} may not be the same as that used in section 10.5 for the abbreviated Theil method. For equally spaced x_i for reasons given below the pairing of i and j are best made working inward from the extremes, e.g., for $n = 9$ the pairings should be $a_{1,9}$, $a_{2,8}$, $a_{3,7}$ and $a_{4,6}$. There appears to be no obvious advantage in other pairings for unequal spacing when $\sum x_i = 0$, although there may be an advantage in using the pairings proposed in section 10.5 for estimating β by the abbreviated Theil method if an intercept for an x value outside the interval containing the x_i is involved, e.g. for determining α (corresponding to $x = 0$) when all x_i are positive.

For any appropriate subset of a_{ij} chosen to be independent, denoting the analogue of (10.10) for that subset by $V_1(a)$ it is easily seen that under $H_0: \alpha = a$ the mean and variance of $V_1(a)$ are $E[V_1(a)] = 0$ and $\mathrm{Var}[V_1(a)] = \frac{1}{2}n$ or $\frac{1}{2}(n-1)$ depending upon whether n is even or odd. The statistic $V_1(a)$ is a linear function $V_1(a) = 2U - 1$ of a statistic U that has a binomial distribution with $\pi = \frac{1}{2}$. What this means in practice is that we may apply sign test inference methods to the variates $\mathrm{sgn}(a_{ij} - a)$.

Appropriate subsets comprising more than one a_{ij} can only be formed if $n \geq 4$. The median of the chosen a_{ij} is an appropriate point estimator of α and confidence interval at or near conventional levels can only be obtained if $n \geq 12$.

If the x_i are equally spaced and $\sum x_i = 0$, if the a_{ij} are chosen so that $x_i = -x_j$ it follows that $w_{ij} = \frac{1}{2}(e_j + e_i) = \frac{1}{2}(y_j + y_i) - a$. Thus the w_{ij} are independent and identically distributed and can be used directly for inferences about α. In particular, inferences about α may now be based on the Wilcoxon signed rank statistic applied to the relevant $u_{ij} = \frac{1}{2}(y_j + y_i)$. The median of the Walsh averages provides the Hodges–Lehmann estimator of α and confidence limits may be obtained using the method described in section 5.3 for sample size $n \geq 10$. Also, the Pitman efficiency is the same as that for the Wilcoxon signed rank test and for the normality case, for example, is $3/\pi$ (compared to $2/\pi$ for the sign test).

Example 10.9 For the data in Example 10.7 relevant values of u_{ij} are $u_{1,7} = \frac{1}{2}(11.1 + 2.5) = 6.8$, $u_{2,6} = \frac{1}{2}(3.1 + 5.1) = 4.1$, $u_{3,5} = \frac{1}{2}(4.6 + 3.4) = 4.0$. The Hodges–Lehmann estimator is easily found (using Walsh averages) to be 4.0, equal in this trivial case to the median of the u_{ij}, the estimator relevant to the sign-based test described above.

10.7 Some general considerations

Several methods described in this chapter show appreciable robustness when there are outliers. However, if there is little or no evidence that assumptions relevant to classic least square break down, because of lower efficiency some of these methods may then not perform as well as least squares. There are intermediate situations where some features of the data show peculiarities that, while not indicating blatant abnormalities, suggest minor violation of least squares assumptions. Then bootstrapping, discussed in section 10.8, or a distribution-free method of analysis might be appropriate. It may be wise to use both least squares and an unconditional distribution-free or other robust method such as an M-estimator of the type described in section 11.4. If both methods lead to similar inferences one may accept these as reasonable. If the analyses lead to different conclusions a comparison based on data characteristics is often illuminating.

Example 10.10 Gat and Nissenbaum (1976) give ammonia concentrations (y mg l⁻¹) at various depths (x m) in the Dead Sea. I fit a linear regression of y on x by classic least squares and by the Theil–Kendall methods and compare and comment upon the results.

x	25	50	100	150	155	187	200	237	287	290	300
y	6.13	5.51	6.18	6.70	7.22	7.28	7.22	7.48	7.38	7.38	7.64

Any standard program gives the least squares fit:

$$y=5.745+0.00653x.$$

The 95% confidence intervals are $(0.00413, 0.00893)$ for β and $(5.261, 6.229)$ for α.

For the Theil method the values of b_{ij} are obtainable from Table 10.4 where values of $b_{ij} \times 10^5$ are given to facilitate reading.

The med(b_{ij}) $= 0.00627$. To estimate α I use the method proposed in section 10.6, analogous to the abbreviated Theil–Kendall procedure. Because $x = 0$ lies outside the range of observed x_i I pair x_7 with x_1, x_8 with x_2, etc., and finally x_{11} with x_5. These give (Exercise 10.7)

$$a_{1,7} = 5.974, \quad a_{2,8} = 4.983, \quad a_{3,9} = 5.538, \quad a_{4,10} = 5.971, \quad a_{5,11} = 6.771$$

and med(a_{ij}) $= 5.971$

Table 10.4 Values of $b_{ij} \times 10^5$ for data in Example 10.10

-2480	67	456	838	710	623	637	477	468	549
	1340	1190	1629	1292	1140	1053	789	779	852
		1040	1891	1264	1040	949	642	632	730
			1040	1568	1040	897	496	486	627
				188	0	317	121	119	290
					-462	400	100	97	319
						703	184	178	420
							-200	-189	254
								0	2000
									2600

The estimating equation is therefore

$$y = 5.971 + 0.00627x$$

broadly in line with the least squares equation. A confidence interval for β may be obtained by the method outlined in section 10.4. If the exact distribution of $T_T(b)$ cannot be obtained an approximate nominal 95% confidence interval may be obtained using the asymptotic result given in that section. In Exercise 10.8 I ask you to verify that this is obtained by excluding the 15 largest and 15 smallest b_{ij} and that it is (0.00255, 0.01039). The sample is too small to give a 95% interval for α. However, by analogy with the sign test situation where 5 plus or 5 minus signs in a sample of 5 correspond to a two-tail $p=1/16=0.0625$ it follows that the interval between the extreme values of the a_{ij} computed above, i.e., (4.983, 6.771) is a 93.75% confidence interval for α.

Clearly, for the data in this example the Theil–Kendall method gives reasonable point estimators of α, β but the confidence intervals are appreciably wider than those for classic least squares. A plot of the data indicates a fairly substantial scatter about the fitted line in either case but with no blatant outliers like that in the data in Example 10.1.

These wider limits are surprising both in the light of efficiency studies and Monte Carlo simulations of the type performed by Hussain and Sprent (1983). Inspecting the data suggests some rounding of the y_i values. Although these are given to two decimal places repeated entries at 7.22 and 7.38 together with the fact that 7 of the 11 entries have a final digit 2 or 8 suggests there may be a limitation in the accuracy of recording of ammonia concentrations. The effect of any such rounding is likely to impact upon individual b_{ij}, tending perhaps to make their dispersion about their median sometimes greater than it would be for more accurate recording. While this may have some effect on least squares confidence intervals it may well have a greater effect on confidence intervals for some distribution-free methods because it lowers the concentration of values near the med(b_{ij}) and this may well be the situation here.

This example highlights the need to compare results of different methods of estimation when they lead to not entirely similar inferences.

A problem arises with the Theil–Kendall method if there are tied values of x_i because the b_{ij} corresponding to such a pair becomes infinite. For only a small proportion of ties there will be little loss of efficiency if such points are replaced by one data entry with the y value set equal to the mean of the y values for all the points with that tied x value. This is generally preferable to artificial tie-breaking devices such as splitting the ties by making arbitrarily small changes to the tied x_i to separate them, for that process may lead to large or even bizarre b_{ij} associated with such splits in view of the small denominators. Another alternative which has greater appeal is to exclude all comparisons between points with a common x_i value but to consider all joins of each of these points to points with other values x_j, i.e. where $x_j \neq x_i$. The extreme case occurs when there are only two x values, x_1 and x_2, and repeated y-values at each. The problem is then equivalent to a two-sample problem and if we consider the slopes of all pairwise joins between the y-values associated with different x it is easy to see that these are similar to the differences computed for the Hodges-Lehmann estimator in section 4.4; indeed they are simply these differences each divided by the constant $x_2 - x_1$. Indeed our method is now equivalent to that of the WMW test.

10.8 The bootstrap in regression

I consider the least squares and the Theil–Kendall methods of estimating β when fitting a straight line to the data in Example 10.1 in the context of bootstrapping. In that example least squares led to the unsatisfactory point estimator $b = 1.107$ while the Theil–Kendall method gave a point estimator $b = 0.567$. Approximate 95% confidence intervals were:

Classic least squares (Example 10.1)	(0.230, 1.984)
Permutation least squares (Example 10.1)	(0.495, 1.984)
Theil–Kendall (Example 10.3)	(0.500, 1.925)

Classic and permutation least squares both give (as they must) the same point estimate and a fit that should please nobody, but the Theil–Kendall point estimate is sensible if the point (6, 11.1) is regarded as aberrant. It is interesting, and perhaps surprising, that although the least squares and Theil–Kendall point estimators differ, the confidence intervals are broadly similar.

I took 1000 bootstrap samples of the 7 observations in Example 1.1 and for each computed the Theil–Kendall estimator and the least squares estimator, and determined approximate 95% confidence intervals by the quantile-based method, omitting the 25 smallest and 25 largest values of the estimators $b*$. As in most bootstrap estimation problems further light is thrown on the nature of the problem by a more detailed study of the bootstrap distribution of $b*$; in particular by a studying the histogram of the sampling distribution.

Example 10.11 From the $B = 1000$ bootstrap samples I obtained and ordered the least squares estimators and rejected the 25 smallest and largest $b*$ to give a 95% quantile based confidence interval (0.489, 1.919) for least squares and similarly an interval (0.500, 1.925) for the Theil-Kendall estimator, in good agreement with the permutation results. Remember that another sample of 1000 (Exercise 10.9) will not lead to identical bootstrap estimates. As I pointed out in Example 10.3 the asymmetric nature of confidence intervals with respect to the point estimator in the Theil case reflects the upward pressure on the estimator of the aberrant point (6, 11.1) which however, has little effect on the median of the b_{ij} that provides the point estimate. Inspecting the histograms of the sample of 1000 $b*$ values shows how the bootstrap reflects the 'information' in the sample that one out of seven observations is aberrant. If this observation is accepted as genuine our best estimate of the number of likewise aberrant observations in whatever population our sample came from is also 1 in 7. This in a sense is built into and reflected in any bootstrap estimation procedure. It is of course also built into classical parametric or permutation theory inference. Another typical example of how a characteristic of a sample is used to imply a population characteristic is the use of the sample standard deviation as the estimate of the population analogue in the t-test.

Figures 10.3 and 10.4 show histograms of the 1000 least squares and Theil bootstrap estimators. The most striking feature of Figure 10.3 is the bimodality. A moment's reflection traces the cause of this to a dichotomy of the bootstrap samples into those which do not include the point (6, 11.1) and those that include that point at least once. The high values clearly represent samples where the points involved include one or more replicates of (6, 11.1) probably associated strongly with other points corresponding to relatively higher values of x. The asymmetry shown by these samples would make a nonsense of confidence intervals based on the point estimator for the original data plus or minus two bootstrap standard errors. Confidence in the procedure for rejecting the 25 largest and smallest $b*$ is engendered by the closeness of the resulting interval to that based on the permutation distribution. However, one should not ignore the possibility of bias in the bootstrap estimates, a topic I referred to briefly in sections 2.5 and 2.6. The limits obtained here make sense even though the whole procedure of fitting a straight line to these data by least squares is questionable in the light of a breakdown in key assumptions.

The histogram in Figure 10.4 for Theil–Kendall estimators again shows bimodality clearly associated with presence or absences of the aberrant point (6, 11.1). The break between categories near 0.8 is even sharper than the histogram suggests because the arbitrary division into intervals of the form (0.6, 0.8–) results in that interval including 208 estimates **each of which takes the value 0.6 exactly.** The most striking difference between the histograms in Figures 10.3 and 10.4 is that in the latter the number of estimates in the classes below 0.8 is nearly 2.3 times that in Figure 10.3, reflecting the fact that the aberrant point is less influential upon a point estimator based on the median of the b_{ij} (Theil–Kendall) than it is upon one based on a weighted mean of these (least squares, section 10.5). I showed in section 3.4 that median estimators in general are more tolerant to rogue points than are mean based estimators because the former will tolerate more rogue points with little effect on the estimate obtained.

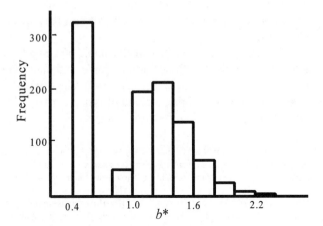

Figure 10.3 Histogram for 1000 bootstrap least squares slope estimators for data in Example 10.11. Five other values of b^* were greater than 2.4.

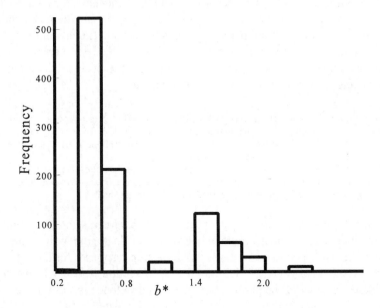

Figure 10.4 Histogram of 1000 Theil-Kendall slope estimators. Example 10.11 data. Thirteen other values of b^* were greater than 2.4.

The above example used a data set where an outlying observation strongly influenced patterns of bootstrap estimators depending on its presence or absence in individual bootstrap samples.

For reasonably well behaved data even for a small sample one might expect less irregularity in histograms associated with bootstrap samples although effects such as skewness or odd outliers may arise if there are even slight departures from classic least squares assumptions and that method of fitting is used. Again such departures are likely to be more marked for small samples.

Example 10.12 Using the data in Table 9.8 I obtained 1000 bootstrap samples of the least squares regression coefficient. The plot of these data given in Figure 10.5 suggests that a straight line fit might not be unreasonable except that the first point looks to lie below what might be an ideal line through the remaining points.

Figure 10.6 is a histogram of $|b*|$ for the 1000 samples. Effectively this is a plot of $-b*$ since all values of $b*$ were negative. There is no evidence of the bimodality so clear in Figure 10.3, but the longer upper tail indicates skewness that reflects the tendency of bootstrap samples that omit the first point to give slightly higher negative values for b*.

If a larger sample had shown a similar tendency for points in the neighbourhood of the one that appears to be below the line in this example, but the proportion of such points remained much the same (i.e. about one is seven) this asymmetry in bootstrap samples is likely to be less marked because a moment's reflection shows that the probability of **all** such slightly aberrant points being excluded from a bootstrap sample is smaller.

Example 10.13. The complete data set given by Hand *et al* (1993) for an uninsulated house for which the data in Table 9.8 comprised a random sample consisted of 26 values. These were plotted on a scatter diagram in Figure 9.4. Again the points with low values of *x* appear to be a little below what might be a reasonable straight line fitted to the rest of the data.

Figure 10.7 shows a histogram of 500 least squares bootstrap values of $|b*|$ for these data. It looks more like the histogram one might expect for a sample for a normal or near normal distribution despite minor reservations one might have about the adequacy of a straight line fit.

In this brief introduction to use of the bootstrap in regression I have described only one approach. Another approach in which residuals from a fit to the original data are bootstrapped to give a new set of bootstrap responses for the fitted model is discussed by Efron and Tibshirani (1993, section 9.4). The method has analogies to some of the permutation methods developed in this chapter. These authors in the same chapter (section 9.5) discussion when each bootstrap method is the more appropriate. These alternative approaches emphasize the point that there is often no uniquely 'correct' or 'best' method of bootstrapping.

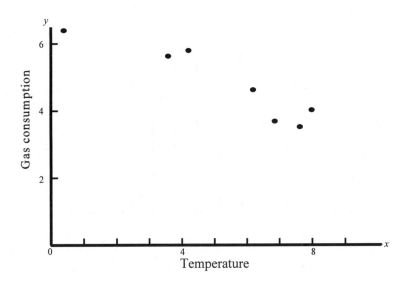

Figure 10.5 A plot of the data in Table 9.8.

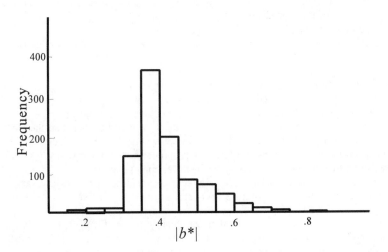

Figure 10.6 Histogram of 1000 bootstrap slope estimators of $|b*|$ for data in Table 9.8.

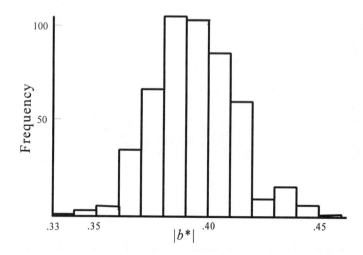

Figure 10.7 Histogram for 500 bootstrap estimators of |b*| for the data in Figure 9.4.

Exercises

10.1 Verify that the values of $T_s(b)$ given in Table 10.1 are correct. Also verify for several values of b_{ij} the correctness of the claim in Example 10.2 that the value of any $T(b_{ij})$ is the mean of the values for b immediately below and above the relevant b_{ij}.

10.2 Verify the values of the b_{ij} recorded in Example 10.5.

10.3 Confirm the values of $T_P(b)$ for each of the several relevant intervals considered in Example 10.5

10.4 Verify in Example 10.6 that when $b = 0.48$ then $T_D = 10$.

10.5 Confirm the confidence interval for b given in Example 10.6.

10.6 Confirm the values of the signed ranks of residuals given in Example 10.7.

10.7 Compute the values of a_{ij} quoted in Example 10.10.

*10.8 In Example 10.10 verify the stated confidence limits for β and confirm that these are obtained when using the asymptotic theory given in section 10.4 by excluding the 15 largest and smallest b_{ij}. If you have access to appropriate computer software you should compare exact theory limits with those given by the asymptotic theory.

*10.9 Using the most suitable available software repeat the studies carried out in Example 10.11 for the same data set. Compare your results with those in Example 10.11 and comment on any differences.

*10.10 The following is a sub-sample of a data set obtained by Lee (1965) [(the full set is given in Hand et al (1994, table 102)] . Plot the data. Do you consider a straight line fit for the regression of mortality index on mean annual temperature may be satisfactory for some or all of the data. Do you think any of the observations might be treated as potential outliers? Use any of the methods developed in this chapter or others you may think appropriate and comment upon their relevance and adequacy in contributing to our understanding of the relationship between mean annual temperature and mortality index for these data. The mortality index refers to a type of breast cancer.

Temperature °F	31.8	34.0	42.3	44.2	45.1	49.2	49.9	51.3
Mortality index	67.3	52.5	65.1	81.7	89.2	95.9	104.5	102.5

*10.11 The following data are given in *Regional Trends,* 24 published by the UK Central Statistical office (1989) and are also given in Hand *et al* (1994, Set 69). The data are for percentage male unemployment and average percent of household expenditure on motoring and travel fares for 11 UK regions. Hand *et al* remark 'the data can be used to illustrate basic correlation and regression' adding that there are potential difficulties with an outlier. Use any methods in this book (or any other methods) that you consider appropriate to explore the characteristics of these data and write a report summarizing your conclusions

Region	Unemployment %	Motoring/fares %
England		
North	14.0	12.8
Yorkshire, Humberside	11.3	14.5
East Midlands	9.0	15.4
East Anglia	6.8	15.0
South East	7.1	15.0
South West	8.2	15.3
West Midlands	11.1	14.6
North West	12.7	14.2
Wales	12.5	14.4
Scotland	13.0	14.1
Northern Ireland	17.6	15.5

*10.12 Gilchrist (1984) and Hand *et al* (1994, Set 115) give distances by road and straight line (linear) distances between different pairs of points in Sheffield, England. Is a linear relationship useful for predicting road distance from linear distance?

Road distance	Linear distance	Road distance	Linear distance
10.7	9.5	11.7	9.8
6.5	5.0	25.6	19.0
29.4	23.0	16.3	14.6
17.2	15.2	9.5	8.3
18.4	11.4	28.8	21.6
19.7	11.8	31.2	26.5
16.6	12.1	6.5	4.8
29.0	22.0	25.7	21.7
40.5	28.2	26.5	18.0
14.2	12.1	33.1	28.0

11

Other regression models and diagnostics

11.1 Multiple regression and the general linear model

The emphasis in earlier chapters has been on permutation tests and other rerandomization procedures such as the bootstrap where these have provided nonparametric, distribution-free or other appropriate procedures based primarily on data. Usually only a few broad distributional or modelling assumptions such as symmetry or identity apart from possible location (or sometimes scale) differences, etc. were made. In this chapter I consider more complicated data structures that may call for different approaches either because randomization or permutation test procedures are not readily available or are cumbersome to use. The models considered are firstly those of multivariate linear regression and the extension to general linear model which, under conventional normality and homogeneity assumptions cover many used in analysis of variance or covariance. I consider the effect of relaxing some standard assumptions.

I showed in Chapter 10 that even for bivariate straight line regression, classic least squares or analogous exact procedures based on raw data permutations may not perform well if an assumption that the error terms are distributed about zero with constant variance break down. Procedures based on tests of correlation between the explanatory variable x and the ranks of the residuals often proved more robust for answering questions about the slope coefficient β. Maritz (1995, chapter 6) extends some of those tests to multiple regression emphasizing the case with two explanatory variables x_1, x_2 where for Y_i corresponding to given x_{1i}, x_{2i} we have $E(Y_i|x_{1i}, x_{2i}) = \alpha + \beta_1 x_{1i} + \beta_2 x_{2i}$. I return to this topic in section 11.2.

The general parametric linear hypothesis model that includes multiple regression and the analysis of variance and analysis of covariance models as particular cases owes much of its relative simplicity in applications to an assumption that errors are identically, and usually normally, distributed. I have already shown in particular cases, e.g. in the discussion on interactions with factorial treatment structures in section 7.6, that this simplicity does not always translate smoothly from a parametric situation to more general nonparametric or distribution free contexts. More generally, Maritz (1995, p.165) comments

It is certainly possible to develop a distribution-free approach to a general linear model along lines somewhat similar to those of the traditional treatment. But it soon becomes clear that exact inferential statements can only be made about rather uninteresting questions, unless one considers generalizations with certain special properties . . .

Diagnostic tests often based upon simple graphical methods to detect unusual or influential observations are useful to detect and deal with situations where classical regression assumptions break down and I consider these in section 11.3. A second and truly distribution-free approach to fitting regression curves or surfaces is one in which closeness of fit to data is played off against the desirability of a smooth fitted function that is not grossly influenced by the quirks of individual data points, an approach I outline briefly in section 11.5.

11.2 Permutation tests in multiple regression

The data in Example 10.1 included the point (6,11.1) which did not lie close to a straight line suggested by the remaining points. Permutation test procedures based on Pearson's product moment correlation coefficient between the x_i and residuals, the analogue of classic least squares estimation, were not robust whereas methods based on rank transformations of the residuals equivalent to using test and estimation procedures based on Spearman's ρ (for equally spaced x) or Kendall's τ gave point estimators that showed some robustness against such an outlier, although the confidence intervals were little different from those given by least squares and reflected an upward thrust giving asymmetry about the point estimator that reflected the influence of the outlier. This gives a warning about limitations to robustness, in that although a rank transformation may improve point estimation it may have limited influence upon a confidence interval. Some other method such as m-estimation, discussed briefly in section 11.4, may then be more appropriate.

When we move from 1 to just 2 regressor or explanatory variables x_1, x_2 it becomes less clear by mere data inspection or a simple data graph (which now becomes three dimensional) which are likely outliers or influential points. Fortunately the regression diagnostics to be described in section 11.3 help with this problem. Further there is a scarcity of appropriate computer software for the multivariate analogues of the techniques discussed in sections 10.2 to 10.5 although the basic theory behind such extensions is given by Maritz (1995, Chapter 6). In this section I sketch the basic idea behind a raw data randomization technique and one rank-based exact procedure for the case of 2 explanatory variables, mainly to show the rationale behind such extensions but also to indicate difficulties in their practical application unless n is large enough to justify asymptotic results. I take an unrealistically small example to illustrate rationale with minimal computation and consider only estimation of β_1, β_2 in a model where $E(Y_i | x_{1i}, x_{2i}) = \alpha + \beta_1 x_{1i} + \beta_2 x_{2i}$ and the Y_i have the same distribution for each $i = 1, 2, 3, \ldots, n$ apart from the implied two-dimensional location differences, differences that do not involve α. As in section 10.2 I assume, without loss of generality when considering only location differences, that $\sum x_{1i} = \sum x_{2i} = 0$. The analogous statistics to (10.1) for estimating β_1, β_2 are now

$$T_1(\beta_1, \beta_2) = \sum x_{1i} d_i$$
$$T_2(\beta_1, \beta_2) = \sum x_{2i} d_i \qquad (11.1)$$

where $d_i = y_i - \beta_1 x_{1i} - \beta_2 x_{2i}$. With the constraints that $\sum x_{1i} = \sum x_{2i} = 0$ it is easily verified that equating these statistics to zero and solving leads to the usual least squares point estimators of β_1 and β_2. Hypothesis tests about these parameters are based on the joint permutation distribution for (11.1) with the d_i computed for values $\beta_1 = \beta_{01}$, $\beta_2 = \beta_{02}$ specified in H_0. This is a bivariate distribution and care is needed to obtain appropriate critical regions for rejection of H_0.

Example 11.1 Given the data

y_i	−13	8	16	4
x_{1i}	−1	0	3	−2
x_{2i}	−5	0	3	2

believed to satisfy a relationship of the form $E(Y_i | x_{1i}, x_{2i}) = \alpha + \beta_1 x_{1i} + \beta_2 x_{2i}$ investigate the acceptability of the hypothesis $H_0: \beta_1 = 1$, $\beta_2 = 3$.

If the test is based on the statistics (11.1) the joint permutation distribution of T_1, T_2 is obtained by considering all 24 permutations of the relevant d_i in much the way we do in obtaining, for example, the reference sets for the Pearson correlation coefficient.

Under H_0 we have $d_1 = -13 - (-1 \times 1 - 5 \times 3) = 3$ and similarly $d_2 = 8$, $d_3 = 4$ and $d_4 = 0$, whence for the permutation giving the observed data $T_1 = (-1) \times 3 + 0 \times 8 + 3 \times 4 + (-2) \times 0 = 9$ and $T_2 = -3$. Table 11.1 gives the values of T_1, T_2 corresponding to this and five other permutations of the d_i for the given x_{ji}, $j = 1, 2$.

There are 24 equiprobable values for (T_1, T_2) and each such pair with its associated probability ($\times 24$) is given in Table 11.2. The reader should verify these values (Exercise 11.1).

To find a critical region for rejection of H_0 we have first the familiar small-sample problem that with only 24 equally likely permutations we are restricted in the choice of significance level effectively to $p = 1/24$ or for a two-tail test perhaps to $p = 1/12$. Further, I have not yet discussed the choice of a critical region in a bivariate case although this is required in both parametric and distribution-free analyses. Clearly, we want to reject H_0 for values of (T_1, T_2) that are in some sense extreme or distant from the mean or median of the joint distribution. Since

Table 11.1 The statistics T_1, T_2 for various permutations of d_i in Example 11.1

x_{1i}	−1	0	3	−2		
x_{2i}	−5	0	3	2	T_1	T_2
d_i	3	8	4	0	9	−3
	3	4	8	0	21	9
	8	3	4	0	4	−28
	4	3	8	0	20	4
	4	8	3	0	5	−11
	0	3	4	8	−4	28

Table 11.2 Joint permutation distribution of (T_1, T_2), probabilities (×24)

T_1 / T_2	−20	−19	−16	−14	−11	−10	−7	−4	−2	1	4	5	6	9	13	14	16	18	20	21
−34	0	0	0	1	0	0	0	0	0	0	0	0	0	0	0	0	0	0	0	0
−32	0	0	1	0	0	0	0	0	0	0	0	0	0	0	0	0	0	0	0	0
−31	0	0	0	0	0	0	0	0	0	1	0	0	0	0	0	0	0	0	0	0
−28	0	0	0	0	0	0	0	0	0	0	1	0	0	0	0	0	0	0	0	0
−23	0	0	0	0	0	0	1	0	0	0	0	0	0	0	0	0	0	0	0	0
−22	0	0	0	0	0	0	0	0	1	0	0	0	0	0	0	0	0	0	0	0
−14	0	0	0	0	0	1	0	0	0	0	0	0	0	0	0	0	0	0	0	0
−11	0	0	0	0	0	0	0	0	0	0	0	1	0	0	0	0	0	0	0	0
−7	0	0	0	0	1	0	0	0	0	0	0	0	0	0	0	0	0	0	0	0
−4	1	0	0	0	0	0	0	0	0	0	0	0	0	0	0	0	0	0	0	0
−3	0	0	0	0	0	0	0	0	0	0	0	0	0	1	0	0	0	0	0	0
1	0	1	0	0	0	0	0	0	0	0	0	0	0	0	0	0	0	0	0	0
4	0	0	0	0	0	0	0	0	0	0	0	0	0	0	0	0	0	0	1	0
5	0	0	0	0	1	0	0	0	0	0	0	0	0	0	0	0	0	0	0	0
9	0	0	0	0	0	0	0	0	0	0	0	0	0	0	0	0	0	0	0	1
10	0	0	0	0	0	0	0	0	0	0	0	0	0	0	1	0	0	0	0	0
13	0	0	0	0	0	0	1	0	0	0	0	0	0	0	0	0	0	0	0	0
17	0	0	0	0	0	0	0	0	1	0	0	0	0	1	0	0	0	0	0	0
18	0	0	0	0	0	0	0	0	0	0	0	0	1	0	0	0	0	0	0	0
25	0	0	0	0	0	0	1	0	0	0	0	0	0	0	0	0	0	0	0	0
28	0	0	0	0	0	0	0	1	0	0	0	0	0	0	0	0	0	0	0	0
30	0	0	0	0	0	0	0	0	0	0	0	0	0	0	0	0	0	1	0	0
32	0	0	0	0	0	0	0	0	0	0	0	0	0	0	0	1	0	0	0	0

T_1, T_2 are correlated and clearly have different variances care is needed in selecting an appropriate measure of distance. In these circumstances a widely used measure is the Mahalanobis distance described below, but to use it we require the mean and covariance matrix of the statistics T_1, T_2. These are obtained by arguments similar to those used in section 10.2 for the statistic T. As in that section, if we apply the usual sampling theory results regarding d_i as the observed value of a random variable D we estimate $E(D)$ by

$$\bar{d} = \Sigma(d_i/n)$$

and var(D) by

$$s_d^2 = \Sigma[(d_i - \bar{d})^2/(n-1)].$$

Since the x_{1i}, x_{2i} are fixed and have means zero it follows that

$$E(T_r) = \Sigma x_{ri}\bar{d} = 0, \;\; r = 1, 2$$

and

$$\mathrm{Var}(T_r) = s_d^2 \Sigma x_{ri}^2 \tag{11.2}$$

and additionally here that

$$\mathrm{Cov}(T_1, T_2) = s_d^2 \Sigma x_{1i}x_{2i}.$$

Thus the covariance matrix of T_1, T_2 may be written

$$V_T = s_d^2(X^TX)$$

where

$$X^T = \begin{Bmatrix} x_{11}, x_{12}, & \cdots & x_{1n} \\ x_{21}, x_{21}, & \cdots & x_{2n} \end{Bmatrix}$$

is the transpose of X.

The Mahalanobis distance from the origin $(0,0)$ to any $t' = (T_1, T_2)$ obtained for any permutation in the reference set is given by

$$Q = t'V_T^{-1}t \tag{11.3}$$

where V_T^{-1} is the inverse of V_T.

If Q is computed for t corresponding to each permutation in the reference set, high values of Q determine a critical region for rejection of H_0. I do not establish the result here, but under H_0 the distribution of Q is asymptotically chi-squared with 2 degrees of freedom, a result that may be used in hypothesis testing for large n.

Example 11.1 (continued). Inspection of Table 11.2 shows that the observed $t' = (9,-3)$ is among the more central values of the statistic that have nonzero probability, so it does not indicate rejection of H_0. Straightforward calculations from the given data yield $s_d^2 = 10.9166$, $\Sigma x_{1i}^2 = 14$, $\Sigma x_{2i}^2 = 38$, $\Sigma x_{1i}x_{2i} = 18$ whence

$$V_\mathrm{T} = 10.9166 \begin{Bmatrix} 14 & 18 \\ 18 & 38 \end{Bmatrix}$$

and the inverse is

$$V_\mathrm{T}^{-1} = \begin{Bmatrix} 0.0167 & -0.0079 \\ -0.0079 & 0.0062 \end{Bmatrix}$$

Thus, for the given data under H_0 where $t' = (9, -3)$ simple computation (Exercise 11.2) gives $Q = 1.835$. Computation of Q using the relevant t' for other permutations indicates that this is among the smaller Q for this example. As already indicated for this small data set significance at a conventionally acceptable level implies a critical region of one point corresponding to the greatest value of Q. Intuitively it is clear that this value will be associated with some t' with nonzero probability near the edges of Table 11.2 Obvious candidates, together with the computed Q (Exercise 11.2), are given in Table 11.3

From Table 11.3 we see that the permutation giving $t' = (-20, -4)$ has the greatest Q-value and the associated p-value had our data given this t' would have been $p = 1/24 = 0.042$. The sample is too small for a reliable asymptotic test, but with two degrees of freedom $\chi^2 = 5.52$ corresponding to an upper tail $p = 0.063$.

Maritz (1995, section 6.2.1) discusses asymptotic confidence regions for the parameters β_1, β_2 but as in many other situations in this book raw data permutation inference tends not to be robust against outliers and certain influential observations of a kind I discuss in section 11.3 so I do not pursue this approach here, nor do I consider its extension to more general multiple regression model with more explanatory variables.

Most modifications to rank-based inference procedures discussed in section 10.2 to 10.5 for bivariate regression extend (in theory at any rate) to two or more regressors, but software developments for exact permutation tests in these situations seem not at the time of writing to be readily available, probably because the methods developed in section 11.3 and 11.4 have proved so fruitful in dealing with data where widely used parametric assumptions break down. I consider only one such test procedure here for the two variable case because it is useful in demonstrating the link between regression and ranked based analyses for location

Table 11.3 Higher values of Q for permutation reference set under H_0

t'		Q
-14	-34	2.92
-17	-32	2.53
-20	-4	5.52
16	32	2.53
18	30	2.46
21	9	4.88

differences between several samples as formalized, for instance, in the Kruskal–Wallis test in Section 7.3.

The procedure I describe parallels that given above for raw data except that the d_i are replaced by their ranks and the statistics T_1, T_2 by

$$T_{S1} = \sum x_{1i} \, \text{rank}(d_i),$$
$$T_{S2} = \sum x_{2i} \, \text{rank}(d_i).$$

The only change in V_T is that s_d^2 is replaced by the corresponding rank variance $s_r^2 = n(n+1)/12$ and in forming Q, $t' = (T_1, T_2)$ is replaced by $t_S' = (T_{S1}, T_{S2})$.

Example 11.2. For the data in Example 11.1 to test the hypothesis $H_0: \beta_1 = 1$, $\beta_2 = 3$, we proceed as in that example except that the d_i are replaced by the corresponding ranks. For the observed data since $d_1 = 3$, $d_2 = 8$, $d_3 = 4$ and $d_4 = 0$ these are $r_1 = \text{rank}(d_1) = 2$, $r_2 = 4$, $r_3 = 3$ and $r_4 = 1$. Proceeding as in Example 11.1 (see Exercise 11.3) we find that for the given data under H_0 the statistics take the value $T_{S1} = 5$ and $T_{S2} = 1$. Table 11.4 corresponds to Table 11.2 for the joint distribution of T_{S1}, T_{S2}.

Clearly this is a more patterned joint distribution than that depicted in Table 11.2 and reflects the uniformity introduced by using ranks. Further, given the x_{ij}, the distribution is clearly no longer conditional upon the observed y_i providing these give rise to the same rankings under H_0. Testing the null hypothesis follows lines analogous to those in Example 11.1 after making the necessary modification to s_d^2 given above.

Pioneers of parametric ANOVA soon realized that that method could be formulated as a multiple regression model. For the one-way ANOVA with three

Table 11.4 Joint distribution of T_{S1}, T_{S2} in Example 11.2. Nonzero entries are $p \times 24$

T_{S2} \ T_{S1}	−8	−7	−5	−4	−3	−1	0	1	3	4	5	7	8
−13	0	0	1	0	0	0	0	0	0	0	0	0	0
−12	0	0	0	0	0	0	1	0	0	0	0	0	0
−11	0	1	0	0	0	0	0	0	0	0	0	0	0
−9	0	0	0	0	0	0	0	1	0	0	0	0	0
−8	0	0	0	2	0	0	0	0	0	0	0	0	0
−7	0	0	0	0	0	0	0	2	0	0	0	0	0
−4	1	0	0	0	0	0	0	0	0	0	0	0	0
−1	0	0	2	0	0	0	0	0	0	0	0	1	0
1	0	1	0	0	0	0	0	0	0	0	2	0	0
4	0	0	0	0	0	0	0	0	0	0	0	0	1
7	0	0	0	0	0	2	0	0	0	0	0	0	0
8	0	0	0	0	0	0	0	0	0	2	0	0	0
9	0	0	0	0	1	0	0	0	0	0	0	0	0
11	0	0	0	0	0	0	0	0	0	0	0	1	0
12	0	0	0	0	0	0	1	0	0	0	0	0	0
13	0	0	0	0	0	0	0	0	0	0	1	0	0

groups or treatments this can be represented as a regression model with two explanatory variables only. If the usual model for the expected yield for treatment group i, $i = 1, 2, 3$, is written $E(Y_i) = \mu + \tau_i$ it is well known that for unique estimation of μ, τ_1, τ_2, τ_3 a constraint must be placed on the parameters. If the number of replicates of the treatments are n_1, n_2, n_3 a commonly used constraint is $\sum_i n_i \tau_i = 0$. It is obvious from the expression for $E(Y_i)$ above that if I denote the yield for the jth unit for treatment i by y_{ij} then $E(y_{ij}) = \mu + \tau_i$ and that this could also be written as a regression model in the form

$$E(y_{ij}) = \mu + \tau_1 x_{1j} + \tau_2 x_{2j} + \tau_3 x_{3j} \qquad (11.4)$$

where $x_{ij} = 1$ for any unit receiving treatment i and $x_{ij} = 0$ otherwise. The constraint $\sum_i n_i \tau_i = 0$ gives $\tau_3 = (-n_1/n_3)\tau_1 + (-n_2/n_3)\tau_2$, reducing our model to one with two explanatory variables with the conventional β_1, β_2 replaced by τ_1, τ_2. Tedious elementary algebra (Exercise 11.4) shows that for this model the distance Q given by (11.3) is identical to the statistic T given in (7.2). If an overall test of location difference using the raw data permutation test is required one may use the method described in section 7.2 which exploits fully the special features of the analysis of variance. It is instructive to apply the methods of this section to a simple example to demonstrate the equivalence.

Example 11.3 In Example 7.1 I applied a raw data permutation test to the following data:

Sample I	0 1 3 4
Sample II	6 8 12
Sample III	10 11 61 65

and although for these data I used there a simpler statistic it is easy to show (Exercise 11.5) that (7.2) gives $T = 4.883$.

Considering (11.4) after making the substitution $\tau_3 = (-n_1/n_3)\tau_1 + (-n_2/n_3)\tau_2$. which in this case gives $\tau_3 = -\tau_1 - 0.75\tau_2$ and writing e_{ij} for the appropriate error terms we get

$$0 = \mu + \tau_1.1 + \tau_2.0 + e_{11}$$
$$1 = \mu + \tau_1.1 + \tau_2.0 + e_{12}$$
$$3 = \mu + \tau_1.1 + \tau_2.0 + e_{13}$$
$$4 = \mu + \tau_1.1 + \tau_2.0 + e_{14}$$
$$6 = \mu + \tau_1.0 + \tau_2.1 + e_{21}$$
$$8 = \mu + \tau_1.0 + \tau_2.1 + e_{22}$$
$$12 = \mu + \tau_1.0 + \tau_2.1 + e_{23}$$
$$10 = \mu + \tau_1.(-1) + \tau_2.(-0.75) + e_{31}$$
$$11 = \mu + \tau_1.(-1) + \tau_2.(-0.75) + e_{32}$$
$$61 = \mu + \tau_1.(-1) + \tau_2.(-0.75) + e_{33}$$
$$65 = \mu + \tau_1.(-1) + \tau_2.(-0.75) + e_{34}$$

It is easily verified that the coefficients of both τ_1 and τ_2 sum to zero. The overall test of the hypothesis that the samples do not differ in location is equivalent to testing $H_0: \tau_1 = \tau_2 = 0$ and this implies that all d_i are simply the corresponding observed y_{ij} and that s_d^2 is then the usual total sum of squares of deviations about the mean in the analysis of variance, i.e. the denominator of T given in (7.2) divided by $N - 1 = 10$. Direct computation gives $s_d^2 = 545.872$. It is easy to verify that $X^T X$ is

$$\begin{Bmatrix} 8 & 3 \\ 3 & 5.25 \end{Bmatrix}$$

Also direct computation gives $T_1 = -139$, $T_2 = -84.25$. The inverse of $X^T X$ is

$$\begin{Bmatrix} 0.159091 & -0.090909 \\ -0.090909 & 0.242424 \end{Bmatrix}$$

and computation using (11.3) gives $Q = 4.883$, identical with the value of T given by (7.2).

In this model where $\tau_1 = \tau_2 = 0$, if we replace the d_i by their ranks this is equivalent to replacing the observations by their ranks leading to the Kruskal-Wallis test and the corresponding Q is now equivalent in the no-tie case to T given by (7.3). See Exercise 11.6.

11.3 Regression diagnostics

Section 11.2 indicated some difficulties that arise with permutation inference in multiple regression even when there are only two explanatory (x) variables. Classic least squares regression is model-based rather than data-based. It usually, but as simple examples have shown not always, perform satisfactorily and prior to the advent of modern computers, it was often difficult to detect model inadequacy. Since the 1960s there have been rapid developments in **regression diagnostics**. Diagnostics are essentially data-based and while many tests require adequate computing facilities, the procedures are similar to those needed to fit a model by least squares and the additional computation for straightforward diagnostic tests is often considerably less than that required for the original fit, because many of the calculations needed for diagnostic tests use by-products of the least squares process, or quantities easily produced by modest extensions.

There are both graphical and analytic approaches to regression diagnostics – the latter based largely on studies of the effect of deleting small amounts of data. Diagnostic techniques have proliferated rapidly and I only outline a few key ideas. Excellent accounts of the theory and practice are given by Atkinson (1985) and by Cook and Wiseberg (1982).

For bivariate regression a scatter plot like that in Figure 10.1 highlights data points that are in some way strange. For multiple regression scatter plots are in essence multidimensional and may be hard to interpret even if one could produce, say, a 5-dimensional graph. Graphical diagnostic methods concentrate on two-dimensional plots of residuals against fitted values or other two-dimensional plots described below.

Well-known inadequacies that may cause difficulty with least squares include:

- **Aberrant observations**. These may arise from incorrect measurement, faulty data recording or copying, or some unusual characteristic of the unit being measured that leads to an outlying value of a measured variable.
- **Model inadequacies.** The chosen model may not be appropriate for describing the systematic structure of the data, or the error distribution may not be normal and homogeneous. Model inadequacies may sometimes be overcome by data transformations or by incorporating additional explanatory variables.

Recording errors may be minimized by careful checking and by using methods that reduce the likelihood of misplacing decimal points or misreading a 5 for an 8 and so on – this latter an ever-lurking danger with hand-written records. In small data sets a visual scan often detects glaring recording errors. The height of an adult male is unlikely to be 1783 cm.

Data that suggest peculiarities in an experimental unit may sometimes be easy, at other times hard, to detect or explain. I was once given a set of data on sizes and chemical content of apples from a number of trees. Fruits from one tree were some 20% larger and had a much higher calcium content than those from other trees. The differences were clear, but only later did I learn that in the previous year a cow had been buried close to the tree giving the atypical fruit!

Different diagnostic tests pick out some peculiarities more easily than others. Some also give a better indication than others of how serious a particular inadequacy is likely to be. Even if strange observations are detected the often difficult and sometimes controversial task may be to decide what to do about them. When there is no obvious reason for an anomaly the choice may be between collecting more data, if that is possible, rejecting an anomalous datum, or trying a different model.

I first outline some graphical and closely related methods. Readers intending to use these or the analytic deletion techniques I introduce later, many of which are included in general statistical software packages, should consult Atkinson (1985) or Cook and Weisberg (1982) for more detail. I discuss the methods using examples mostly for bivariate regression and in many cases the generalization to multiple regression is clear in principle but computation calls for good statistical software rather than a pocket calculator because the necessary matrix operations involve appreciable number crunching.

Two-dimensional plots involving residuals are relevant for linear regression with one or more explanatory (x) variables, but careful interpretation is needed. Contrary to intuition a very small residual may indicate a potential difficulty.

Most standard statistical packages allow many or all of the following plots:

- Scatter plots of response y against each of the p explanatory variable x_k, $k = 1$, $2, \ldots, p$. These detect any close linear relationship between y and a particular x_k, but often they are of limited usefulness except in the bivariate case ($p = 1$).
- A plot of residuals $r_i = y_i - \hat{y}_i$ (where $\hat{y}_i$ is the predicted value using the fitted regression) $i = 1, 2, \ldots n$ against each x_k, $k = 1, 2, \ldots, p$. Care in interpretation is needed because small residuals may imply leverage, a concept discussed later in this section. However, a curvilinear relationship suggests that adding, say, a quadratic term in the particular explanatory variable might be helpful.
- A plot of the r_i against $\hat{y}_i$. If the variance of the r_i appears to increase as $\hat{y}_i$ increases this suggests some transformation of the response variable y, e.g. to logarithms, may be appropriate.
- A plot of the ordered residuals against normal scores, i.e. the expectations of the order statistics of a normal distribution, (called a *normal plot*) gives a near-straight line scatter if the residuals are nearly normally distributed.
- Plots of responses or of residuals against explanatory variables **not** in the model. If either of these plots suggest a relationship it indicates that that variable should be included in the model. In particular, if observations have been collected over time, then time should be regarded as a potential explanatory variable for such plots. Unexpected variables in this category are sometimes called *lurking variables*. See, e.g. Atkinson (1985, section 1.2).

Example 11.4   I illustrate some of these ideas using the data in Example 10.1 and modifications thereof. Data Set I below is that used in Example 10.1. Data Set II is the same except that for the final point $x = 6$ is replaced by $x = 16$. This may seem an unlikely x-value (an outlier in the sense that it causes surprise in the light of the other x values) but if it were a correct value then missing a digit and entering it as $x = 6$ is a common type of mistake. When a sequence of observations have a pattern such as 0, 1, 2, 3, 4, 5 it is easy to record a following correct 16 as 6 because one's thinking has become attuned to the pattern and intuitively expects the value 6. Data Set III differs from Set II only by replacing $y = 4.0$ by $y = 5.0$. Thus the sets are

Set I	x	0	1	2	3	4	5	6
	y	2.5	3.1	3.4	4.0	4.6	5.1	11.1
Set II	x	0	1	2	3	4	5	16
	y	2.5	3.1	3.4	4.0	4.6	5.1	11.1
Set III	x	0	1	2	3	4	5	16
	y	2.5	3.1	3.4	5.0	4.6	5.1	11.1

Figure 11.1 shows scatter diagrams for each set together with the fitted least squares lines. For Set I the fit, as noted in section 10.1, is unsatisfactory, the line being 'pulled' upward by

the outlier (6, 11.1). For Set II the line is a good fit to all points. So good a fit would be rare for most data in the life, medical or social sciences, although perhaps less so in some physical sciences or engineering contexts.   The fact that the x value for the point (16, 11.1) is very different from the remaining x values does not greatly affect the fit here as the point clearly follows the straight-line trend inherent in the other points.   In that sense the point is not influential, but we see below that the fit will be changed to accommodate that point if the x value remains the same but the y value changes.   Such changes are associated with a property known as leverage.   For Set III the fit is good for all points except (3, 5.0) which might in this sense be described as an outlier.   The term outlier here is relevant to the pair (x, y) relative to other pairs.   Neither the values of x nor of y when considered alone are outliers (in the sense of being extreme) with respect to other values of that same variable.

Figure 11.2 gives plots of residuals against x. For Set I the residuals for the first six points show a near-linear decreasing trend then a sudden upward jump for the last residual, a pattern indicating inadequacy of the model. For Set II the residuals are small (note the different scales on the residual axis for Sets I and II) with no appreciable pattern and indicate no model inadequacy. For Set III the residuals are small except for $x = 3$ which is large relative to the others. This is different from the situation for Set I, where although the point (6, 11.1) is well away from the trend suggested by the remaining points, the residual associated with that point is not appreciably larger in magnitude than a couple of other residuals. Yet, as the scatter diagrams with the fitted lines in Figure 11.1 clearly shows, the outlier in Set I is more influential (so far as estimation of β is concerned) than is the outlier in Set III. Outliers in responses y may or may not have a great influence on the estimate of β, but they generally increase the residual sum of squares.

Figure 11.3 shows normal plots for the residuals for each data set. Evidence of non-normality is indicated by departures from a straight line and such a departure is only shown strongly for the very large residual [that associated with the point (3, 5.0)] for Set III, and there is a less strong indication of nonlinearity for the final point in Set I.

For the record the equations of the fitted lines are:

Set I	$y = 1.51 + 1.11x$,
Set II	$y = 2.44 + 0.54x$,
Set III	$y = 2.62 + 0.53x$.

Atkinson (1985, Chapter 1) gives a detailed discussion of diagnostic plots.

The general multivariate linear regression model is often written in matrix notation. If there are p explanatory variables $x_1, x_2, \ldots, x_p$ and a response variable y, the ith data point is written $(y_i, x_{1i}, x_{2i}, \ldots, x_{pi})$ where $i = 1, 2, \ldots, n$. It is customary to write the explanatory variables as an $n \times p$ matrix X, the responses as a column vector y of n-components $y_1, y_2, \ldots, y_n$, the unknown regression coefficients as a p-component vector β. A superscript ^{T} represents a row vector or the transpose of a matrix. A model that includes a constant is incorporated by setting all $x_{1i} = 1$. In this notation the least squares estimates $\hat{β}$ of the parameters β are obtained by choosing $\hat{β}$ to minimize

$$R(β) = (y - Xβ)^T(y - Xβ).$$

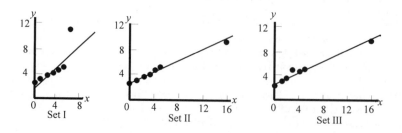

Figure 11.1 Scatter diagrams and fitted lines for data in Example 11.4.

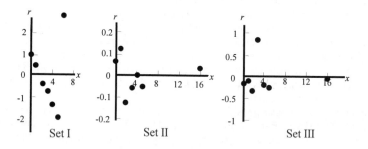

Figure 11.2 Plot of residuals (r) against x for data in Example 11.4. Note the different scales for the residuals in each plot.

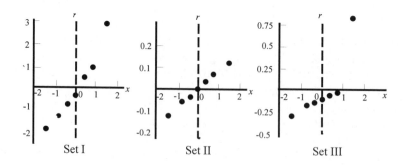

Figure 11.3 Normal plots of residuals (r) for data in Example 11.4 Note the different scales for the residuals for each data set.

Standard textbooks on regression such as Draper and Smith (1981) or Weisberg (1980) discuss the estimation procedure showing in particular that

$$\hat{\beta} = (X^TX)^{-1}X^Ty \qquad (11.5)$$

and that the vector of predicted y values is

$$\hat{y} = X\hat{\beta} = X(X^TX)^{-1}X^Ty. \qquad (11.6)$$

The matrix $H = X(X^TX)^{-1}X^T$ in (11.6) is called the **hat** matrix because in speech y is often referred to as 'y hat', so the $n \times n$ matrix H is the matrix that 'puts the hat on y'. The residual sum of squares is $R(\hat{\beta})$; the n-vector of residuals is $r = y - \hat{y}$.

The hat matrix plays a key role in regression diagnostics. In this elementary treatment I consider it only for bivariate regression where if we write the constant α as β_1 and the slope β as β_2 this comes into line with the notations in (11.5) and (11.6) if, for n data points (x_i, y_i), we set

$$X^T = \begin{pmatrix} 1 & 1 & 1 & . & . & . & 1 \\ x_1 & x_2 & x_3 & . & . & . & x_n \end{pmatrix}$$

In this simple case the hat matrix can be obtained for small n in minutes using a pocket calculator, or in microseconds with good statistical software. For this case the diagonal elements h_{ii}, a notation commonly abbreviated to h_i, which have an important role in diagnostics, take the simple form

$$h_i = 1/n + (x_i - \bar{x})^2/[\sum(x_i - \bar{x})^2] \qquad (11.7)$$

From (11.7) it is clear that the minimum possible value of h_i is $1/n$, a value only attained if some $x_i = \bar{x}$, and also that h_i increases as x moves away from $\bar{x}$.

An example shows some important properties of H.

Example 11.5 For Set I in Example 11.4 (Exercise 11.7) the hat matrix is

$$\begin{matrix}
0.464 & 0.357 & 0.250 & 0.143 & 0.036 & -0.071 & -0.179 \\
0.357 & 0.286 & 0.214 & 0.143 & 0.071 & 0.000 & -0.071 \\
0.250 & 0.214 & 0.179 & 9.143 & 0.107 & 0.071 & 0.036 \\
0.143 & 0.143 & 0.143 & 0.143 & 0.143 & 0.143 & 0.143 \\
0.036 & 0.071 & 0.107 & 0.143 & 0.179 & 0.214 & 0.250 \\
-0.071 & 0.000 & 0.071 & 0.143 & 0.214 & 0.286 & 0.357 \\
-0.179 & -0.071 & 0.036 & 0.143 & 0.250 & 0.357 & 0.464
\end{matrix}$$

The following general properties of a hat matrix H are easily verified for this example

- the matrix is symmetric, i.e. $h_{ij} = h_{ji}$ for all i, j.
- the sum of the elements of any row (or column) is (apart from round-off effects) 1.
- the diagonal elements increase as we move away from the mean $\bar{x}$, which here equals 3 and gives the minimum $h_3 = 1/7 = 0.143$. In a row and column corresponding to $\bar{x}$ all elements of H are equal.

From (11.6) it follows that each $\hat{y}_i$ is a weighted mean of all y_i with weights given by the elements of the ith row of H, e.g., for this example

$$\hat{y}_2 = 0.357y_1 + 0.286y_2 + 0.214y_3 + 0.143y_4 + 0.071y_5 + 0.000y_6 - 0.071y_7$$
$$\hat{y}_4 = 0.143y_1 + 0.143y_2 + 0.143y_3 + 0.143y_4 + 0.143y_5 + 0.143y_6 - 0.143y_7 \quad (11.8)$$
$$\hat{y}_7 = -0.179y_1 - 0.071y_2 + 0.036y_3 + 0.143y_4 + 0.250y_5 + 0.357y_6 - 0.464y_7$$

In practice one obtains the $\hat{y}$ directly using $\hat{y} = X\hat{\beta}$, but the weighted mean approach shows how the pattern of x values influences the predicted y irrespective of whether a point is an outlier. In this sense the hat matrix alone tells us nothing about the *influence* of an outlier. Had the final point been (6, 5.6) instead of (6. 11.1) the point would not have been an outlier, but H would have remained unaltered although our estimates $\hat{y}$ would have been altered by the change in y_7. In passing note that the estimate of $\hat{y}_4$ given above equals, as it should $\bar{y}$, because the least squares line passes through the mean $(\bar{x}, \bar{y})$.

Here is the hat matrix for Set II in Example 11.4:

$$
\begin{array}{rrrrrrr}
0.256 & 0.230 & 0.205 & 0.179 & 0.154 & 0.128 & -0.152 \\
0.230 & 0.211 & 0.191 & 0.171 & 0.151 & 0.132 & -0.086 \\
0.205 & 0.191 & 0.177 & 0.163 & 0.149 & 0.135 & -0.019 \\
0.179 & 0.171 & 0.163 & 0.155 & 0.146 & 0.138 & 0.048 \\
0.154 & 0.151 & 0.149 & 0.146 & 0.144 & 0.141 & 0.114 \\
0.128 & 0.132 & 0.135 & 0.138 & 0.141 & 0.145 & 0.181 \\
-0.152 & -0.086 & -0.019 & 0.048 & 0.114 & 0.181 & 0.914
\end{array}
$$

giving rise to estimates that include

$$\hat{y}_2 = 0.230y_1 + 0.211y_2 + 0.191y_3 + 0.171y_4 + 0.151y_5 + 0.132y_6 - 0.086y_7$$
$$\hat{y}_4 = 0.179y_1 + 0.171y_2 + 0.163y_3 + 0.155y_4 + 0.146y_5 + 0.138y_6 - 0.048y_7 \quad (11.9)$$
$$\hat{y}_7 = -0.152y_1 - 0.086y_2 - 0.019y_3 + 0.048y_4 + 0.114y_5 + 0.181y_6 + 0.914y_7$$

Comparing corresponding lines in (11.8) and (11.9) for $\hat{y}_2$ the main difference is that in (11.9) the weights are more evenly distributed than they are in (11.8) although once again the extreme observation corresponding to x_7 has little weight. Since x_4 for the second data set is no longer equal to $\bar{x}$, the weights for $\hat{y}_4$ are no longer all equal as they were in (11.8) and again the observation corresponding to x_7 has the least weight. For $\hat{y}_7$ the situation has now changed dramatically. Whereas in (11.8) three points had associated weights exceeding 0.2 in magnitude, in (11.9) only the point corresponding to y_7 itself now has a weight exceeding 0.2. Indeed this last point is crucial in determining $\hat{y}_7$. It is clear from Figure 11.1 that this point is not an outlier for data set II in the sense that the point it replaces in data set I was an outlier. However clearly the point is influential in determining $\hat{y}_7$ in the sense that any change in y_7 will greatly change $\hat{y}_7$. Points with this characteristic are said to have **high leverage**.

High values of h_i indicate high leverage. Remembering that the sum of the elements in reach row of H is 1, it follows that for a high leverage point the remaining elements in that row will have small values, or else any others that are not small will tend to cancel out by having opposite signs. For the bivariate case considered here it is clear from (11.7) that high leverage implies a point is situated at an appreciably greater distance from the mean of x than most other points. It is of course possible, especially with larger data sets, to have several points with high leverage.

Since the ith residual is $r_i = y_i - \hat{y}_i$ it follows that the vector r of residuals is given by $r = (I - H)y$ where I is the unit matrix with all diagonal elements unity and all other elements zero. The matrix $I - H$ has the property of idempotency, i.e. $(I - H)(I - H) = I - H$. Please accept this result or consult a book on matrix algebra if you are not familiar with the way it is established. This implies that the covariance matrix of r is

$$\text{var}(r) = \text{var}[(I - H)y] = (I-H)\sigma^2 \tag{11.10}$$

where σ^2 is the variance of Y. This reflects the well-known fact that the residuals are correlated. Further, since (11.10) gives

$$\text{var}(r_i) = \sigma^2(1 - h_i)$$

it follows that in general each r_i has a different variance. In particular for a point of high leverage since h_i is large the residual will have a lower variance than the residuals at points of lesser leverage. The general effect of this is that we get a better fit at points of high leverage. This is true whether or not the point is an outlier in the sense of not conforming to the linear trend indicated by the remaining points. Differences in variance between the r_i often result in peculiarities in normal plots and may sometimes suggest heterogeneity of variance for the y_i when this is not the situation. These difficulties may be overcome by using the standardized residuals given by (11.11) below. These all have the same variance if the least squares model with homogeneous error variance holds. Since we do not know σ^2 we replace it by the usual estimate, i.e. $s^2 = R(\beta)/(n - p)$ and the **standardized residuals** are defined as

$$r_{si} = r_i /[s\sqrt{(1 - h_i)}] \tag{11.11}$$

These standardized residuals, as well as the r_i are given by many statistical software packages in their general least squares linear regression program with p explanatory variables. These residuals are also sometimes referred to as studentized residuals but because that name is used for another kind of residual given below, I follow the terminology recommended by Atkinson (1985).

Example 11.6. For the data sets in Example 11.4 I give in Table 11.5 both residuals and standardized residuals. Despite the fact that the aim of standardization is to render residuals that have something like a t-distribution, in view of the correlations it is not appropriate to submit them to a simple formal test like a t-test. However, high values of a standardized residual, particularly when associated with a relatively low value of an unstandardized residual point to model inadequacy, as do residuals that remain high relative to others after standardization. This

Table 11.5 Residuals, r_i, and standardized residuals, r_{s_i}. for data sets in Example 11.4

Observation no.		1	2	3	4	5	6	7
Set I	r_i	0.993	0.486	–0.321	–0.829	–1.335	–1.942	2.950
	r_{s_i}	0.752	0.318	–0.196	–0.496	–0.817	–1.274	2.234
Set II	r_i	0.060	0.121	–0.118	–0.057	0.003	–0.037	0.029
	r_{s_i}	0.806	1.565	–1.409	–0.723	0.033	–0.458	1.132
Set III	r	–0.119	–0.050	–0.281	0.788	–0.144	–0.175	–0.019
	r_{s_i}	–0.352	–0.144	–0.791	2.186	–0.397	–0.483	–0.163

is the situation for observation 4 in Set III. The high value for the standardized residual for observation 7 in data Set I also suggests further investigation of that point is desirable.

Although the diagonal elements h_i of the hat matrix are used to detect points with high leverage, this is little help in isolating influential outliers. For the bivariate case a scatter diagram like that for Set I in Figure 11.1 may do this satisfactorily. For multivariate data detection of outliers and influential points is harder, for with several explanatory variables the values of each for a particular point may not be unusual or extreme, but their combined influence on least squares regression may be dramatic. Here a technique reminiscent of jackknifing is useful. The idea is, as in the jackknife, to obtain estimates of the parameters omitting one point at a time and looking at the resulting coefficients. If a point is influential, omitting it is likely to have a dramatic effect on the estimates of coefficients, but when less influential points are omitted the effect will be less striking. If there are several high influence points the situation is more complicated. One might perform analyses deleting single points, pairs of points, triplets and so on, but the computation then is involved and time consuming.

Example 11.7 Data Set I in Example 11.4 houses the outlier (6, 11.1) and clearly its presence increases the slope estimator, β, upward to accommodate that point. The one-point deletion least squares fits for α and β obtained by omitting each point in turn are easily computed by any standard regression program and using the Jackknife notation in section 2.5 may be written $\alpha_{(i)}$, $\beta_{(i)}$ when the ith data point is omitted. For this example the relevant values are

Point deleted	Constants	
i	$\alpha_{(i)}$	$\beta_{(i)}$
1	0.647	1.3057
2	1.264	1.1557
3	1.605	1.0932
4	1.645	1.1071
5	1.565	1.1652
6	1.313	1.3014
7	2.490	0.5171

Clearly the most, indeed the only, dramatic effect on the estimation of slope comes from omitting the last observation.

Atkinson (1985, section 3.1) pursues this approach via the concept of deletion residuals. I outline the method here but please refer to Atkinson for more detail. An obvious question of interest is how good is the equation that omits the point corresponding to y_i at predicting that value? Denoting the prediction for y_i obtained when the ith observation is deleted by $\hat{y}_{(i)}$, I define the corresponding deletion residual by $r_i^\dagger = y_i - \hat{y}_{(i)}$. Since the data point involving y_i has been deleted it follows that if we estimate the variance of $r_i^\dagger$ using a deletion estimate $s_{(i)}^2$ for σ^2 then if there is agreement between predicted and observed values under the least squares model that $r_i^* = r_i^\dagger/(\sqrt{[\text{estimated var}(r_i^\dagger)]})$ will have a t-distribution with $n - p - 1$ degrees of freedom. I omit the derivation, but reasonably straightforward matrix algebra leads to

$$r_i^* = \frac{r_i}{s_{(i)}\sqrt{(1-h_i)}} \tag{11.12}$$

where r_i and h_i are computed for the complete data. Thus (11.12) is the same as (11.11) except that s is replaced by $s_{(i)}$. Several names have been proposed for this statistic including *RSTUDENT*, *TRESIDUAL*, *externally studentized residual*, *cross-validatory* or *jackknife residual* and *deletion residual*. All have merit, but I follow Atkinson and use the term deletion residual. Many standard statistical packages including MINITAB compute this statistic routinely if requested.

Example 11.8 For data Set I in Example 11.4 we saw from the deletion fits in Example 11.7 that the observation (6, 11.1) appeared to be influential. One might therefore expect the corresponding deletion residual to give a significant *t*-value. MINITAB gives each deletion residual; in order these are

0.714   0.288   −0.177   −0.455   −0.785   −1.387   48.005

clearly only the last of these gives a significant t and the value $r_7^* = 48.005$ with 4 degrees of freedom leaves no doubt about the influence of that observation.

While a high value of a deletion residual pinpoints an aberrant observation these residuals do not give direct information about changes in parameter estimates due to deletion. i.e., they do not give direct information on the differences $\hat{\beta}_{(i)} - \hat{\beta}$. Several measures that do this have been proposed and details of their rationale and derivation are given by Atkinson (1985, section 3.2). One in common use is the Cook statistic proposed by Cook (1977) for which the rationale becomes clear if it is written in the form

$$D_i = (\hat{\beta}_{(i)} - \hat{\beta})^T X^T X (\hat{\beta}_{(i)} - \hat{\beta})/ps^2.$$

Clearly large values of D_i indicate that the ith observation is influential. For computational purposes an equivalent more useful form of the Cook statistic is an expression in terms of the standardised residuals and diagonal elements of the hat matrix for the full data set, namely

$$D_i = \frac{r_{s_i}^2 h_i}{p(1 - h_i)}$$ (11.13)

An alternative statistic proposed by Belsey, Kuh and Welsch (1980) is equivalent to the square root of D_i except that $r_{s_i}^2/p$ is replaced by the squared deletion residual $(r_i^*)^2$. These latter statistics are sometimes called *DFITS* or *DDFITS* and are widely available in regression software. High values draw attention to influential observations and more formal tests based on these and related statistics are available, (see, e.g. Atkinson, 1985, Chapter 4). As a rule of thumb the MINITAB manual suggests that *DFITS* greater than $2\sqrt{(p/n)}$ are worthy of attention.

Example 11.9. For the data set considered in Example 11.8 Minitab give the following values for the Cook statistics and the *DFITS*.

Cook statistic	0.245	0.020	0.004	0.021	0.073	0.325	2.163
DFITS	0.665	0.182	–0.082	–0.186	–0.366	–0.877	44.690

Not surprisingly, the highest values of both statistics occur for the final observation. Since $2\sqrt{(p/n)} = 1.069$ this last *DFITS* value is the only one calling for attention.

Here is an example involving three explanatory variables.

Example 11.10 Fit a least squares linear regression to the following data and explore the fit for evidence of high leverage or influential data points.

x_1	3	7	5	1	6	4	5	8	5	2	7	4
x_2	5	2	8	32	11	9	30	12	5	7	4	8
x_3	4	3	7	34	5	10	3	12	12	2	1	8
y	13.6	13.8	21.6	35.6	38.5	17.8	97.0	26.0	23.4	26.0	21.8	20.0

The least squares regression line is

$$y = 7.09 + 1.28x_1 + 2.82x_2 - 1.76x_3.$$

For a complete diagnostic analysis for this fit I strongly urge you to use the facilities (including plotting facilities) in any available major statistical package. I summarize here only some of the key features recovered from the diagnostics generated by MINITAB.

● The only standardized residual of magnitude greater than 2 was 2.749 for observation number 9. (the corresponding raw residual was 16.926, the magnitude of all other such residuals being less than 5).

- The deletion residual for observation number 9 was 10.909 and that for observation number 7 was 2.120.
- Observation 4 and 7 showed high leverage, $h_4 = 0.880$ and $h_7 = 0.878$.
- Highest Cook statistic values were 2.997 and 5.619 for observations 4 and 7 and *DFITS* of greatest magnitude were −3.630, 5.683 and 5.590 for observations 4, 7, 9 respectively.
- The residual sum of squares was 383.0

The above values were all given by output without having to compute the actual fits with deletions. The suggestion from these statistics is that observation 9 might be an outlier and that observations 4 and 7 have high leverage.

Visual inspection of the data does not suggest any obvious peculiarity about observation 9. Observations 4 and 7 on the other hand have high values in at least one explanatory variable and this may in itself indicate potential high leverage although it must be emphasized that it is not always easy to pick out high leverage points by visual inspection. The indication that observation 9 has high influence and therefore might well be an outlier suggests further examination of a fit with this point omitted. The resulting least squares fit is

$$y = 5.73 + 1.08x_1 + 3.08x_2 - 2.04x_3.$$

The new diagnostic statistics for this fit are generally more stable as a result of omitting observation 9. The standardized and deletion residuals give no cause for alarm. There is a dramatic reduction in the residual sum of squares from 383.0 to 21.3. Not surprisingly observations 4 and 7 remain high leverage points ($h_4 = 0.900$ and $h_7 = 0.938$). The Cook and *DFIT* values for observation 7 are respectively 3.226 and −3.548. These are a little high for comfort and make it tempting to see what happens if this observation is also omitted. However, caution is needed. The degrees of freedom for estimating error would be reduced to 6 and some loss of precision in inferences based on standard least squares statistics will result. With larger data sets one may feel happier with further deletions of this kind. However, bearing the difficulties of data reduction in mind, it is worth seeing what happens if observations 7 and 9 are both omitted.

The fit now is

$$y = 4.87 + 1.14x_1 + 3.32x_2 - 2.24x_3.$$

Residuals are now all satisfactory and not unexpectedly the only remaining high leverage point is observation 4 which also has a high, but hardly alarmingly high, *DFIT* value. The residual sum of squares is reduced only to 18.7 and because of loss of degrees of freedom there is indeed a small decrease in precision.

The data in this example were artificially generated by choosing the explanatory variables so as to ensure that at least observations 4 and 7 had high leverage because they represented points in the 3 dimensional x-space remote from the ellipsoid of concentration of the other x values. The true y values were generated from the model $y = 7 + x_1 + 3x_2 - 2x_3$ and the observed values of y given in the example were formed by adding errors generated as a random sample from a $N(0, 4)$ distribution, **except** for observation 9 for which the gross error 20.4 was incorporated.

In the light of this explanation of how the data were generated one may feel satisfied with the performance of the diagnostic process. Apart from the constant term, the best fit to the 'true' model is obtained by omitting observation number 9. The fit to the known true model is less good if the influential observation 7 is also omitted. This should come as no surprise for omission of a high-leverage observation, unless it is suspect on other grounds generally reduces precision of estimates. Poor estimation of the constant should come as no surprise because it is effectively an estimate of $E(Y)$ at a point where all x_i are zero. This is well away from any observed x_i triplets and hence will have a high standard error. Indeed its standard error for the fit omitting points 9 is 1.852 implying that the true value of 7 is well within the relevant 95% least squares confidence interval.

   Atkinson (1985) and Cook and Weisberg (1982) include many diagnostic studies using real data. One set discussed by Atkinson is a classic data set widely used in the regression literature that was first given by Brownlee (1965) and is commonly referred to as Brownlee's stack loss data (Exercise 11.8).

11.4 *M*-estimators in regression

In principle *m*-estimators like those described briefly in section 3.5 extend to linear regression problems including multiple regression. A definitive early expository paper is that by Andrews (1974) and this and a number of later papers have drawn attention to potential difficulties. The procedure is an iterative one and it is important to use robust starting estimates of the regression coefficients and to select the function to be minimized with care. Failure to do so may result in what is basically a least squares solution when that is not appropriate. I give here only a trivial example after a few introductory remarks and this is only indicative of how the concept is applied rather than a demonstration of its practical use. The topic is discussed briefly by Atkinson (1985, section 12.1) and is mentioned in various places by Barnett and Lewis (1995). Both give a number of references to papers dealing with specific theoretical developments and practical applications. Important references, some to more general aspects of robust estimation, include Huber (1981, 1983), Carroll (1980, 1982) and Schrader and Hettmansperger (1980).

   The illustration here is based on the Huber-type *m*-estimator for a mean like that considered in section 3.5, only looked at from a slightly different angle. The key to the approach is to regard the differences from the mean, i.e. $r_i = x_i - \mu$ as residuals which I now replace by regression residuals which, in the simple bivariate case, are $r_i = y_i - a - bx_i$ where a, b are, at the various stages, some current estimates of parameters α, β. I also use a modification mentioned in section 3.5 replacing r_i by r_i/s, where s is some robust measure of spread. A measure suggested by Andrews (1974) is the median of the $|r_i|$ computed with some appropriate initial a, b. In section 3.5 I indicated that in the situation there (where a different estimator of s was proposed) a choice of $k = 1.5$ was usually satisfactory. That choice is often also satisfactory in a regression context. The estimating equation analogous to (3.10) is now $\sum \psi(r_i/s) = 0$ and for practical purposes we use an analogue of (3.12) where now $w_i = \psi(r_i/s)/(r_i/s)$ and differentiating with respect to the parameters leads (Exercise 11.9) to the estimating equations appropriate to weighted least squares with solutions of the form

$$\alpha^* = (\sum w_i y_i)/(\sum w_i) - \beta^*(\sum w_i x_i)/(\sum w_i) \tag{11.14}$$

$$\beta^* = [\sum w_i x_i y_i - (\sum w_i x_i)(\sum w_i y_i)/\sum w_i]/[\sum w_i x_i^2 - (\sum w_i x_i)^2/\sum w_i)]. \tag{11.15}$$

The value of s is determined using the residuals from initial robust estimates a_1, b_1 for the coefficients. This value is not changed between iterations. The weights for the first iteration are computed using the residuals based on the same a_1, b_1 and new estimates $a_2 = \alpha^*$, $b_2 = \beta^*$ are computed from (11.14) and (11.15). The cycle is repeated until convergence.

I indicated at the start of this section that Andrews pointed out that if there are suspect observations least squares estimators of α, β may not be satisfactory starters. In the bivariate case one could use a robust estimator such as the Theil–Kendall estimator described in section 10.4 but had one computed that estimator there may be little point, except perhaps for confirmatory purposes, in also computing an *m*-estimator. Indeed, in practice *m*-estimators are more likely to be used with several explanatory variables because the robust methods based on ranks do not extend readily to more than two explanatory variables. Andrews (1974) suggests one approach to getting suitable preliminary estimates of coefficients in these circumstances. but for illustrative purposes here I use a simple robust procedure for bivariate regression due to Bartlett (1949) who introduced it for a rather different problem (that in which the explanatory variables are measured with error).

Example 11.11. I use again the now familiar Set I from Example 11.4 with the influential observation (6, 11.1). Bartlett's method for estimating β is to divide the data into two groups (ideally of equal size) on the basis of the ordered x values. For an odd number of observations that at the median x value is omitted. For each group the medians of x and y in that group are computed. Using suffixes h, l for the higher and lower groups and writing the relevant medians as m_{xh}, m_{yh}, m_{xl}, m_{yl}, the estimate of β is $b = (m_{yh} - m_{yl})/(m_{xh} - m_{xl})$. An appropriate estimate of α is the median a of all $a_i = y_i - bx_i$. For data Set I you should verify (Exercise 11.10) that $b = 0.5$ and $a = 2.5$. These are close to the various robust estimators obtained in this and the previous chapter by a variety of methods. The residuals, computed as $r_i = y_i - 2.5 - 0.5x_i$ are easily found to be 0, 0.1, –0.1, 0, 0.1, 0.1,5.6. Clearly med$|r_i| = 0.1$. Setting $s = 0.1$ and $k = 1.5$ and remembering that $\psi(r_i/s) = 1.5$ if $r_i/s > 1.5$ it follows that $w_i = 1$, $1 \le i \le 6$. When $i = 7$ we get $r_7/s = 5.6/0.1 = 56$, thus $w_7 = 1.5/56 = 0.0268$, whence (11.14) and (11.15) (see Exercise 11.10) give $\alpha^* = 2.442$ and $\beta^* = 0.546$. Using these new estimates of the coefficients it is easily verified that the residuals are such that once again all $w_i = 1$, expect for w_7. The new value of $r_7 = 5.382$ whence $w_{7(1)} = 1.5/53.82 = 0.0279$.   Using these estimators the new estimates of intercept and slope are $\alpha* = 2.441$ and $\beta* = 0.547$.   It is easily verified that the weights calculated with these estimates are unchanged, so this is the accepted fit.

The above computations may be carried out with any software that fits a weighted linear regression after one forms robust initial estimates a, b and one has also to compute new r_i at each step to determine any changes in the w_i.   I emphasize that the above is a trivial example to show how the method works.   In practice there are a range of alternative m-estimators that use different ways to downweigh the influence of tail or other outlying values.   Several are suggested and their properties discussed in Andrews *et al.* (1972), Huber (1977) and elsewhere.

11.5 Penalty functions in regression

A modification of least squares gives a distribution-free way to fit a smooth curve to indicate trends.   Figure 11.4 is a plot of the height in metres achieved by winners of the pole vault in the Olympic games between 1896 and 1976. The data are from Hand *et al* (1994, Set 300).   The games were not held in 1916, 1940, 1944 for obvious reasons. Clearly achievement has increased over the period but a least squares straight line fit to the data would not be an adequate summary of trend either for descriptive or predictive purposes, for it fails to indicate that while there was rapid improvement in achievement between 1896 and 1912 and between 1960 and 1972, the improvement is less marked between 1912 and 1960.

The classic solution to improving matching in these circumstances is to fit a polynomial. A quadratic would not be satisfactory because it has the property that the second derivative is constant and geometrically it represents a parabola which has no point of inflection, i.e. at no point does the curve cross its tangent at that point and this is clearly needed for a smooth curve that follows the general trend in Figure 11.4. A cubic curve allows one point of inflection, since its second derivative is a linear function of x and the point of inflection is found by setting $x = 0$ in this second derivative.

Figure 11.5 shows the least squares cubic fitted to these data. It fits well in the early years, but slightly less well in later years and clearly prediction by extrapolation would not be meaningful. We get better fits with polynomials of higher degree and indeed it is well-known that a polynomial of degree $n - 1$ may be found that passes through any set of n points. Especially when there are outliers present the shape of a polynomial, even at points remote from an outlier may be strongly influenced by the existence of the outlier. We met a trivial example of this with a straight line fit to data Set I in Example 11.4. A further

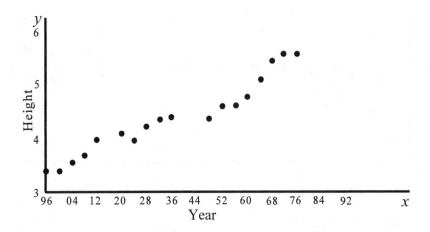

Figure 11.4 Heights achieved by winners of Olympic pole vault, 1896–1976.

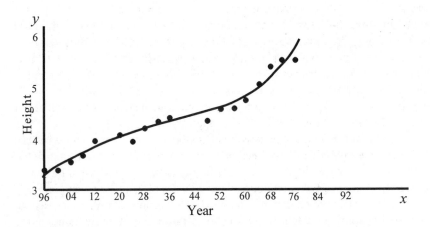

Figure 11.5 Cubic fit to heights achieved by winners of Olympic pole vault, 1896–1976.

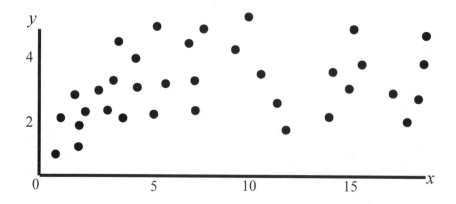

Figure 11.6   A noisy data set.

criticism of polynomials for smoothing is that altering the degree of a polynomial even by 1 may lead to dramatic changes in the curve of best fit. For the data in Figure 11.4 for example, while there is little change between the straight line of best fit and the best quadratic fit, the cubic of best fit is quite different from the best quadratic.

Indeed, if a smooth trend curve is required for data like that in Figure 11.4 we may get a more satisfactory fit than a polynomial by tracing a curve with judicious use of drawing aids like a set of French curves. The resulting curve is unlikely to be a polynomial or any curve that is easily specified analytically. Although most skilled draughtsman would probably draw much the same curve for the data in Figure 11.4 if we had a noisier data set (i.e. one for which the trend is less obvious, as in Figure 11.6) or a less skilled draughtsman, agreement about what is a satisfactory fit may be harder to attain.

In recent years computer-intensive mathematical procedures using what are called **roughness penalties** have been developed that enable us to formalize the drawing-board approach without the constraint of adhering to polynomial or any other family of curves, while retaining many good features of least squares.

It is relatively easy to explain the rationale of the proposal, rather less easy to recognise a simplification that enables us to carry out the fitting without having to specify the precise form of the function we are fitting (thus making the method both nonparametric and distribution free). Once this is done one needs only appropriate software to implement the procedure. More sophisticated uses of the approach include developing objective methods to compare trend curves for several data sets.

I confine discussion to a brief outline of the rationale. A full treatment of the theory and practical implementation of the technique is given by Green and Silverman (1993).

Writing the model

$$y = f(x) + \text{error}$$

without specifying $f(x)$ is a generalization from polynomial regression that allows the systematic part of the model to be any function. Generally for curve fitting we apply the constraint that the function is to be smooth. With a further restriction to homogeneity of error, least squares becomes a sensible method of fitting. If I choose a family of curves with sufficient flexibility (in general this means with a large number of parameters) I can, using least squares, fit a model that passes through all of a finite number of data points. For some models such fits might be bizarre in the sense that although the model fits the data perfectly in the sense that every data point lies on the fitted curve one would not be happy to use the fit for interpolation between data points. Figure 11.7 illustrates a situation of this sort for the data in Figure 11.4. I do not really believe that if Olympic Games had been held in the early 1940s that somebody would have won the event a vault of about 4.7 metres!

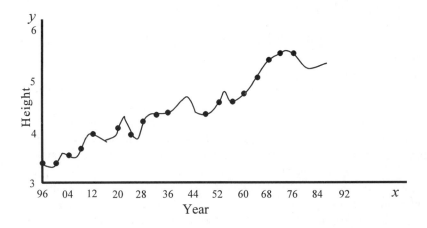

Figure 11.7 Heights achieved by winners of Olympic pole vault, 1896–1976 and an arbitrary curve through all points.

Applying least squares to arbitrary functions when we have no clear idea what arbitrary functions are appropriate does not solve the smoothing problem!

The key idea behind curve fitting with a smoothness penalty, sometimes called **penalized least-squares regression** is that a term is added to the sum of squares, $\sum_i[(y_i - f(x_i)]^2$ that is minimized in least squares. Least squares effectively leads to choice of some specific minimizing function which may be denoted by f. The aim of the roughness penalty is to reduce the effect of roughness by introducing a term that will influence the function we are trying to minimize if the fitted curve is too rough. For a smooth curve a measure of such roughness in a broad sense is the amount of change in the second derivative. If our curve is to be fitted over the interval (a, b), where typically $a = x_{(1)}$ and $b = x_{(n)}$, then one measure of roughness of this curve is the integrated value of the square of the second derivative values over that range. This is intuitively obvious for polynomials because for a straight line the second derivative is zero so the roughness is always zero. There is no smoother curve than a straight line! For a quadratic the second derivative is a constant. For higher order polynomials the second derivative is a lower order polynomial and thus the proposed roughness measure tends to increase as the degree of the polynomial increases.

In a more general context improved fit tends to induce more roughness so the idea of a roughness penalty is that there is a pay-off between the fit at individual points and the smoothness of the fitted curve overall. The function that is minimized is of the form

$$U(f) = \sum_i[(y_i - f(x_i)]^2 + \alpha \int_a^b [f''(x)] \tag{11.16}$$

where α is a positive constant and f'' represents the second derivative of f. I denote by f^* the function that minimizes $U(f)$. The value of α has to be chosen and clearly the greater the value of α the greater the influence of the roughness penalty on the solution. When $\alpha = 0$ (11.16) reduces to least squares. As $\alpha \to \infty$, for minimization of $U(f)$ the squared integral of the second derivative is forced to zero, which implies the fitted curve does not change in slope and thus will approach linear regression for any f. Green and Silverman (1994, p.5) aptly describe α as representing 'the rate of exchange between residual error and local variation and it gives the amount in terms of summed squared residual error that corresponds to one unit of integrated squared second derivative'.

A remarkable property of the method is that once α is chosen we do not need to specify the family f to estimate f^*. This is because the latter can always be expressed in terms of a special sort of piecewise polynomials called cubic splines. A description of these or the fitting procedure for (11.16) is beyond the scope of this book, but cubic splines are widely used in numerical analysis.

Green and Silverman (1994, Chapters 2, 3) explain how cubic splines are fitted to data, giving algorithms based on subtle but relatively straightforward matrix operations easily carried out with appropriate software. They also consider in detail such matters as a suitable choice of α for particular problems.

A different approach to smooth curve-fitting is described by Efron and Tibshirani (1993, section 7.3), using what they call the **loess** method.

11.6 A case study

In events like figure skating or diving competitions a panel of judges customarily score independently each competitor's performance either for each of a number of attributes (e.g. artistic merit, technical skill) or for execution of each of several set routines. Scores are typically on a scale 0.0 to 10.0; e.g. in a diving competition a judge may award a competitor scores of 9.3, 8.6 and 9.1 for each of 3 types of dive. In general, scores of this kind are awarded by each of n judges to each of k competitors. Conventionally individual judges scores are combined in some way for each competitor to produce an ordering to reflect the relative merit of competitors' performances. Ideally, the method of combination should allow for the following factors:

- (i) although attributes are scored on the same scale they may not be of equal importance (e.g. each of 3 diving routines are unlikely to be equally difficult).
- (ii) even skilled judges will not agree in their assessment of individuals for any attribute or in any routine. Some may use a wider or narrower range of scores than others, or may consistently award lower scores than their colleagues and yet reflect similar overall orderings.
- (iii) some judges may show inconsistency over time. A common complaint is that judges tend not to give high scores to early performers on the grounds that to do so leaves little scope to reward adequately a later superior performance, but that judges become more relaxed when only a few competitors remain to be assessed. Other factors that may lead to inconsistency include, for example, a run of poor competitors followed by a good competitor when there may be a subconscious tendency to compare the latter with recent poor performers rather than with earlier good competitors. Because all judges assess the competitors in the order in which they perform, if a majority, or even a substantial minority, of the judges show similar inconsistencies this may affect the rankings. Competitors order of performance is usually randomly determined but this does not overcome this problem. It could be overcome, at least partially, if competitors were presented to each judge in a different random order, but this is impractical if each competitor performs only once.

The problem in (i) is overcome by taking a weighted mean of the separate attribute scores, the weights representing an agreed assessment of the relative difficulty or importance of each. If this is done for each of, say, 5 judges it is not usually satisfactory to simply add these 5 weighted means and rank the resultant sum because of differences between the distributions of scores for individual judges mentioned in (ii). To overcome this a trimmed mean is sometimes recommended where, if there are 5 judges then the highest and lowest scores among the 5 awarded to each competitor is trimmed out. This has two drawbacks. The first is that because it is unlikely that the scores trimmed out will be those of the same two judges for every competitor there are in effect different subsets of 3 judges determining individual scores. The second objection is that a low or high score that is trimmed out may be awarded by a judge who is actually highly skilled in assessing the relative abilities of competitors but is simply using a different part of the scoring scale than some of the other judges who may not necessarily be discriminating so clearly or accurately.

One may remove such distributional differences by converting each judges scores to ranks 1 to k and adding judges' ranks for each competitor but this tends to downweigh large differences between scores awarded and to increase small differences and it.does not cope with inconsistencies of the type outlined in (iii). It is hard to deal with such inconsistencies although there are tests that will detect them. If inconsistencies are confined to, say, just one of 4 or 5 judges it might be wise to disregard that judge's scores; a riskier, and in practice difficult, alternative is to try and correct for inconsistencies if these show any clear pattern.

Assessing properties of scoring systems both for differences between judges and for inconsistencies within individual judges assessments calls for use of many techniques introduced in this book together with intelligent application both of exploratory data analysis and of standard model-based analyses when appropriate.

Example 11.12 Fligner and Verducci (1988) and Hand *et al* (1994, Set 442) give the data in Table 11.6 based on assessments by each of 5 judges for 40 competitors in synchronised swimming events at the 1986 National Olympic Festival at Houston, Texas. Each of two routines was scored on a scale 0.0 to 10.0 and if the awards for a competitor in the routines are R_1, R_2 what is recorded for that competitor is a weighted score $X = 1.7R_1 + 2.2R_2$ where the weights 1.7 and 2.2 reflect the relative difficulty of the routines.

Fligner and Verducci were interested in testing whether there was any overall tendency for inconsistency that took the form of scoring late competitors more generously. These authors rightly point out that tests such as the Jonckheere–Terpstra or Page test are inappropriate for detecting such ordered effects because, despite the superficial resemblance to a randomized block experiment with judges as blocks and competitors as treatments there is no randomization within blocks. Effectively this means that one must take account of strong correlations between judges scores for individuals in any measure of a possible effect. The reader should refer to their paper for details but they find no evidence of any competitor-order dependent effect using

Table 11.6 Judges scores in a swimming contest, Fligner and Verducci (1988)

Competitor	Judge					Competitor	Judge				
	1	2	3	4	5		1	2	3	4	5
1	33.1	32.0	31.2	31.2	31.4	21	27.3	28.1	28.4	27.5	26.4
2	26.2	29.2	28.4	27.3	25.3	22	29.5	28.1	27.3	28.4	26.4
3	31.2	30.1	30.1	31.2	29.2	23	28.4	29.5	28.4	28.6	27.5
4	27.0	27.9	27.3	24.7	28.1	24	31.2	29.5	29.2	31.2	27.3
5	28.4	25.3	25.6	26.7	26.2	25	30.1	31.2	28.1	31.2	29.2
6	28.1	28.1	28.1	32.0	28.4	26	31.2	31.2	31.2	31.2	30.3
7	27.0	28.1	28.1	28.1	27.0	27	26.2	28.1	26.2	25.9	26.2
8	25.1	27.3	26.2	27.5	27.3	28	27.3	27.3	27.0	28.1	28.1
9	31.2	29.2	31.2	32.0	30.1	29	29.2	26.4	27.3	27.3	27.3
10	30.1	30.1	28.1	28.6	30.1	30	29.5	27.3	29.2	28.4	28.1
11	29.0	28.1	29.2	29.0	27.0	31	28.1	27.3	29.2	28.1	29.2
12	27.0	27.0	27.3	26.4	25.3	32	31.2	31.2	31.2	31.2	28.4
13	31.2	33.1	31.2	30.3	29.2	33	28.1	27.3	27.3	28.4	28.4
14	32.3	31.2	32.3	31.2	31.2	34	24.0	28.1	26.4	25.1	25.3
15	29.5	28.4	30.3	30.3	28.4	35	27.0	29.0	27.3	26.4	28.1
16	29.2	29.2	29.2	30.9	28.1	36	27.5	27.5	24.5	25.6	25.3
17	32.3	31.2	29.2	29.5	31.2	37	27.3	29.5	26.2	27.5	28.1
18	27.3	30.1	29.2	29.2	29.2	38	31.2	30.1	27.3	30.1	29.2
19	26.4	27.3	27.3	28.1	26.4	39	27.0	27.5	27.3	27.0	27.3
20	27.3	26.7	26.4	26.4	26.4	40	31.2	29.5	30.1	28.4	28.4

a test that allows for these correlations, although, perhaps rather surprisingly, an incorrect application of the Jonkheere–Terpstra test suggests some such effect.

Even if such an effect is indicated, because the order of appearance is the same for all judges there is no way of telling whether the effect is due to judges' bias or to the particular randomization putting a preponderance of good performers late in the competition. The latter possibility is effectively confounded with one of scoring bias. The Fligner–Verducci test may be modified to test for other systematic biases such as a tendency for the judges to be more generous in their scoring of early competitors or other ordered patterns, but what seems a more likely problem is that these or other oddities may show up for perhaps only one or two judges. Table 11.6 lends itself to a variety of largely data-driven tests for such oddities.

If the judges are competent and consistent and there are real differences in competitors' performances one expects high values of Kendall's coefficient of concordance, W. Although formally equivalent to the statistic used in the Friedman test use of W is not invalidated by failure to randomize within blocks (time order of making assessments) because we use W only as a measure of association. Our only worry is if W is near zero. This could imply either (i) that some at least of the judges are not very competent, or (ii) there is little difference between the achievement of competitors. For these data $W = 0.7626$ indicating a good measure of agreement. This statistic does not tell us anything about possible biases relating to order in which the competitors perform, for it does not take that ordering into account.

For a preliminary comparison of judges scoring I look at 5-number summaries of the distribution of the 40 scores allotted by each judge and also at the mean and standard deviation of the scores. The 5-number summary is the minimum, 1st quartile, median, 3rd quartile and maximum in that order. These are:

Judge 1	24.0	27.075	28.40	31.20	33.1
Judge 2	25.3	27.350	28.25	30.10	33.1
Judge 3	24.5	27.300	28.10	29.20	32.3
Judge 4	24.7	27.300	28.40	30.75	32.0
Judge 5	25.3	26.550	28.10	29.20	31.4

One normally expects some diversity in the extreme values even between 'samples' from the same distribution so a more interesting feature here is the low values of the 3rd quartiles for Judges 3 and 5 and of the 1st quartile for Judge 5. The differences implicit in these summaries are often more easily digested in adjacent box and whisker plots. The means and standard deviations of scores for the judges also indicate differences in distribution patterns, particularly in dispersion as measured by standard deviations. These statistics are:

	Judge 1	*Judge 2*	*Judge 3*	*Judge 4*	*Judge 5*
Mean	28.785	28.833	28.375	28.655	28.000
Standard dev.	2.156	1.707	1.805	2.010	1.647

Judge 5 scores give the lowest mean and the smallest standard deviation, suggesting that judge tends to give lower than average scores and that these are more concentrated about the mean, although we have not established from these summaries that any differences are statistically significant. One might use parametric tests based on the normal distribution to explore the differences as these should be reasonably robust for samples of 40 that do not show clear indications of nonnormality. If in any doubt nonparametric tests might be preferred. For illustrative purposes I have used the Wilcoxon signed rank test for differences in location and the Conover test for differences in dispersion applied to the 10 different pairings of judges to indicate differences that are likely to be real. Because such pairwise tests are clearly not independent, p-values should not be interpreted literally as indicators of significance. However a high p-value is a clear indicator of a lack of evidence of population score differences between judges, whereas if all or nearly all comparisons between a particular judge and all fellow judges are low this is indicative of some difference in that judge's scoring methods. A more sophisticated approach would then be to look at specific comparisons in, say, an overall Friedman test, although here again one must have reservations about validity because there is no randomization within blocks (judges). Table 11.7 gives exact p-values for the pairwise Wilcoxon signed ranks above the diagonal and asymptotic p-values for the Conover squared-rank test below the diagonal. For such large samples there is little difference between exact and asymptotic levels. It is clear from the entries above the diagonal and the sample means that Judge 5 scores show a downward shift in location compared to those for the other judges, with a less strong indication of a slight downward shift for judge 3 relative to judges 1, 2 and 4. The Conover test suggests lower dispersion for judges 2 and 5 relative to judge 1 with less strong suggestions that dispersion for judge 3 is less than that for judge 4 and that for judge 5 is lower than that for judge 4. In summary, there is strong evidence of some location and dispersion differences in judges scores which strengthens a case for transformation to ranks before combining scores. Some thought is needed about the validity of what I have just done because the Wilcoxon signed rank test is a paired sample test whereas the Conover test introduced in section 6.5 applies to independent samples. In applying the latter to these data we are taking

Table 11.7 One-tail *p*-values for the Wilcoxon signed-rank test (above diagonal) and for the Conover test for dispersion (below diagonal) for all pairs of judges.

	Judge 1	Judge 2	Judge 3	Judge 4	Judge 5
Judge 1		0.480	0.065	0.233	0.001
Judge 2	0.021		0.031	0.243	<0.001
Judge 3	0.108	0.333		0.107	<0.001
Judge 4	0.405	0.066	0.163		<0.001
Judge 5	0.032	0.385	0.270	0.078	

into account the fact that the judges make their assessments independently but we cannot ignore the fact that they are making them on the same units. Thus any inferences we make about differences in dispersion apply only to the judges assessments of these particular units. The same considerations apply to parametric tests of dispersion differences for such data. For example, under an assumption of normality one might use Bartlett's test (described in most standard statistical textbooks) as an over-all test for variance homogeneity over all 5 judges.

In section 9.7 I pointed out that the coefficient of concordance is a linear function of the mean of the pairwise Spearman coefficients r_s in the no-tie case. Modification is needed with ties but a close relationship still holds. It is therefore of interest to examine the pairwise Spearman coefficients between judges. If those between one judge and most of his/her fellows is well below the average value of the remaining coefficients it implies that that judge is producing rankings out of line with those of colleagues. For the data in Table 11.6 the Spearman rank correlation matrix is:

	Judge 1	Judge 2	Judge 3	Judge 4	Judge 5
Judge 1	1.000	0.636	0.707	0.785	0.725
Judge 2	0.636	1.000	0.656	0.676	0.670
Judge 3	0.707	0.656	1.000	0.775	0.666
Judge 4	0.785	0.676	0.775	1.000	0.737
Judge 5	0.725	0.670	0.666	0.737	1.000

This matrix indicates that the correlations in ranks between judge 2 and the remaining judges are slightly lower than other correlations between judges, an exception being that between judges 3 and 5. If judge 2 is eliminated W increases from 0.763 to 0.778, not a dramatic increase although the omission results in minor adjustments to final rankings. In a decision to eliminate one judge on the grounds of lower correlations we are assuming that it is his/her rankings that are erratic; it may be of course that that judge is actually making a better assessment of merit than any of the other judges! There is no way of knowing if this is so; any decision can only be made on a balance of probabilities based on an implicit belief that good judges are more likely to agree than bad ones.

Even though the Fligner and Verducci test provides no evidence in these data for common inconsistencies among all judges in the shape of more or less generous scores to later competitors there may be a tendency for individual judges to do so. If they do this should be reflected in the rank correlations between the order of performing (ranked 1 to 40) of competitors and the ranking for a particular judges score. It is therefore of interest to compute the Spearman or Kendall coefficients (r_S and t_K) between order of competing and scores for each judge. Values and the associated asymptotic one-tail p-values for each are:

	r_S	p	t_K	p
Judge 1	−0.056	0.366	−0.040	0.374
Judge 2	−0.085	0.300	−0.044	0.335
Judge 3	−0.251	0.059	−0.173	0.063
Judge 4	−0.235	0.072	−0.163	0.081
Judge 5	−0.080	0.311	−0.049	0.376

Features of these correlations are that all are negative but none produce p-values that we would normally regard as indicating significance. The implication is that there is a slight tendency to give lower scores to later competitors and that this tendency is more marked for judges 3 and 4. There is certainly no evidence for any claim that judges tend to be more generous in assessing later competitors. The consistency of slight but not significant negative correlations probably reflects a genuine slight overall inferiority of later competitors as a consequence of the particular random ordering. It is indeed comforting that all judges seem to agree about this, although it could be argued that there is a slight indication that judges 3 and 4 have been rather more severe than the others on later competitors.

In competitions where this method of judging is used there is almost inevitably a subjective element in the scoring. This is usually smoothed out by using competent judges and sensible scoring systems. If it is important to eliminate anomalies that might arise in final rank orderings a number of statistical techniques might be used to look for outlying scores. For example, regression diagnostics might be used on a fit of judge 3 scores to competitor order to see whether the nearly significant regression implied by the negative correlation is due to just one or two aberrant points and if so one might see if, for those points, the other judges gave very different scores. I urge readers to perform these and other diagnostic tests and also to consider any other appropriate tools in their statistical armoury relevant to assessing characteristics of scoring systems of this type. Plenty of other tests are possible. To explore further indications of possible lower scoring for later competitors an *ad hoc* procedure would be to arbitrarily divide the results for each judge into two samples, the first consisting of competitors 1 to 20 and the second of competitors 21 to 40 and to examine for each judge whether there were indications of a lower mean score in the latter group using a test such as the *t*-test if it is thought appropriate, or a WMW or some other nonparametric test.

Converting scores to ranks is just one way to attain a common distribution for all judges' scores. One might transform to normal scores. Another transformation widely used for combining examination marks or results in psychological tests is effectively equivalent to a studentization of scores, although it is usually executed by transforming all scores linearly so that they generate a distribution with a common sample mean and variance but these are not necessarily the mean 0 and variance 1 implicit in Studentization. For example, if the scores X awarded by a judge have a mean 28.1 and a standard deviation 1.9 it is easily verified that the transformation $Y = (2X - 3)/1.9$ give a sample with mean 28 and standard deviation 2. Similarly, scores for other samples may be converted to scores with mean 28 and standard deviation 2 before they are added to produce a final ranking. Transformations like this have a less drastic effect on outliers than transformation to ranks. In these circumstances choice of the more appropriate transformation should depend, for example, upon whether a high outlier represents a valid high score showing a competitor's marked superiority or is just an aberration on the part of a judge.

The crucial comparative test of different scoring systems that attempt to arrange competitors in order of merit is whether they achieve the same end and if not to find in what way they differ. If several scoring systems, e.g. addition of individual judges raw scores; addition of ranks based on these scores; trimmed means of judges scores, etc. all lead to exactly the same rankings there is no obvious problem although there could still be an injustice if all judges had been inconsistent in a similar way (e.g. all had unjustly given more generous scores to later competitors). When the ordering of competitors differs between various schemes for combining individual judges' scores there is a problem. Unless tests like those suggested in Example 11.12 or others suggest that one or more judges should be excluded on the grounds of evident inconsistency, etc. the most profitable line is likely to be a close examination of scores awarded to competitors who attain different rankings under alternative methods of combining scores and attempting to assess whether these may be a consequence of an apparently aberrant score awarded by a particular judge (Exercise 11.15).

Exercises

11.1 Confirm the probabilities given in Table 11.2 for the data in Example 11.1.

11.2 Verify the calculated value of the distance Q for the t' values in Table 11.3.

11.3 Verify all calculations in Example 11.2 and confirm the entries in Table 11.4.

11.4 Verify for 3 treatments in the one-way classification ANOVA described in section 11.2 that the distance measure Q given by (11.3) is equivalent to the statistic T given by (7.2).

11.5 Verify the calculations in Example 11.3.

11.6 Confirm the assertion made following Example 11.3 that if observations are replaced by ranks in that example that this leads to the Kruskal–Wallis test and that here the calculated Q is equal to the statistic T given by (7.3).

*11.7 Verify the calculated hat matrices for data Sets I and II of Example 11.4 that were obtained in Example 11.5. This is a tedious (but possible) exercise with a pocket calculator if no suitable computer program is available.

*11.8 The Brownlee stack loss data (Brownlee, 1965) has been widely analysed to illustrate the possible influence of outliers in multiple regression. I give the data below and if you have suitable software try your hand at analysing it. A detailed diagnostic analysis is given by Atkinson (1985, pp. 129–36). There are three explanatory variables, *air flow, cooling water temperature* and *acid concentration* and a response *stack loss* which was a measure of ingoing ammonia escaping unconverted. There are 21 observations. Some changes in units have been made for simplicity but this should not affect your conclusions.

Air flow	Water Temperature	Acid concentration	Stack loss
80	27	89	42
80	27	88	37
75	25	90	37
62	24	87	28
62	22	87	18
62	23	87	18
62	24	93	19
62	24	93	20
58	23	87	15
58	18	80	14
58	18	89	14
58	17	88	13
58	18	82	11
58	19	93	12
50	18	89	8
50	18	86	7
50	19	72	8
50	19	79	8
50	20	80	9
56	20	82	15
70	20	91	15

11.9 Verify the form of the robust estimators given in (11.14) and (11.15).

11.10 Verify the initial estimates of the coefficients obtained using Bartlett's method for data Set I of Example 11.4 that are used in Example 11.11 and confirm the remaining calculations in the latter example.

*11.11 Mazess, Peppler and Gibbons (1984) and Hand el al (1994, Set 17) give the following data for age, percentage fat and sex for 18 normal adults. Interest was in a study of the relationship between age and %fat and whether there is a difference between males and females. In studying the fit there is scope for using regression diagnostics.

Age	23	23	27	27	39	41	45	49	50	53
% fat	9.5	27.9	7.8	17.8	31.4	25.9	27.4	25.2	31.1	34.7
Sex	M	F	M	M	F	F	M	F	F	F

Age	53	54	56	57	58	58	60	61
% fat	42.0	29.1	32.5	30.3	33.0	33.8	41.1	34.5
Sex	F	F	F	F	F	F	F	F

*11.12 Mapes and Dajda (1976) and Hand et al (1994, Set 97) give data on child's age and the percentage of illnesses reported to a doctor for children of that age. How is reporting related to age? An inspection of the data suggests that a straight line fit will not be adequate. How might regression diagnostics be used to highlight such inadequacies and suggest a possible model to overcome them?

Child's age	0	1	2	3	4	5	6	7	8	9	10	11	12	13	14
% illness reported	70	76	51	62	67	48	50	51	65	70	60	40	55	45	38

*11.13 Weiner (1977) and Hand *et al* (1994, Set 346) give the following data on the average oral vocabulary of children of various ages. Explore the relationship between age and vocabulary.

Age (years)	1.0	1.5	2.0	2.5	3.00	3.5	4.0	4.5	5.0	6.0
Words	3	22	272	446	896	1222	1540	1870	2072	2562

*11.14 The following data on numbers of divorces (degrees nisi granted in thousands) in the UK in various years is given in *Social Trends* **21**, (1991) published by the UK Central statistical office. Explore any trends with time.

Year 19..	61	71	76	81	84	85	86	87	88	89
Decrees	27	89	132	148	148	162	153	150	155	152

*11.15 In Example 11.12 for the data in Table 11.6 I suggested that final ordering of competitors might be based on (i) the sum of the scores awarded by each judge as given in Table 11.6 or (ii) the mean of such scores after trimming the highest and lowest of the 5 judges' scores for each competitor or (iii) replacing each judges scores by the corresponding ranks over all competitors and calculating the sum of ranks awarded by all 5 judges to each competitor or (iv) omitting the results for Judge 2 on the grounds that the Spearman coefficient between his scores and those of the other judges is below average and then using either method (i) or (iii). Compare the orderings given by these methods. You will find that all give orderings with a broad measure of agreement but there are a few anomalies or differences. By careful examination of the data in Table 11.6 and these final orderings attempt to explain why such differences occur. In the light of your explanations which of the above methods of ordering competitors do you consider to be the fairest?

12

Categorical data analysis

12.1 Types of categorical data

In section 1.6 I showed that the Fisher–Pitman two-sample permutation test may be put in a contingency table format with each of two rows corresponding to a sample and each column corresponding to a recorded measurement. The cell entries were counts of units allocated to a specific sample (row category) with a given measurement (column category). In the no-tie case in Table 1.1 all counts were 1 and all column totals were also 1, the row totals being 4 and 5, the sample sizes. The Wilcoxon–Mann–Whitney test in section 4.2 provides a special case of that formulation with each column corresponding to a rank. Other examples of test and estimation procedures developed for measurement data that are expressible in contingency table format include the median tests of section 4.6 and 7.4 and some correlation tests in section 9.1. For the median test in Example 4.11 the table of counts of numbers of observations above and below the median in each sample gave a conventional 2×2 contingency table. Actual measurements, even if available, are absorbed or telescoped into these groupings.

The above tests were usually concerned with data presented as measurements or with orderings based upon quantities that could in theory at least be measured. The contingency table formulation shifts the emphasis from **measurements** made on units to **counts** of numbers of units in some category (e.g. a sample) having each measurement, or, in the case of grouped data, as in Example 4.8, counts of numbers of units having measurements within each group. While the column categorization in the above examples was by measurements or scores (e.g. ranks) or intervals (e.g. below or above a median) the other grouping (rows) could either by a similar type of categorization (e.g. in correlation) or by a qualitative characteristic (e.g. Group I and Group II in Table 1.1; one of k groups or samples in a Kruskal–Wallis or a Jonckheere–Terpstra test). For Kruskal–Wallis the qualitative categories, samples, were not ordered, whereas for Jonckheere–Terpstra they were.

Recognition of this duality between measurements and counts of numbers of units having each measurement has stimulated interest in *categorical data analysis* or *the analysis of contingency tables*, but more importantly counts often occur as primary data with no associated physical measurement.

Categorical data analysis is receiving ever-increasing attention. Books on the subject include Bishop, Fienberg and Holland (1975), Breslaw and Day (1980), Agresti (1984,1990, 1996),Christensen (1990) and Everitt (1992). Both Agresti (1990, 1996) and Everitt (1992) give comprehensive treatments of topics covered in this and the next chapter. In broad terms Agresti's approach is model-driven and Everitt's data-driven. It is clear from a study of both that the two approaches converge. Agresti (1996) is less technical than Agresti (1990) and puts more emphasis on generalized linear models which I discuss briefly in Chapter 14.

A shift in emphasis from measurements to counts unifies the treatment of topics that at first sight look unrelated and consolidates links between data- and model-driven approaches. When the mechanisms of generation, collection and presentation of data are clear these may lead to a specific model, and a model-driven analysis then becomes appropriate. When an appropriate model is less clear interest shifts to patterns in data that are not highly model dependent. In the rest of this book I mix data-driven and model-driven approaches to highlight the interface between them.

Historically, early categorical data analyses asked only whether counts indicated an association between characteristics. Do people with fair hair tend to have blue eyes and people with dark hair tend to have brown eyes? Are people who smoke more likely to develop lung cancer than those who do not? The well-known chi-squared test and Fisher's exact test indicate whether or not there may be association, but one often wants information about the nature of any possible association. This has led to new developments over several decades. .

I consider first an example where a routine classical data-driven approach leads to a certain conclusion, but an examination of the way the data were collected shows why this simple analysis is misleading.

Example 12.1. The data in this example are artificial, but illustrate an anomaly that often occurs with real data and shows how the way data are presented may lead to that anomaly.

To test a new drug for treating a respiratory disorder 3000 patients are given either a standard drug or the new drug. Only 1700 doses of the new drug are available, so the remaining 1300 patients are treated with the standard drug. At each centre in urban and rural areas the patients to receive the new drug are chosen at random. Originally the only data given to a statistician is the drug received by each person, and whether or not a cure has followed. The results are summarized in Table 12.1

The classic tests for no association between drug and outcome (i.e. of H_0:*drugs are equally effective*) are the chi-squared test and the Fisher exact test and no matter whether an exact or asymptotic test is used both lead to similar conclusions. The asymptotic results give two-tail $p = 0.0466$ and 0.0467 respectively and the exact tests both give, as is usual for a 2×2 table, identical p-values, here $p = 0.0495$. From the data the percentage cures are easily computed

Table 12.1 Numbers cured or not cured by each of two drugs

	No effect	Cure	Row totals
Standard drug	850	450	1300
New drug	1170	530	1700
Column totals	2020	980	3000

as 34.62% for the standard and 31.18% for the new drug, indicating that the new drug is less successful.

What does equal effectiveness of the drugs imply from a modelling viewpoint? A simplistic model would be that no matter which drug is supplied, each individual has a probability π of being cured. If, for the standard drug, this probability had remained constant over many years and was, say, π_0, we would be unlikely to do the experiment outlined above, for the hypothesis of interest would be *Does $\pi = \pi_0$ for the new drug?* In that situation the experiment could sensibly concentrate on the new drug to test this hypothesis and probably to obtain a confidence interval for π for that new drug. This is standard parametric binomial theory testing. The experimental set-up and test outlined above makes no assumption about a common π, perhaps because old records are not available or because there is a year to year variation in π. The latter is likely for a respiratory disease as incidence and severity (and therefore response to treatment) may vary from year to year due to changes in weather, atmospheric pollution, etc. Further, the assumption that each patient has a fixed, even if unknown, probability of a cure with a given treatment is unrealistic because the probability of cure for an individual will almost certainly depend not only upon that treatment but upon other factors including severity of the respiratory condition, the patient's general health and also their diet, housing and working conditions. In these circumstances the best we might hope for modelwise, is that under H_0 in a reasonably large random sample these disturbing factors might balance out to give a proportion of cures that is fairly constant in large samples. Trying to achieve this is a key reason for random assignment of patients to treatments. This justifies the above tests under this looser model.

In my description of the way the data were obtained one sentence suggests a potential for verifying an assumption of a common 'average' π value over all patients and also a potential problem that might arise if there is evidence against such homogeneity. That sentence is:

At each centre in urban and rural areas the patients to receive the new drug are chosen at random.

If we had separate data for each centre we could test other questions potentially of interest. These include:

- If there is no differential effect of drugs is the probability of cure under H_0 the same at all centres?
- If H_0 does not hold is the differential effect the same at all centres?

Common sense suggests that if we have appropriate data we should use them to answer these questions before we combine the data for an overall test, because if the probability of cure differs between centres and/or between drugs at different centres, tests on the combined data are not justified and any interpretation of the pooled results is questionable.

Table 12.2 Responses to two drugs in urban and rural areas

| | Urban | | Rural | | Row |
	No effect	Cure	No effect	Cure	totals
Standard drug	500	100	350	350	1300
New drug	1050	350	120	180	1700
Column totals	1550	450	470	530	3000

Data for individual centres were not available but the data in Table 12.1 were a combination of data from two sources for which separate records were kept. These were *urban* and *rural* areas; a sensible split on medical grounds, for respiratory complaints are often more prevalent in one or other of these environments, depending upon factors like atmospheric pollution, conditions under which people work (urban populations may include a higher proportion of factory workers exposed to toxic fumes; rural workers may be more exposed to inclement weather). The data in Table 12.1 split into urban and rural groups is given in Table 12.2.

From that table the percentage success rates for the two drugs in each area are:

	Urban	Rural
Standard drug	16.67	50
New drug	25	60

Clearly the new drug does better in both city and country and equally clearly the cure rate for either drug is more than twice as high in rural compared to urban areas. The chi-squared test establishes in both urban and rural areas an association between drugs and cure rate and the success rates indicate that the new drug does better in each. Yet for the pooled data in Table 12.1 the standard drug gave a higher cure rate!

This example highlights the danger of combining data where the proportions in corresponding categories in two tables differ appreciably. It is tempting to combine the data in this example because of the intuitive feeling that if two sets of data show association in the same direction then combining the data should enhance the evidence. Broadly speaking this is only so if the separate data comply with the same model. Here clearly the proportion of cures in city and country are very different. That such effects may sometimes, as here, imply for the combined data an effect in the opposite direction is well known and is called Simpson's paradox. I explain why it arises in section 12.6. If the data in Example 12.1 were further broken down by individual clinics within urban and rural areas an even more complex pattern of responses might emerge.

The normal distribution has a pivotal role in model-driven analyses for continuous data, but for categorical data a more important distribution is the multinomial distribution with the special case of the binomial distribution also prominent. The Poisson distribution, too, is central in data involving counts, but as Agresti (1990, section 3.1.1 and 3.1.2) shows, when we condition on fixed total numbers, as we usually do in categorical data analysis, the cell counts generated by Poisson processes have a multinomial distribution. If you did not know this, you should refer to Agresti or to a standard text on distribution theory to confirm it. In many situations when analysing counts in categorical data the same tests apply no matter whether the data generating system is a multinomial distribution or several independent binomial or Poisson distributions.

Except for some cases in sections 13.3 and 13.4, I consider counts of individuals or units (e.g. machine components or manufactured items) categorized by two or more attributes, considering the individuals or units as independent samples. In some situations in section 13.4 observations are made on the same individual to categorize each before and after some event, or we may be dealing with matched pairs. These are the sort of situation that gave rise to McNemar's test described in section 6.3. Questions of interest then are different from those of departure from independence that I deal with in the rest of this chapter.

Counts may be presented in one-, two- three- or higher-dimensional tables, commonly referred to as one-, two-, three- etc. way contingency tables. I only mention the one-way case in passing. Essentially it covers the case of univariate discrete data in the form of counts with specific attributes.

Two-way tables with counts in r rows and c columns are the best known and most widely explored type of contingency table and many basic ideas in categorical data analysis are easily illustrated for $r \times c$ tables, some by very special cases such as the 2×2 table. Each *dimension* or *way* (i.e. rows or columns) corresponds to an attribute split into categories that may be either **explanatory** (e.g. dose levels of a drug; the names of several different drugs; gender; religious affiliations; age groups like <30, 31–40, 41–50, over 50; income levels; countries of residence, etc.) or **responsive** (e.g. side effects of drugs given as none, moderate or severe; effects of a treatment measured as deterioration, no change, or improvement; or measurements, scores or ranks for some characteristic). Attributes may be **qualitative** or **quantitative**. Qualitative attributes that have no natural ordering are described as nominal (e.g. religious affiliations, matrimonial status, country of birth, eye colour). Qualitative attributes that have a natural ordering and all quantitative attributes are **ordinal** (e.g. reactions of a drug described as slight, moderate or severe; grouping of individuals by age as under 50 or 50 and over; grading manufactured items as, poor, average, good).

Typically, in two-way tables each of the r rows represents either the level of an explanatory variable or if there are no explanatory variables, a level of a response variable, and each of the c columns the level of a response. In the case of an explanatory and a response categorization it is a common and useful, though not a universal convention, to assign the explanatory categories to rows and the response categories to columns. The entry at the intersection of the ith row and

Table 12.3 Number of patients exhibiting side-effects for three drugs

	None	Slight	Moderate	Severe	Row totals
		Side-effect level			
Drug A	35	6	2	1	44
Drug B	17	8	0	4	29
Drug C	43	6	1	1	51
Column totals	95	20	3	6	124

*j*th column, referred to as cell (i, j) is the number of items exhibiting that combination of row and column attributes. Table 12.3 is an example with $r = 3$ and $c = 4$. The row categories (drugs) are explanatory and the columns represent ordinal responses. Is there evidence that the incidence rate of side effects depends on the drug administered? The null hypothesis is that side effect incidence proportions are independent of the drug administered. A classic contingency table approach to this problem uses the chi-squared test. The asymptotic form of that test, assuming the statistic has approximately a chi-squared distribution with (here) 6 degrees of freedom is suspect for Table 12.3 because of low expected numbers in many cells under the hypothesis of no association. An exact test considering all permutations of the usual chi-squared statistic, subject to fixed marginal totals, is available. Alternatively one might use an extension of the Fisher exact test, the Fisher–Freeman–Halton test, described for a special case in section 7.4. Neither of these tests takes account of the ordering of the responses from none to severe, a matter clearly relevant to clinicians.

More effective approaches take the ordering of column categories into account. The data in Table 12.3 may be looked on as highly tied measurement data because the response categories of none, slight, moderate and severe are ordered. If appropriate tied rank scores are given to these data a suitable test is the Kruskal–Wallis (KW) test. The 95 patients showing *no effect* are all given the mid-rank 48, the 20 giving the response *slight* receive the average of ranks 96 to 115, i.e. 105.5, the 3 with *moderate* symptoms are each ranked 117 and the six with *severe* are each ranked 121.5. This approach takes account of the ordering of column categories. This is important because a clinician is interested in the relative frequency of side effects, attaching more importance to differences that manifest themselves through more serious side effects. An alternative to mid-rank scores is to assign scores to the relative seriousness of side effects based on a clinician's assessment of the importance of each. For example, if a clinician considers slight side effects to be of little clinical importance, moderate ones to be 10 times more serious, and severe ones 5 times more serious than moderate ones he might use a scoring system of 0, 1, 10, 50 for none, slight, moderate and severe side effects respectively and then perform a nonparametric one way ANOVA using these scores (or other chosen scores) as data. StatXact and some other programs enable these tests to be carried out either as exact or asymptotic tests. In StatXact Monte Carlo approximations to the exact test are also available. These are useful when data sets are too large for an exact test, but there is doubt about an asymptotic test because of many low cell frequencies. Another possibility in situations like this is a bootstrap approach.

Example 12.2 For the data in Table 12.3 StatXact gave the following *p*-values for the tests discussed above.

- Exact chi-squared test $p = 0.0398$ (asymptotic approximation 0.0490).
- Fisher–Freeman–Halton test $p = 0.0539$ (asymptotic approximation 0.1029).
- Kruskal–Wallis test $p = 0.0230$.
- ANOVA with scores 0, 1, 10, 50 $p = 0.0393$.

Because they take ordering of columns into account, the Kruskal–Wallis and ANOVA tests give different results if we interchange column data. For example, if the second and fourth column data (but not the column labels or scores) are interchanged, in effect, providing different numbers for the slight and severe side effects, the *p* values for these tests are respectively $p = 0.0367$ and $p = 0.1688$. Because they ignore column ordering the chi-squared test and the Fisher–Freeman–Halton test are unaffected by permutation of rows or columns.

Example 12.2 highlight differences in analyses that depend upon whether or not categories in rows and columns are ordered. More generally, different analyses are appropriate depending upon whether the row and column categories are:

- both unordered;
- row categories are not ordered, but column categories are, or *vice versa;*
- both row and column categories are ordered.

Analyses may also be modified depending upon whether rows represent explanatory or response categories. The nature of the data suggests different models for each situation. In this sense analysis of much categorical data is data-driven, although the background models for the chosen analysis are often quite specific. While various hypothesis tests are available about how data fit some prespecified model (goodness-of-fit-tests), the problems of wider interest are whether there is independence between row and column classifications, and if independence is rejected, what is the nature of the association? The results of the tests in Example 12.2 suggested an association between drugs and side effects, i.e. that the side-effect incidence and level varied between drugs. Both the Kruskal–Wallis test and the nonparametric ANOVA results suggest a more specific relationship, i.e. not only a difference in some or all side effect levels between drugs, but a difference related to degree of severity, since these tests take ordering of side-effects into account when scores are allocated.

In a general $r \times c$ table the count in cell (i, j) is written n_{ij}, the total for row i is n_{i+}, i.e. $n_{i+} = \sum_j n_{ij}$, and that for column j is n_{+j} and $n = \sum_i n_{i+} = \sum_j n_{+j} = \sum_{ij} n_{ij}$ is the total for all cells.

The models that arise in connection with different data structures often depend upon specific values of certain parameters, but these values are often irrelevant to questions of association, in that their values are not specified in our null hypotheses (or indeed in the alternative hypotheses), and are in that sense nuisance parameters. A device known as conditional inference avoids the need to specify these parameters. This makes use of the properties of sufficient and ancillary statistics, notions not described here, but which are well-known statistical concepts defined in textbooks which cover general theory of statistical inference at any but the most elementary level. For those familiar with the basic ideas of maximum likelihood estimation, excellent accounts of inference theory for categorical data are those by Agresti (1990, Chapter 3 and 1996, Chapter 2).

The conditional feature is to make inferences **conditional** upon row and column totals being fixed. In sections 1.6, 7.5 and 9.1 where I presented well established tests in a contingency table format, the row and column totals were fixed. Later in this chapter I give other examples where again these marginal totals are fixed. The validity of this conditional approach has been debated for some 50 years. There are situations where this restriction may validly be lifted and others where it may seem at odds with the way data have been obtained, but its validity in most situations is now widely accepted, even when it is not a consequence of the experimental set-up giving rise to the data. Difficulties may arise with conditioning on fixed marginal totals if these depend on the process under study. I illustrate this in Section 13.6 for the data from Example 1.2 and detailed comments on the situation arising there are given in Gastwirth (1997).

The conditional argument is easily described for a 2×2 table. Are two drugs equally effective in relieving pain? With 25 patients the following results are obtained

	Relief	No relief	Total	
Drug A	8	4	12	
Drug B	3	10	13	(12.1)
Total	11	14	25	

Here row totals are fixed by the pre-trial decision to administer Drug A to 12 and Drug B to 13 patients. Under the hypothesis of no association, i.e. that the drugs are equally effective, one is in essence saying it is immaterial to the outcome which drug is administered and in effect that whatever drug we give individuals we have a fixed number (i.e. 11) who will obtain relief, whilst the remaining 14 will not. Clearly, the question we ask is similar to that raised in the discussion of the Fisher–Pitman test in section 1.6, i.e. given the observed outcome is this an unlikely one under H_0: *the drugs are equally effective*. In that example the scoring system (there a physical measurement on each unit) differed from that here where the response is binary (*relief* or *no relief*). However, if we arbitrarily score relief as 1 and no relief as 0 we are in an equivalent situation to that in Example 1.3 except that we have heavily tied measurements, all either 0 or 1. Our permutation reference set is all permutations of those scores in samples of 12 and 13. All possible permutations in the reference set give rise to tables with marginal totals the same as those in (12.1).

An argument against a conditional test is that if we performed the above experiment by taking a random sample of 25 from a large population and allocated 12 at random to *Drug A* and the remainder to *Drug B*, then under the null hypothesis we might well get the above result, but if we repeated the experiment with another sample of 25 from that larger population we might, because of sampling variation get, say, 9 cases getting relief and 16 not, even if the null hypothesis were true. This implies that in repeated sampling our column totals may not remain fixed. However, in making inferences in nearly all other situations in this book they have been conditional on the samples we have! The only exceptions have involved the bootstrap where we form new samples by sampling *with replacement*.

The above argument used a data-driven approach. What about a model driven approach? Here we might assume that the probability of relief for any individual given drug A is π_a and for an individual given drug B it is π_b. The null hypothesis of no association is $H_0: \pi_a = \pi_b$. The column totals, as do the cell entries in each row, give information about the common value, π, of π_a and π_b when H_0 is true. Indeed, the column totals combine the information about this parameter that is contained in the two rows, but the precise value of π is irrelevant to whether H_0 is true. It is the individual cell entries in each row that carry the information about differences between π_a and π_b, so the conditional test based on the fixed marginals is again relevant.

There are mechanisms that generate data like that in (12.1) without regarding either row or column totals as pre-fixed. This happens, for instance, in what are called retrospective studies. For example, hospital records might have given the information in the cells of (12.1) for all patients so treated in the past month. Again, however, our method of testing for no association using conditional inference would be the same, i.e. conditional upon observed marginal totals, but the validity of any statistical inferences is restricted by the nonrandomness of the method of data selection. Strictly the inferences apply to that data only.

12.2   Inferences in 2×2 tables

The 2×2 table is one of the simplest data formats in statistics. Simple indicators and measures of association for these tables often extend readily to $r \times c$ tables and also to comparisons of sets of k tables, each 2×2, which may be looked upon as three-way $2 \times 2 \times k$ tables. Many also extend with modification or added complexity to three-way $r \times c \times k$ tables and also to tables in more than 3 dimensions.

Even in a 2×2 table there are basic differences between appropriate models for the situation where one category (rows) is explanatory and the other (columns) is a response and the situation where both categories are responses. An example of the former was given by (12.1) while (12.2) gives an example of the latter and records the numbers of people in a random sample of 100 who, in the previous 12 months, have or have not used two forms of public transport at least once:

| | | \multicolumn{2}{c}{*Train*} | |
		Yes	*No*	
Bus	*Yes*	43	27	(12.2)
	No	19	11	

The distinction between explanatory and response variables is not always clear cut. For example if people are classified into two groups according to salary earned (e.g. £20 000 per annum or less, more than £20 000 per annum) and by job satisfaction (high or low) the salary division may be regarded by some as a variable that explains a response *job satisfaction* or by others as a response to an explanatory variable *job satisfaction* in that satisfaction is likely to motivate workers in a way that leads to promotion and better prospects of higher salary. Fortunately for some inferences we need not make such distinctions, for certain inferences about association will not be affected. The distinction may be more important in general $r \times c$ tables. What is then often of more impact (section 12.3) is whether explanatory or response categories are nominal or ordinal. With only two categories for an attribute order is immaterial except in the sense that difference implies an inequality and is in that sense an ordering. We may of course be interested in the direction or magnitude of any such inequality.

I first consider a model for situations like that in (12.2). If the sample is a random sample from some large population and the proportions in that population with the characteristic corresponding to that of row i and column j $(i, j = 1,2)$ is π_{ij}, then in a random sample of n, standard statistical theory tells us that the numbers in each cell have a multinomial distribution with parameters $(n, \pi_{11}, \pi_{12}, \pi_{21}, \pi_{22})$. Since $\pi_{11} + \pi_{12} + \pi_{21} + \pi_{22} = 1$, one of these probability parameters is expressible in terms of the remaining three. Because the data are bivariate (rows corresponding to a variate X, say, and columns to a variate Y) we have a bivariate multinomial distribution. The marginal distributions in the 2×2 case are binomial. Writing

$$\pi_{i1} + \pi_{i2} = \pi_{i+}, \; i = 1,2 \text{ and } \pi_{1j} + \pi_{2j} = \pi_{+j}, \; j = 1,2$$

the marginal distribution of X (rows) for a sample of n is $B(n, \pi_{1+})$ and for Y (columns) is $B(n, \pi_{+1})$. If X, Y are independent it follows that $\Pr(X = i, Y = j) = \pi_{ij} = \Pr(X = i).\Pr(Y = j) = \pi_{i+}\pi_{+j}$. It is easy to see that independence implies $\pi_{11}\pi_{22} = \pi_{21}\pi_{12}$.

If the π_{ij} take a set of prespecified values, then given a sample of n the expected count in cell (i, j) is $m_{ij} = n\pi_{ij}$. If the observed count in cell (i, j) is n_{ij} the chi-squared statistic for testing goodness of fit (section 6.7) is $X^2 = \sum_{i,j}(n_{ij} - m_{ij})^2/m_{ij}$. Under H_0 this has asymptotically a chi-squared distribution with 3 degrees of freedom. For small samples an exact permutation test is possible. It is seldom that one uses this particular test (a goodness of fit test) in a contingency table framework because tests about prespecified values of the π_{ij} are seldom wanted.

A more usual test is that for independence of X and Y, (row and column classifications) and when there is not independence one may want to explore the nature or strength of the association. Under $H_0:X$ and Y are independent, if we make no assumption about the π_{ij} we might estimate them from the data. Since independence implies $\pi_{ij}=\pi_{i+}\pi_{+j}$ a sensible approach is to estimate the binomial marginal probabilities using the usual binomial estimators. For example, that for π_{i+} is $p_{i+} = n_{i+}/n$ and that for π_{+j} is $p_{+j} = n_{+j}/n$, giving $p_{ij} = p_{i+}p_{+j} = (n_{i+}n_{+j})/n^2$ as an estimator of π_{ij}. Thus, under independence the expected cell counts are $m_{ij} = np_{ij} = (n_{i+}n_{+j})/n$. The chi-squared statistic $X^2 = \sum_{i,j}(n_{ij} - m_{ij})^2/m_{ij}$ may be used in an exact permutation test or in an asymptotic chi-squared test. The degrees of freedom are now 1, since we have estimated two independent parameters π_{1+} and π_{+1} from the data. Once these are known the remaining estimated marginal probabilities are fixed as $p_{2+} = 1 - p_{1+}$ and $p_{+2} = 1 - p_{+1}$.

Independence implies $\pi_{11}\pi_{22} = \pi_{21}\pi_{12}$ and it is easily verified (Exercise 12.1) that the expected counts $m_{ij} = np_{ij}$ satisfy the relationship $m_{11}m_{22} = m_{21}m_{12}$.

If independence is not assumed, for row 1, the conditional probabilities associated with columns 1 and 2 are respectively $\pi_{(1)1} = \pi_{11}/\pi_{1+}$ and $\pi_{(1)2} = \pi_{12}/\pi_{1+}$ with corresponding results for row 2. I write $\pi_{(i)j}$ for the conditional probability for j given i instead of the more conventional $\pi_{j|i}$ because juxtaposition of row and column suffixes in the conventional notation is confusing in this context. Since $\pi_{11} + \pi_{12} = \pi_{1+}$ it follows that $\pi_{(1)2} = 1 - \pi_{(1)1}$, consistent with the well known fact that this conditional distribution is binomial. The ratio $\Omega_i = \pi_{(i)1}/\pi_{(i)2}$, $i = 1, 2$ is the odds on column 1 for row i. The ratio $\theta = \Omega_1/\Omega_2$ is the odds ratio. It is easily verified (Exercise 12.2) that the independence condition $\pi_{11}\pi_{22} = \pi_{21}\pi_{12}$ also holds for the conditional probabilities, i.e. $\pi_{(1)1}\pi_{(2)2} = \pi_{(2)1}\pi_{(1)2}$, whence it follows that for independence $\theta = \pi_{(1)1}\pi_{(2)2}/\pi_{(2)1}\pi_{(1)2} = 1$. It is also easily seen from the above discussion that if m_{ij} is the cell expectation under the multinomial model, whether the rows and columns are independent or not, that the odds ratio also equals $\theta = m_{11}m_{22}/m_{21}m_{12}$ with, of course $\theta = 1$ if this definition is used when the expectations are computed on the assumption of independence with fixed marginal totals. The corresponding ratio for observed counts, $\theta* = n_{11}n_{22}/n_{21}n_{12}$ is called the empirical odds ratio. A moment's reflection shows that in general $0 < \theta$, $\theta* < \infty$. Throughout this chapter several examples show the importance of the odds ratio in inference. Clearly, for instance, the greater the departure of $\theta*$ from unity the stronger the evidence against independence. Under the hypothesis of independence the distribution of $\theta*$ is not symmetric about $\theta = 1$. Clearly, interchanging rows or columns has the effect of replacing $\theta*$ by $1/\theta*$. A transformation to $\phi = \log \theta$ gives $\phi = 0$ as the condition for independence and $\phi*$ is then distributed symmetrically about zero. The large sample distribution of $\phi*$ approaches normality with mean zero and, although I do not prove it, standard deviation

$$\sigma* = [(1/n_{11}) + (1/n_{12}) + (1/(n_{21}) + (1/n_{22})]^{\frac{1}{2}}. \qquad (12.3)$$

A difficulty occurs if any cell entries are zero, for $\phi*$ then becomes infinite and its variance is no longer finite. The amended estimators

$$\theta^\dagger = (n_{11} + \tfrac{1}{2})(n_{22} + \tfrac{1}{2})/[(n_{21} + \tfrac{1}{2})(n_{12} + \tfrac{1}{2})] \qquad (12.4)$$

and $\phi^\dagger = \log\theta^\dagger$ with corresponding adjustments to the terms in $\sigma*$ are often used to avoid this difficulty and in most practical situations lead to similar inferences.

Example 12.3 For the data for rural patients in Example 12.1, i.e.

	No effect	Cure
Standard drug	350	350
New drug	120	180

$\theta^\dagger = (350.5 \times 180.5)/(120.5 \times 350.5) = 1.498$ and $\phi^\dagger = \log 1.498 = 0.404$. Also the modified (12.3) gives $\sigma* = 0.140$, whence assuming normality, an approximate 95% confidence interval for Φ is $0.404 \pm 1.96 \times 0.140$, i.e. (0.130, 0.678). Back transforming, the interval for θ is (1.139, 1.970). Since $\theta = 1$ is not included, this is consistent with the chi-squared test indication of association.

The bivariate multinomial model is relevant to situation like that in (12.2). In Example 12.3 the row classifications are clearly explanatory, so it makes no sense to regard these as random variables and now only responses, represented by columns are potentially random variables. Clearly with only two columns an appropriate model is one specifying sampling from a binomial distribution separately within each row where for row i, $i=1,2$ there is a probability π_i that a unit is allocated to column 1 and a probability $1-\pi_i$ of allocation to column 2. Lack of association or independence now corresponds to $\pi_1=\pi_2$, but as pointed out in section 12.1 a test for possible association makes no assumptions about the actual value of the nuisance parameters $\pi = \pi_i = \pi_j$, because the null hypothesis that responses are independent of the explanatory (row) categories implies only that $\pi_1 = \pi_2$. Here we proceed on the basis of our discussion of the situation in (12.1) that the column totals provide the best information about the common estimator p of π_1 and π_2. The relevant estimator is $p = n_{+1}/n$, whence $m_{11} = pn_{1+} = n_{1+}n_{+1}/n$. The remaining cell expectations may be obtained *ab initio* in a similar manner, or as is well-known and easily verified, by the relationships $m_{12} = n_{1+} - m_{11}$, etc. The expectations have exactly the same form as those obtained above for the bivariate multinomial distribution relevant to the case of two response categories as in (12.2) and the same test procedures are relevant. The equivalence becomes notationally identical by writing $\pi_i = \pi_{(i)1}$ and $1 - \pi_i = \pi_{(i)2}$, the notation I used for conditional probabilities, although it must be remembered that π_i is no longer a conditional probability in this situation. All the tests for independence and the statistics that measure association carry over with minor notational changes. The odds ratio now becomes $\theta = [\pi_1/(1 - \pi_1)]/[\pi_2/(1 - \pi_2)]$. Independence implies $\theta = 1$ and θ, θ^* and $\theta^\dagger$ and ϕ, ϕ^* and $\phi^\dagger$ may be expressed in terms of m_{ij} or n_{ij} in the way described above. Thus the procedure used in Example 12.3 was appropriate after all.

Fisher's exact test was developed for a situation where both row and column marginal totals are fixed *a priori*. The situation is in many ways analogous to that giving rise to the two sample Fisher–Pitman permutation test. There we had a total of M measurements to be allocated to two groups of $m+n$ subject to the constraint $m + n = M$ and our test examines the property of some statistic, e.g. the sum, for all possible samples of m. The situation envisaged by Fisher for the 2×2 table was that where we have a total of N items of which n_{+1} are labelled A and n_{+2} are labelled B. This is the mechanism that determines column totals. The experimental procedure corresponding to that of the randomization in the Fisher–Pitman procedure is to select at random from these items without replacement a sample of $m = n_{1+}$ to form the first row. By implication those not selected form the second row sample of $n = n_{2+}$. A moment's reflection shows that with these constraints once we know n_{11}, the number of items labelled A in the first row [i.e. in cell (1,1)] all other cell totals are fixed. The number n_{11} in cell (1,1) is an appropriate statistic for a permutation test of a null hypothesis of independence. Extreme values of n_{11} indicate association. It is well known that in sampling without replacement from a population with n_{+1} items labelled A and n_{+2} items labelled B that in a sample of n_{1+} the number of items labelled A has a hypergeometric distribution. There is some economy of notation if we write the general 2×2 table with fixed marginal totals in the form:

$$\begin{array}{cc@{\qquad}c}
a & b & a+b \\
c & d & c+d \\[1ex]
a+c & b+d & n = a+b+c+d
\end{array}$$

If X is the count in cell (1,1), bearing in mind the fixed marginal totals the hypergeometric distribution implies

$$\Pr(X = a) = \frac{{}^{a+b}C_a \times {}^{c+d}C_c}{{}^{n}C_{a+c}}$$

and this reduces to the more usually quoted form (4.5) given on p. 96.

For the hypergeometric distribution $E(X) = (a + b)(a + c)/n$ and with a notational change this is easily seen to equal m_{11} as obtained above for the bivariate multinomial model or the row-wise binomial model. The appropriate test for independence may then be performed in the way described in section 4.6.

For the multinomial model given above I showed that the expected cell count m_{ij} in cell (i, j) is $m_{ij} = n\pi_{ij}$ for all $i, j = 1, 2$. For independence this reduces to $m_{ij} = n\pi_{i+}\pi_{+j}$, whence

$$\log m_{ij} = \log n + \log \pi_{i+} + \log \pi_{+j}.$$

Thus, on a logarithmic scale under independence the expected count in cell (i, j) is a linear function of what may be regarded as an ith row and a jth column effect. The model is called the **loglinear model for independence** and clearly has similarities to the classic linear model associated with randomized blocks. As in that model the 'parameters' are not unique, and constraints are needed to give a unique specification. One possible specification is

$$\log m_{ij} = \mu + \rho_i + \gamma_j$$

where for a 2×2 table

$$\mu = \log n + \tfrac{1}{2}(\log \pi_{1+} + \log \pi_{2+}) + \tfrac{1}{2}(\log \pi_{+1} + \log \pi_{+2})$$

$$\rho_i = \log \pi_{i+} - \tfrac{1}{2}(\log \pi_{1+} + \log \pi_{2+})$$

and

$$\gamma_j = \log \pi_{+j} - \tfrac{1}{2}(\log \pi_{+1} + \log \pi_{+2})$$

It is easily verified that the independence model satisfies the condition that the odds ratio $\theta = 1$ or $\phi = \log \theta = \log m_{11} + \log m_{22} - \log m_{12} - \log m_{21} = 0$. If you are familiar with the concept of interaction in a 2×2 factorial experiment you will recognise this as the analogue of the relationship between factor combination expectations that corresponds to no interaction between the two factors in such an experiment. Values of $\phi \neq 0$ imply association, the categorical data analogue of interaction in factorial experiments. This situation is discussed in more detail with a simple example in Sprent (1993, section 10.2.1). Analogous relationships involving odds ratio hold if the m_{ij} are replaced by π_{ij} providing the independence condition holds.

If there is association in a 2×2 table and all $m_{ij} > 0$, the loglinear model, like its linear analogue, requires an additional term to represent interaction. Writing

$$\eta_{ij} = \log m_{ij}, \quad \eta_{i.} = \tfrac{1}{2}(\eta_{i1} + \eta_{i2}), \quad \eta_{.j} = \tfrac{1}{2}(\eta_{1j} + \eta_{2j}), \quad \eta_{..} = \tfrac{1}{4}(\eta_{11} + \eta_{12} + \eta_{21} + \eta_{22})$$

$$\mu = \eta_{..}, \quad \rho_i = \eta_{i.} - \eta_{..}, \quad \gamma_j = \eta_{.j} - \eta_{..} \text{ and } (\rho\gamma)_{ij} = \eta_{ij} - \eta_{i.} - \eta_{.j} + \eta_{..}$$

it is easily seen by substitution that

$$\log m_{ij} = \mu + \rho_i + \gamma_j + (\rho\gamma)_{ij} \tag{12.5}$$

and also that

$$\rho_1 + \rho_2 = 0, \ \gamma_1 + \gamma_2 = 0, \ (\rho\gamma)_{i1} + (\rho\gamma)_{i2} = 0 \ (i = 1, 2),$$

$$(\rho\gamma)_{1j} + (\rho\gamma)_{2j} = 0, \ (j = 1, 2).$$

The constraints imply there are 4 independent parameters which may conveniently be taken as μ, ρ_1, γ_1, $(\rho\gamma)_{11}$ and these specify the log m_{ij} exactly for any m_{ij} including those equal to the observed counts n_{ij}. If $(\rho\gamma)_{11} = 0$ then (12.5) reduces to a model having row and column effects only and is thus an independence model. For the general model (12.5), using the constraints the log odds ratio is

$$\phi = \log(m_{11}m_{22}/m_{12}m_{21}) = \log m_{11} + \log m_{22} - \log m_{12} - \log m_{21} =$$
$$(\rho\gamma)_{11} + (\rho\gamma)_{22} - (\rho\gamma)_{12} - (\rho\gamma)_{21} = 4(\rho\gamma)_{11}.$$

The constant $(\rho\gamma)_{11}$ may be interpreted, as one would expect by analogy with the linear model for a 2×2 factorial structure, as a measure of interaction or association. Independence corresponds to $(\rho\gamma)_{11} = 0$. This is related to the well-known result for 2×2 tables with fixed marginal totals that if m_{ij} are expected cell counts under the hypothesis of independence then, if we write $n_{11} - m_{11} = d$ it follows that $n_{12} - m_{12} = n_{21} - m_{21} = -d$ and $n_{22} - m_{22} = d$, consistent with the need for only one constant for interaction in the loglinear model. This model is said to be saturated, since with appropriate choice of parameters, it provides an exact fit of a loglinear model to any 2×2 table.

The loglinear model does not distinguish between explanatory and response variables.

Other measures of association are widely used, especially in the context of $r \times c$ tables discussed in the next section. Yule (1900) proposed for 2×2 tables

$$Q = \frac{m_{11}m_{22} - m_{21}m_{12}}{m_{11}m_{22} + m_{21}m_{12}}$$

which has properties like a correlation coefficient in that $|Q| \leq 1$ and that $Q = 0$ implies the odds ratio has the value $\theta = 1$ indicating independence. The relationship between Q and θ is $Q = (\theta - 1)/(\theta + 1)$. Positive values of Q imply a positive association in the sense that if the row and column categories represent in some sense increasing orders or stronger preferences, the responses reflect such preferences. For example if it is thought that a University degree will lead to higher salaries in employment a survey might be taken among employees aged about 40 and the results presented in a 2×2 table as in Example 12.4

Example 12.4 In a sample of 110 employees aged between 40 and 42 in an industrial complex the employees are grouped on the basis of annual salary above and below £15 000 per annum and whether or not they have a University degree or equivalent qualification. The results were

	No degree	degree
<£15 000 pa	15	9
>£15 000 pa	33	53

We estimate Q by its *sample value*, i.e. by taking n_{ij} as an estimate of m_{ij}, giving an estimate $Q^* = (15 \times 53 - 33 \times 9)/(15 \times 53 + 33 \times 9) = 0.456$.

Yule's Q has similarities to Kendall's tau for highly tied data. Here the two X values (ordinal only) may be looked upon as *no degree* (low) and *degree* (high) and a similar classification used for salaries. If the number of concordances is computed by the procedure in section 9.2 we have $C = 15 \times 53$ and the number of discordances (high X with low Y) is $D = 9 \times 33$.

12.3 Association in $r \times c$ tables

Many ideas developed for 2×2 tables generalize with appropriate modifications to $r \times c$ tables, but there are additional considerations. The basic model in section 12.2 when row categories were explanatory assumed independent binomial distributions within each row with parameters n_{i+}, π_i for row i, the criterion for no association being $\pi_1 = \pi_2$. For more than 2 response categories the response distribution within each row becomes multinomial, the parameters for the ith row, $i = 1, 2, \ldots r$ being n_{i+}, π_{ij}, $j = 1, 2, \ldots . c$ subject to constraint $\sum_j \pi_{ij} = \pi_{i+} = 1$. Independence implies for all $1 \le s, t \le r$ that $\pi_{sj} = \pi_{tj}$ for every j.

When row and column categories are both responses the simplest model with fixed marginal totals is that for sampling from a bivariate multinomial distribution with parameters n, π_{ij}, $i = 1, 2, \ldots , r$ and $j = 1, 2, \ldots , c$ subject now to the constraint $\sum_{ij} \pi_{ij} = 1$. The marginal distribution for rows is now multinomial with probabilities $\sum_j \pi_{ij} = \pi_{i+}$ while that for columns has probabilities $\sum_i \pi_{ij} = \pi_{+j}$. The condition for independence between row and column categories is $\pi_{ij} = \pi_{i+}\pi_{+j}$.

For either the *multinomial by rows* model or the *bivariate multinomial* model under the hypothesis of independence arguments similar to those in section 12.2 give the maximum likelihood estimators of the expected numbers in cell (i, j) as $m_{ij} = n_{i+}n_{+j}/n$.

Appropriate test procedures for independence or association should take account of characteristics such as whether categories are nominal or ordinal   The well-known chi-squared test for independence is widely used, but an alternative test based on the concept of the likelihood ratio has advantages in some cases. The derivation of the statistic, called the likelihood ratio statistic, is given by Agresti (1990, section 3.3.2) and it takes the form

$$G^2 = 2\sum_{i,j} n_{ij} \log(n_{ij}/m_{ij}). \qquad (12.6)$$

and is a generalization of (6.4). Under H_0, like the Pearson chi-squared statistic X^2, G^2 has asymptotically a chi-squared distribution with $(r-1)(c-1)$ degrees of freedom. An exact permutation test using G^2 is included in StatXact. Except for 2×2 tables when the p-values are nearly always identical for the chi-squared test, the Fisher exact test and this likelihood ratio test, p-values usually differ between tests but are broadly in line with each other when H_0 holds and usually indicate a similar conclusion when H_0 is rejected even though the exact p-values differ. The asymptotic counterparts of each statistic differ even for 2×2 tables but usually lead to broadly similar conclusions about significance.

The concept of odds ratio extends from 2×2 tables in the following way. An odds ratio is defined for any pair of rows s, t and any pair of columns u, v as

$$\theta_{(st)\,(uv)} = \frac{\pi_{su}\pi_{tv}}{\pi_{tu}\pi_{sv}}$$

For independence all $\theta_{(st)\,(uv)} = 1$. This follows because independence implies that $\pi_{ij} = \pi_{i+}\pi_{+j}$ for all i, j.

Clearly in an $r \times c$ table there are ${}^r C_2 \times {}^c C_2$ such ratios but many are expressible in terms of others, e.g. it is easily verified (Exercise 12.3) that

$$\theta_{(st)\,(st)} \times \theta_{(st)\,(tu)} = \theta_{(st)\,(su)}.$$

Agresti (1990, section 2.2.6) shows that all information about association in the odds ratios is contained in certain nonunique sub-sets of $(r{-}1)(c{-}1)$ odds ratio. One such subset is the odds ratios for adjacent rows and columns, referred to as the **local odds ratios**

$$\theta_{(i,i+1)(j,\,j+1)} = \frac{\pi_{ij}\pi_{i+1,\,j+1}}{\pi_{i,\,j+1}\pi_{i+1,\,j}}.$$

A useful abbreviation for this local odds ratio is θ_{ij} where $i = 1, 2, \ldots, r-1$ and $j = 1, 2, \ldots, c-1$. Agresti shows that providing all π_{ij} are nonzero using local odds ratios in place of cell probabilities causes no loss of information. The odds ratio associated with any model is unchanged if we replace the π_{ij} by the appropriate m_{ij}.

For 2×2 tables the odds ratio is a single measure containing all the information on association. There is generally no single measure for $r \times c$ tables that does not result in loss of information. Nevertheless, several single number quantities are widely used to describe features of association. Which such measure is relevant may depend upon whether categories are nominal or ordinal. In particular, for ordinal categories if we associate row categories with a variable X and column categories with a variable Y one may want to know whether X increase with Y. This is analogous to the concept of correlation for measurement data. For ordinal categories where there is no clearly defined metric the corresponding question is broadly framed as *are X and Y monotonic*? In section 12.2 I introduced the idea of concordance and discordance in 2×2 tables, leading to a correlation-like statistic. The idea extends to $r \times c$ tables.

Example 12.5   The investigation giving the data in Example 12.4 may be developed by expanding the row categories to 4 salary ranges and dividing the degree column category into the categories *first degree only* and *higher degree*, giving for the same sample of 110

	No degree	First Degree only	Higher degree
<£15 000 pa	15	8	1
£15 000 but under £20 000 pa	28	21	3
£20 000 but under £25 000 pa	4	10	6
£25 000 or more	1	5	8

A second unit is concordant with a first unit if its rankings in row and column categories are both higher than that for the first units. Thus all 15 subjects (units) with no degree who earn <£15 000 per annum give rise to a concordance with each of the $21 + 3 + 10 + 6 + 5 + 8$ subjects with better qualifications and higher salary grades, i.e., subjects below and to the right in the table. All concordances are calculated on a similar basis taking as first units all units in the first $r - 1$ rows and $c - 1$ columns. Thus the total number of concordances, C, is

$$C = 15(21 + 3 + 10 + 6 + 5 + 8) + 8(3 + 6 + 8) +$$
$$28(10 + 6 + 5 + 8) + 21(6 + 8) + 4(5 + 8) + 10(8) = 2169.$$

A second unit is discordant with a first unit if the second unit is below and to its left in the table. All possible first units giving rise to discordances lie in the *first* r–1 rows and the *last* c–1 columns. The total number of discordances, D, is

$$D = 1(28 + 21 + 4 + 10 + 1+5) + 8(28 + 4 + 1) +$$
$$3(4 + 10 + 1 + 5) + 21(4 + 1) + 6(1 + 5) + 10(1) = 544.$$

whence $Q^* = (C – D)/(C + D) = (2169 – 544)/(2169 + 544) = 0.599$. The excess of concordances over discordances suggests a more or less monotonic association between income and qualifications if these are classified in the above way.

What do concordances and discordances mean in terms of the joint probability distribution for two ordinal variables? Consider a first unit selected at random. The probability that subject is in cell (i, j) is π_{ij}. The subject will be concordant with a second subject if that second subject is in any cell (h, k) where $h > i$ and $k > j$. Thus the probability of the pair being concordant is $2\pi_{ij}(\sum_{h>i}\sum_{k>j}\pi_{hk})$. The factor 2 reflects the fact that if subject 2 is concordant with subject 1 then subject 1 is concordant with subject 2. Considering all subjects the probability of concordance between all pair is obtained by summing over all cells and is

$$\pi_C = \sum_{i,j}[\pi_{ij}(\sum_{h>i}\sum_{k>j}\pi_{hk})]$$

By a similar argument the probability of discordance is

$$\pi_D = \sum_{i,j}[\pi_{ij}(\sum_{h<i}\sum_{k<j}\pi_{hk})]$$

Association is said to be positive if $\pi_C > \pi_D$ and negative if $\pi_C < \pi_D$, while $\pi_C = \pi_D$ implies independence. The ratio $Q = (\pi_C - \pi_D)/(\pi_C + \pi_D)$ is estimated by Q^* defined in Example 12.4 and clearly has similarities to Kendall's tau for a heavily tied situation.

The above measure of association for ordinal categories was proposed by Goodman and Kruskal (1954). I used the notation Q to indicate its close relationship with the measure proposed by Yule for the 2×2 table, but it is often referred to as the gamma measure of association and denoted by Γ or γ. It is based entirely on the ordinal nature of the categories and not surprisingly that when some measure can be assigned to the magnitude of differences between categories alternative measures of association analogous to Spearman's rho or Pearson's product moment correlation coefficient are also available. These are developed using a specific loglinear model in Section 12.4.

Extension of the loglinear model from 2×2 to r×c tables is basically straightforward, the main difference being that there are many possible models of interaction instead of just one interaction term as in (12.5).

For the multinomial model the expected cell count are $m_{ij} = n\pi_{ij}$.   For independence this becomes $m_{ij} = n\pi_{i+}\pi_{+j}$, whence

$$\log m_{ij} = \log n + \log \pi_{i+} + \log \pi_{+j}.$$

extending in an obvious way from the 2×2 situation as the loglinear model for independence.   Constraints are needed for a unique solution and one possible specification is

$$\log m_{ij} = \mu + \rho_i + \gamma_j$$

where

$$\mu = \log n + (\log \pi_{1+} + \log \pi_{2+} + .. + \log \pi_{r+})/r + (\log \pi_{+1} + \log \pi_{+2} + . . + \log \pi_{+c})/c$$

$$\rho_i = \log \pi_{i+} - (\log \pi_{1+} + \log \pi_{2+} + . . . + \log \pi_{r+})/r$$

and

$$\gamma_j = \log \pi_{+j} - (\log \pi_{+1} + \log \pi_{+2} + . . . + \log \pi_{+c})/c$$

It is easily verified that the independence model implies that all the odds ratios $\theta_{(st)\,(uv)} = 1$ or $\phi_{(st)\,(uv)} = \log \theta_{(st)\,(uv)} = \log m_{su} + \log m_{tv} - \log m_{sv} - \log m_{tu} = 0$. Values of $\phi \neq 0$ imply association, the categorical data analogue of interaction. Analogous relationships involving odds ratios hold if m_{ij} are replaced by π_{ij} providing the independence condition holds.

If interactions are possible, extending arguments used for 2×2 tables we write

$$\eta_{ij} = \log m_{ij}, \quad \eta_{i.} = (\eta_{i1} + \eta_{i2} + . . . + \eta_{ic})/c, \quad \eta_{.j} = (\eta_{1j} + \eta_{2j} + . . . + \eta_{rj})/r,$$

$$\eta_{..} = (\textstyle\sum_{i,j}\eta_{ij})/n,$$

$$\mu = \eta_{..}, \quad \rho_i = \eta_{i.} - \eta_{..}, \quad \gamma_j = \eta_{.j} - \eta_{..} \text{ and } (\rho\gamma)_{ij} = \eta_{ij} - \eta_{i.} - \eta_{.j} + \eta_{..}$$

it is easily seen by substitution that

$$\log m_{ij} = \mu + \rho_i + \gamma_j + (\rho\gamma)_{ij}$$

and also that

$$\textstyle\sum \rho_i = \sum \gamma_j = 0, \sum_i (\rho\gamma)_{ij} = \sum_j (\rho\gamma)_{ij} = 0.$$

These constraints mean that the number of independent parameters are $r - 1$ parameters ρ_i, $c - 1$ parameters γ_j and $(r - 1)(c - 1)$ parameters $(\rho\gamma)_{ij}$ and also μ, so there are $r - 1 + c - 1 + (r - 1)(c - 1) + 1 = rc = n$ such parameters, implying that for any $r \times c$ table of counts parameters can be found to give an exact fit. This is a saturated model. Saturated models are of little interest and most analyses consider models where some interaction terms are set at zero. There are a number of models of interest. A few are particularly relevant to data driven analyses. Their diversity springs from the fact that the loglinear model, as already indicated in section 12.2 makes no distinction between explanatory and response categories. In the case of ordinal categories in both rows and columns a key model is the linear-by-linear association model that contains only one interaction term with a particular form. I have already indicated in earlier chapters that a number of permutation tests are special cases of that model.

12.4 Linear-by-linear association

The linear-by-linear association model uses assigned row and column scores to explore the structure of association. I denote the row scores by u_i and the column scores by v_j and to reflect ordering these are constrained to satisfy the inequalities $u_1 \leq u_2 \leq \ldots \leq u_r$ and $v_1 \leq v_2 \leq \ldots \leq v_c$. The loglinear version of the model is

$$\log m_{ij} = \mu + \rho_i + \gamma_j + \beta u_i v_j \qquad (12.7)$$

If $\beta = 0$ (12.7) reduces to the independence model. For fixed u_i, v_j there is only one interaction parameter β. The choice of scores affects the interpretation of β. Important cases include equal interval scores $u_2 - u_1 = u_3 - u_2 = \ldots = u_r - u_{r-1}$ and $v_2 - v_1 = v_3 - v_2 = \ldots = v_c - v_{c-1}$. A special case of such scores is each $u_i = i$ and each $v_j = j$, implying rank scores for the ordered row and column categories. Centered scores are sometimes used, replacing u_i, v_j by $u_i' = u_i - \bar{u}$ and $v_j' = v_j - \bar{v}$. For rank scores this implies $\sum_i \beta u_i' v_j' = \sum_j \beta u_i' v_j' = 0$ and more generally $\beta u_i' v_j'$ is a deviation from independence measurable as a monotonic trend towards the four corner cells, the direction depending on the sign of β.

Example 12.6 When $r = 5$ and $c = 7$, for rank scores clearly $u_i' = i - 3$ and $v_j' = j - 4$. The interaction terms (deviations from independence) $\beta u_i' v_j'$ are given in Table 12.4.

If $\beta > 0$, $\log m_{ij}$ is greater than its value under independence for cells at the top left and bottom right and less for cells at the top right and bottom left. If $\beta < 0$ this situation is reversed. For general scores u_i, v_j and for other values of r, c there will be broadly similar patterns of monotone association as one moves towards the corners of the table.

In terms of log odds ratios if $s < t$ and $u < v$ then

$$\phi_{(st)(uv)} = \log\,[(m_{su}m_{tv})/(m_{sv}m_{tu})] = \beta(u_t - u_s)(v_v - v_u)$$

implying that for linear-by-linear association the magnitude of the log odds ratio is larger for rows and columns that are further apart. Agresti (1990, section 8.1.2) points out that we may interpret β as the log odds ratio per unit distance on a scale determined by our choice of row and column scores. For *equal interval* scores, all local odds ratios as defined in section 12.3, i.e.

Table 12.4 Interaction terms in log m_{ij} using centered rank scores in a 5×7 contingency table.

6β	4β	2β	0	-2β	-4β	-6β
3β	2β	β	0	$-\beta$	-2β	-3β
0	0	0	0	0	0	0
-3β	-2β	$-\beta$	0	β	2β	3β
-6β	-4β	-2β	0	2β	4β	6β

$$\theta_{ij} = \theta_{(i,i+1)(j,j+1)} = \frac{m_{ij}\, m_{i+1,j+1}}{m_{i,j+1} m_{i+1,j}}$$

are equal since

$$\log \theta_{ij} = \beta(u_{i+1} - u_i)(v_{j+1} - v_j)$$

is constant for all choices of i, j. This is sometimes called *uniform association*. In particular, if the scores are unit interval scores, $\log \theta_{ij} = \beta$, whence $\theta_{ij} = e^b$.

The name linear-by-linear association reflects the fact that for any u_i, v_j the interaction contribution within a row, say row i, across columns is a linear function of the v_j with slope $\beta(u_i - \bar{u})$ with an analogous property for columns.

The relevance of this model to permutation theory is that many permutation tests depend upon a statistic of the form

$$S = \sum_{i,j} n_{ij} u_i v_j \qquad (12.8)$$

with appropriate choices of u_i, v_j. Independence ($\beta = 0$) corresponds to the null hypothesis and we reject H_0 if the observed outcome indicates a general pattern of the form exemplified for a particular case in Table 12.4 when $\beta \neq 0$. This corresponds to extreme values of S in (12.8), where intermediate values indicate independence. Statistics expressible in this form include that for the Fisher–Pitman two-sample permutation test (Section 1.6) and the Pearson product-moment permutation test (Section 9.1). The WMW test and that for Spearman's rho, being rank analogues, are special cases of linear-by-linear association.

The general nature of scores u_i, v_j in the model opens the door to *ad hoc* tests based on arbitrarily scores. This arbitrary choice carries dangers, unless the choice is clearly relevant to the problem at hand.

StatXact includes a program to test for linear-by-linear association for any chosen scores in $r \times c$ tables with ordered row and column categories. This is a powerful tool if one chooses scores to reflect appropriate distances between categories. The tests reduce essentially to applying a correlation coefficient permutation test with those chosen scores to, often heavily tied, data.

Example 12.7   Side effect of a drug, if any, exhibited by patients are classified as *none, mild, moderate, severe*. Do the following results indicate that side effects increase with dose?

Dose	Side effect level			
	None	Mild	Moderate	Severe
100 mg	49	0	1	0
200 mg	59	1	0	0
300 mg	38	1	1	0
400 mg	32	1	0	2

Row and column classifications are clearly ordinal. Many tests are available to answer the query. Some are linear-by-linear association tests with specific scores. Others are more general tests of correlation applied to a highly tied situation where 185 observations are allocated to only 4 dose levels and 4 ordinal response levels. One might use a Jonckheere–Terpstra test (section 7.5) since the dose levels are ordered. As Terpstra realized in his original paper, and as I pointed out in section 9.2 this is equivalent to a test using Kendall's tau allowing for ties and using a scoring system 1 for concordance, 0 for ties and -1 for discordance. It is similar to the counts in the Kruskal Γ statistic using the method described in the previous section. For these data it is easily verified that the number of concordances, C, is $49 \times 6 + 1 \times 2 + 59 \times 5 + 1 \times 3 + 38 \times 3 + 1 \times 2 + 1 \times 2 = 712$ and discordances, D, is $1 \times 132 + 1 \times 70 + 1 \times 32 + 1 \times 33 = 267$ whence $\Gamma = 0.4545$. StatXact provides an exact permutation test using this coefficient and this indicates an exact $p = 0.0520$. An exact test based on the closely related Kendall's tau gives $p = 0.051$, only marginally different than that for the test based on Γ. As indicated above the Kendall test allowing for ties is identical with the Jonckheere–Terpstra test.

For a linear-by-linear association test there is a wide choice of potential scores. If we base a test on Spearman's rho the scores used are the mid-rank scores for rows and columns, these being heavily influenced by ties. The midranks scores for rows are based on row totals, e.g. since the first row total is 50 the mid-rank score is 25.5. In essence these are the 50 patients 'tied' at the 100 mg. dose. Continuing, there are 60 patients with a 'tied' dose of 200 mg. so each is scored as the mid-rank of ranks 51 to 110, i.e. ½(51 + 110) = 80.5 and similarly the remaining mid-rank scores for rows are ½(111 + 150) = 130.5 and ½(151 + 185) = 168. Similarly, for columns (Exercise 12.4) the scores are 89.5, 180, 182.5, 184.5. The one-tail significance level for this test may be got using either a test for the Spearman coefficient or the linear-by-linear association test with these row and column scores. Using either in StatXact gives $p = 0.0523$, in good agreement with the result for the Goodman–Kruskal Γ or the Johnckheere–Terpstra test. A valid criticism of mid-ranks scores in the highly tied situation common with categorical data is that the mid-ranks depend heavily on the number of subjects allotted to each treatment. There must be unease about uncritical use of this test in highly tied situations, especially if the numbers allocated differ markedly between treatments. There is a better case for basing scores on the nature of treatments or responses rather than upon the numbers of patients allocated to each treatment. The unsatisfactory nature of mid-rank scores with heavy tying and unbalanced row and column totals has been pointed out by a number of writers including Graubard and Korn (1987).

An appealing choice for row scores is 1, 2, 3, 4 corresponding to doses in units of 100 mg. In the absence of more specific information on relative importance of side effects a 'default' scoring for columns might also be 1,2,3,4. These scores in a linear-by-linear test give exact $p=0.038$ in a one-sided test. However, a clinician may suggest a scoring system for side effects similar to that suggested in Example 12.2 allocating 1 to no side effect, 2 to a slight side effect and 20 and 200, say, to moderate and severe side effects. With these scores for columns and the scores 1,2,3,4 for rows the exact one-sided $p = 0.0198$.

Example 12.7 demonstrated virtually unlimited flexibility in choosing scores for a linear-by-linear association test, a freedom that needs a health warning but not just because it is a medical application. In a group of plausible tests one-tail p values ranged from 0.0198 to 0.0520, tending to decrease as more information was built into the tests. The important caveat is that such information be chosen for its relevance to the problem at hand and not on a hint inspired by a data inspection that by using a particular scoring system we may enhance the prospects of getting a significant result! The speed of modern computers tempts one to try lots of analyses of the same data. This temptation should be avoided in linear-by-linear association tests by choosing a scoring system before data are obtained.

12.5 Stratified 2×2 tables

In Example 12.1 I considered data in each of two 2×2 tables separately and indicated the consequences, often misleading, of combining data from two tables in the way that I did for the earlier analysis in that example. Each separate table had the same row and column classifications, but one referred to urban and another to rural data and sample sizes differed in each, as did the proportions responding favourably to each drug. Stratification of data into categories like *rural* and *urban* is common. Table 12.5 is another example given in Sprent (1993). Data are counts of the numbers of satisfactory and faulty items produced by each of two types of machine using raw material from 3 sources.

Here there are three *strata* corresponding to different sources of raw materials and obvious questions of interest are (i) whether the proportion of unsatisfactory items differs between machine types and (ii) whether any such differences between proportions varies between sources of raw material.

Table 12.5 Output quality for two machines using raw material from 3 different sources

Material source	Machine	Output status	
		Satisfactory	Unsatisfactory
A	Type I	42	2
	Type II	33	9
B	Type I	23	2
	Type II	18	7
C	Type I	41	4
	Type II	29	12

We may also look at k strata of 2×2 tables as one three-way 2×2×k table. Indeed, one useful way of presenting data in three-way tables is as a set ot two-way cross-sectional tables. This applies to general r×c×k tables. That is what I did in Table 12.2. The partitioning into cross-sectional tables is not unique. Other choices of strata are possible, for example, rather than giving Table 2.2 the same data could be presented stratified by drugs, i.e.

| | Standard drug | | New drug | |
	No effect	Cure	No effect	Cure
Urban	500	100	1050	350
Rural	350	350	120	180

These subtables focus attention on the difference between urban and rural cure rates for each drug separately and a chi-squared or Fisher exact test immediately establishes that for either drug the cure rate is associated with locality (i.e. differs between urban and rural). However, the stratification used in Table 12.2 (repeated below) is more appropriate for answering questions about whether the new drug is superior to the standard.

I established in Example 12.3 that cure rate and type of drug are associated for the rural data with a significantly higher cure rate for the new drug. A similar result holds for urban areas (Exercise 12.5). Having established an association between the drug administered and the cure rate and between environments (urban or rural) and cure rate we may then ask whether the association between drug administered and cure rate is similar in urban and rural areas. To answer we must be more specific about what is meant by similar. Clearly from Table 12.2 the estimated probabilities of cure for standard and new drugs in an urban environment are $p_{su} = 1/6$ and $p_{nu} = 350/1400 = 1/4$, whence $p_{nu} - p_{su} = 1/12 = 0.0833$ and for the rural environment the corresponding difference is $p_{nr} - p_{sr} = 0.1$. Standard binomial theory may be used to test whether these differences in probabilities are themselves significantly different. However, differences in probabilities may not be of prime interest, especially if the pairs of probabilities being compared are on different parts of the probability scale. What is often more

Table 12.2 (repeated) Responses to two drugs in urban and rural areas

| | Urban | | Rural | | Row |
	No effect	Cure	No effect	Cure	totals
Standard drug	500	100	350	350	1300
New drug	1050	350	120	180	1700
Column totals	1550	450	470	530	3000

interesting is whether the degree of association as measured by the odds-ratio is the same in both strata, the strata here being *urban* and *rural*.

Writing n_{ijs}, m_{ijs} for observed and expected numbers in cell (i, j) in stratum s, $s = 1, 2$ the hypothesis of equal association asks whether the odds ratio is the same for each stratum, i.e. tests $H_0:(m_{111}m_{221}/m_{211}m_{121}) = (m_{112}m_{222}/m_{212}m_{122})$ against the alternative that they are unequal. Several asymptotic tests are available Two are described and applied to the data in Table 12.2 in Examples 12.8 and 12.9.

Example 12.8 In Example 12.3 I established for the rural data that using the empirical odds ratio modified by adding 0.5 to each n_{ij} was $\theta_r = 1.498$ and that $\phi_r = \log \theta_r = 0.404$ and that the standard deviation of ϕ_r was $\sigma_r = 0.140$. A similar analysis for the urban data (Exercise 12.6) gives $\theta_u = 1.662$, $\phi_u = 0.508$ and $\sigma_u = 0.126$. The large samples involved makes it reasonable to assume that under H_0 the difference $d = \phi_r - \phi_u$ is normally distributed with mean zero and standard deviation $\sigma = (\sigma_r^2 + \sigma_u^2)^{1/2}$. Thus $Z = d/\sigma$ has a standard normal distribution under H_0 and substitution of the above values gives $Z = 0.553$, so clearly one accepts the hypothesis of equality of odds ratios.

Suppose, however, it was discovered that the row labels for the rural stratum had been accidentally interchanged and that in fact the data given for the standard drug really applied to the new drug and vice versa, so that in the country the standard drug is more effective, opposite to the result for the urban strata. As already pointed out, interchange of rows does not affect the chi-squared test because this is essentially a two-tail test. However the effect on the odds ratio of interchanging rows in a 2×2 table is easily seen to be to replace θ_r by its reciprocal and to change the sign of ϕ_r while σ_r is unaltered. Now $Z = d/\sigma = (-0.404 - 0.508)/0.188 = 4.85$. Not surprisingly this indicates a highly significant difference between the urban and rural odds ratios. A disadvantage of this method is that it does not generalize to $k > 2$ strata.

Example 12.9 An alternative method for testing whether there is a common odds ratio does extend, but only at a considerable increase in computational effort, to $k > 2$. It makes use of the loglinear model, but it has a direct analytic solution only in the case $k = 2$, requiring an iterative algorithm for $k > 2$ because $k - 1$ expected values have to be estimated subject to the constraints implied by fixed marginal totals in the three-way table and the further restriction that in each of the k stratum that the odds ratio for the expected frequencies are equal. Several algorithms are available for these procedures including one known as the iterative scaling procedure. For $k = 2$ we only need to determine one entry, e.g. that in cell (1, 1, 1) and the other entries are then all fixed by the marginal totals. In effect there is one degree of freedom in a 2×2×2 table with fixed marginal totals. To test for equality of odds ratios in the urban and rural data in Table 12.2 I denote the expected value under $H_0: \theta_u = \theta_r$ in cell (1, 1, 1) by x. Subject to fixed marginal totals it is easy to compute the remaining cell expectations in terms of x. For example, that for cell (1, 1, 2) is $n_{11+} - x = (500 + 350) - x = 850 - x$. Corresponding to Table 12.2 the cell entries corresponding to x in cell (1,1,1) conforming to the fixed marginal totals are those in Table 12.6.

For equal odds ratios x must satisfy the condition:

$$\frac{x(x - 150)}{(1550 - x)(600 - x)} = \frac{(850 - x)(680 - x)}{(x - 380)(x - 150)}$$

Table 12.6 Expected responses to two drugs in urban and rural areas in terms of an expected response x in cell $(1,1,1)$.

| | Urban | | Rural | | Row |
	No effect	Cure	No effect	Cure	totals
Standard drug	x	$600-x$	$850-x$	$x-150$	1300
New drug	$1550-x$	$x-150$	$x-380$	$680-x$	1700
Column totals	1550	450	470	530	3000

This is a cubic equation in x. If there is more than one real solution there will usually be only one that gives all expected values positive, and this is the required solution and turns out to be $x = 497.0$. A chi-squared test could be used to test the hypothesis of equal odds ratios, but for more general tests in loglinear models that I do not describe here the likelihood ratio statistic given by G^2 in (12.6) is a preferred test statistic. For these data $G^2 = 0.150$, well below the value 3.84 required for significance in the asymptotic test based on the chi-squared distribution.

A situation often arising with k strata of 2×2 tables is one where, because of small numbers in some cells, an exact or asymptotic test (chi-squared, Fisher exact or likelihood ratio) shows no evidence of association within each stratum, but the data indicates that if there were indeed an association it appears to be similar in all strata. The discussion of Example 12.1 showed the danger of combining tables with different total counts and different proportional reponses in different categories. We need a way to pool **information**, rather than data, from different strata in a manner that might demonstrate an association that is common across all strata. I discussed in Examples 12.8 and 12.9 tests for a common odds ratio when $k = 2$. How then do we test whether there is a common odds ratio for all k strata, and if we accept a hypothesis of a common odds ratio, how do we then test whether it is consistent with a hypothesis $\theta = 1$, or equivalently $\phi = 0$, that implies independence? Asymptotic theory and permutation test theory have developed along different lines for this problem. I develop first some asymptotic results and then turn to exact results based on permutation theory.

How does one estimate a common odds ratio for several strata? With various stratum sizes and perhaps different allocations to rows or columns within strata a simple average of all stratum odds ratios is too naïve. Mantel and Haenszel (1959) proposed an estimator that is widely used, namely

$$\theta^* = \frac{\sum_s (n_{11s}n_{22s}/N_s)}{\sum_s (n_{12s}n_{21s}/N_s)}. \tag{12.9}$$

This is a weighted mean of the strata odds ratios θ_s with weights $(n_{12s}n_{21s}/N_s)$ where N_s is the stratum s total. The motivation for choice of these weights is that when θ is near unity, they are close to the reciprocal of the variances of the corresponding θ_s. Breslow and Day (1980) proposed that to test whether the data across strata were consistent with this estimated common odds ratios we compute the expected frequency m_{11s} in cell(1,1,s) for each stratum conditional upon the common odds ratio being θ^*. This amounts to choosing for each stratum $x = m_{11s}$ which is a positive root not exceeding n_{1+s} or n_{+1s} of the equation

$$\frac{x(N_s - n_{1+s} - n_{+1s} + x)}{(n_{1+s} - x)(n_{+1s} - x)} = \theta^*.$$

If odds ratios are equal across strata the differences $|m_{11s} - n_{11s}|$, $s = 1, 2, \ldots, k$ should all be small and Breslow and Day (1980, chapters 4 and 5) discuss appropriate tests for homogeneity of odds ratios in detail. A statistic they propose is

$$M_{BD}^2 = \sum_s \{(m_{11s} - n_{11s})^2 / [\text{var}(n_{11s})]\} \tag{12.10}$$

where

$$\text{var}(n_{11s}) = [(1/m_{11s}) + (1/m_{12s}) + (1/m_{21s}) + (1/m_{22s})]^{-1}.$$

This statistic has asymptotically a chi-squared distribution with 1 degree of freedom under H_0. In the expression for $\text{var}(n_{11s})$ the expected frequencies are subject to fixed marginal totals given m_{11s}.

If H_0 is accepted, confidence limits for the common value may be of interest. There is a strong case for basing these on the log odds ratio since this converges to normality more rapidly than does θ^*. StatXact, which has a program for the above test also computes a confidence interval based on a method of Robins, Breslow and Greenland (1986). I omit details, but there are brief discussions in Sprent (1993, section 10.3.1) and in Everitt (1992, section 2.8.1) The former bases an interval on the log-odds ratio (with the possibility of back transformation), the latter on θ directly. The main complexity in either case is the formulae for variance and a computer program such as that in StatXact is desirable for computation.

Example 12.10 For the data in Table 12.2 we find $\theta^* = 1.593$ and after tedious but elementary arithmetic and algebra that $M_{BD}^2 = 0.3138$. If I change the labels for the rural stratum as I did in Example 12.8 and by doing so make the odds ratios very different for each stratum the Breslow–Day test picks this up clearly, and $M_{BD}^2 = 23.93$, indicating very high significance.

If, in this test we accept the hypothesis of a common odds ratio we may want to know if this is compatible with a hypothesis of independence ($\theta = 1$). A useful asymptotic test was proposed by Mantel and Haenszel (1959) and is widely referred to as the **Mantel–Haenszel** test. A nearly identical test proposed by Cochran (1954) is described by Everitt (1992, section 2.7.2). I find the Mantel–Haenszel formulation more appealing intuitively, especially as it has similarities to the test given above for homogeneity of odds ratios. Basically, the test requires the calculation in each stratum of the cell (1,1) expectation under independence for that stratum, i.e. $m_{11s} = n_{1+s}n_{+1s}/N_s$. The statistic is

$$M^2 = \frac{(m_{11+} - n_{11+})^2}{\sum_s \mathrm{var}(n_{11s})} \qquad (12.11)$$

where $m_{11+} = \sum_s m_{11s}$, $n_{11+} = \sum_s n_{11s}$ and

$$\mathrm{var}(n_{11s}) = n_{1+s}n_{+1s}n_{2+s}n_{+2s}/[N_s^2(N_s - 1)]$$

Asymptotically under H_0, M^2 has a chi-squared distribution with 1 degree of freedom. If the expected and observed values in the cell of each table differ appreciably and these differences are all in the same direction this suggests odds ratios consistently above or consistently below 1 and an increased value of M^2 and so large values of the statistic indicate rejection of H_0.

Example 12.11 For the data in Table 12.2 straightforward calculation (Exercise 12.7) gives $M^2 = 24.96$ clearly indicating rejection of the hypothesis of unit common odds ratio. In Example 12.8 I established acceptability of the hypothesis of a common odds ratio, but this test clearly rejects the possibility that that ratio is $\theta^* = 1$. The prior acceptability of a common-odds ratio is a precondition for this test to be meaningful. If we interchange rows in the rural data as in Example 12.8 it is easily verified that $M^2 = 1.560$ which is considerably less than 3.84 required for significance at the 5% level. However, in this case, as we saw in Example 12.8, the hypothesis of equal odds ratios has been rejected, so the test is not appropriate. The effect of heterogeneity on M^2 is that deviations from expectation in cell (1,1) of each stratum tend to be in opposite directions and so largely cancel out in forming the numerator of (12.11).

I have illustrated the use of the Breslow–Day and Mantel–Haenszel statistics only for $k = 2$, but they are clearly valid for all k.

A more realistic situation where the Breslow–Day and Mantel–Haenszel tests may usefully be applied is to the data in Table 12.7. Inspection suggests that side effects are slightly more common with drug B, the proportions showing side effects being higher with drug B in all age groups. A test for independence separately in each age group using a chi-square or a Fisher exact test does not reject the hypothesis of independence in any age group. However, the trend being in the same direction in each group tempts one to pool the information in the hope that we may then get a more powerful test. Clearly we might use the Breslow–Day test for homogeneity of odds ratios and then the Mantel–Haenszel test for independence. However, one may justifiably have doubts about an asymptotic tests for data that is sparse in the side-effects cells. The flimsy but

Table 12.7 Side effect incidence of two drugs on subjects in various age groups.

		No side effect	Side effect
Age under 30	Drug A	8	0
	Drug B	11	5
Age 30–39	Drug A	14	2
	Drug B	18	7
Age 40–49	Drug A	25	3
	Drug B	42	10
Age over 50	Drug A	49	2
	Drug B	22	4

best hope for reliable asymptotic results must then lie in the overall number of observations being fairly large. Fortunately there are exact tests to answer these questions. Unlike most exact tests met earlier in this book where the asymptotics are, or could be, based on the statistic used in the exact tests, here the exact tests use a different approach. I use a trivial data set to demonstrate the ideas behind the tests used in StatXact for this problem. A program like that in StatXact is needed even for quite small real data sets. Without such a program one might consider bootstrap estimation. I do not pursue that possibility here, but care is needed in selecting the correct sampling distributions for bootstrapping.

Zelen (1971) proposed an exact test for equality of common odds-ratios across strata. His test is based on hypergeometric probabilities occurring in the Fisher exact test for 2×2 tables. Subject to the condition that any set of possible n_{11s} sums over s to the observed n_{11+} commonsense suggests that if there is a common odds ratio the product of the hypergeometric probabilities will attain a maximum when the observations strongly support a common odds ratio, θ, irrespective of the value of θ. This is indeed the case and the value of this product decrease steadily as this support weakens. A simple example indicates this tendency.

Example 12.12 Consider the three 2×2 tables

3	5		4	6		4	5
2	10		2	9		1	4

The empirical odds ratios are 3, 3 and 3.2 and so these configurations look to strongly support a common odds ratios not very different from 3. The hypergeometric probabilities for each of these tables are easily computed using (4.5) and are 0.2384, 0.2128 and 0.3147 with a product 0.0160. The following set of tables involve minimal disturbance to the above tables subject to retaining the $2 \times 2 \times k$ marginal totals:

2	6		5	5		4	5
3	9		1	10		1	4

and the hypergeometric probabilities are now 0.3973, 0.0511 and 0.3147 with product 0.0064. For this set the empirical odds ratios are 1, 10, 3.2. A stronger indication of heterogeneity of odds ratios is illustrated by the following strata with the same marginal totals

5	3		1	9		5	4
0	12		5	6		0	5

giving probabilities 0.0036, 0.0851 and 0.0624 with product 0.00002 and empirical odds ratios ∞, 0.133, ∞.

The test proposed by Zelen is based on the reference sets of all possible strata subject to the given marginal totals. The probability associated with each is expressed as the product of the hypergeometric probabilities for that set divided by the sum of such products over all members of the reference set. The exact p-value is the sum of the probability associated with the observed pattern and all members of the reference set with the same or lower probabilities. Only for trivial sets is it possible to compute the relevant probabilities without a suitable computer program.

Example 12.13 Consider the two-strata data

1	6		5	2
3	6		1	4

It is easy to see that there are only 4 other possible arrangements; these, together with the above data are given in Table 12.8 which also gives the hypergeometric probability associated with each, and also the product of those hypergeometric probabilities each divided by the sum of these products. The latter is the statistic used in the Zelen test. They are also the probabilities associated with each partitioning under the hypothesis of a common odds ratio. There are two configurations with lower probability than that observed, namely configurations II and V, whence $p = 0.1399 + 0.0020 + 0.0083 = 0.1502$. Since this is greater than 0.05 we do not reject the hypothesis of homogeneity. The statistic, which is the probability associated with the observed configuration, can only take 5 possible values. In more realistic samples with more strata and larger data sets the number of possible values rapidly increases. StatXact confirms (apart from a minor rounding difference) the values given here and also gives for comparison the $p = 0.0615$ for the asymptotic Breslow–Day test. For so small a sample and with the exact test being highly discrete this discrepancy should cause no surprise.

The strongest support comes for configuration III, and it is easily verified that this is the configuration with the least difference between stratum odds ratios.

Example 12.14 For the more realistic data in Table 12.7 I used StatXact to test for homogeneity of odds ratios using the exact test. This gave $p = 0.7382$. The asymptotic Breslow–Day test gave $p = 0.6292$. Although there is a noticeable difference in these p-values both lead to the conclusion that there is no evidence against the hypothesis of homogeneity of odds ratios.

Table 12.8 Illustration of Zelen's test for homogeneity in stratified 2×2 tables

Configuration	Stratum I		Prob(P_1)	Stratum II		Prob(P_2)	$Z=P_1P_2/\sum(P_1P_2)$
I (Observed)	1 3	6 6	0.3231	5 1	2 4	0.1136	0.1399
II	0 4	7 5	0.0692	6 0	1 5	0.0076	0.0020
III	2 2	5 7	0.4154	4 2	3 3	0.3788	0.5998
IV	3 1	4 8	0.1731	3 3	4 2	0.3788	0.2500
V	4 0	3 9	0.0192	2 4	5 1	0.1136	0.0083

When a common odds ratio is accepted interest centres on whether or not this ratio might be $\theta = 1$, for if it does not there is evidence of association between drug and side effects. We might risk the asymptotic Mantel–Haenszel test for the Table 12.7 data despite the relatively small counts in some cells, but fortunately an exact test is also available. The approach, suggested by Gart (1970) is somewhat different from that in the Zelen test for homogeneity.

Gart gave not only a test for the hypothesis $H_0:\theta = 1$ but also a method for calculating confidence intervals for θ. The latter are based on noncentral hypergeometric distributions and involve mathematics and statistics beyond the scope of this book. StatXact however provides a program for such intervals. I indicate the approach for testing $H_0:\theta = 1$ (independence).

The reference set for the permutation test of $H_0:\theta = 1$ is the set of all possible combinations of k 2×2 tables where for each stratum the tables included are all possible tables with the fixed marginal totals for that stratum. In general, if for any table in stratum s we denote the cell (1,1) entry by m_{11s} then $T = \sum_s m_{11s} = m_{11+}$ will not equal n_{11+}. For given data put $t = n_{11+}$ and if there is independence this should be near some unknown expected value that would imply independence. What Gart proposed was that we base the test of $H_0:\theta = 1$ on the position of the observed t among the T for all possible permutations. For, under the assumption of a common odds ratio, if this ratio is unity then the data should in each stratum give outcomes among those of higher probabilities in a Fisher exact test in which case the observed t should be closer to $E(T|\theta = 1)$ than would be the case if θ took some other value. Example 12.15 demonstrates the basics of the test using a trivial data set.

Example 12.15 Consider the two strata

1	6		2	5
3	6		4	1

There are 5 possible 2×2 data sets with the marginal totals for stratum 1 and these have n_{111} values given in the first column of Table 12.9. The second column of Table 12.9 gives the hypergeometric probability associated with that configuration computed from (4.5). The third column gives the possible values of n_{112} for all possible stratum 2 tables with marginal totals the same as those for the observed table and column 4 gives the associated hypergeometric probabilities.

The statistic T corresponding to any combination of tables determined by the values of the cell (1,1) entries in each stratum is the sum of those cell (1,1) entries and the associated probability is the product of the corresponding hypergeometric probabilities. Clearly several of the 5×6 = 30 combinations give the same value of T. For example, $n_{111} = 4$ and $n_{112} = 5$ gives $T = 9$ as does $n_{111} = 3$ and $n_{112} = 6$. The associated probabilities are, from Table 12.9, $0.0192 \times 0.1136 = 0.00218$ and $0.1731 \times 0.0076 = 0.0132$ so $\Pr(T = 9) = 0.00218 + 0.0132 = 0.00350$. By considering all 30 combinations the probabilities for all possible values of T given in Table 12.10 may be derived from Table 12.9

Table 12.9 Fisher exact test hypergeometric probabilities for tables with fixed marginals relevant to Example 12.15

First stratum cell (1,1)	Probability	Second stratum cell (1,1)	Probability
0	0.0692	1	0.0076
1	0.3231	2	0.1136
2	0.4154	3	0.3788
3	0.1731	4	0.3788
4	0.0192	5	0.1136
		6	0.0076

Table 12.10 Probabilities associated with each T value in Gart's test for independence for data in Example 12.15.

T-value	Probability
1	0.00053
2	0.01032
3	0.06607
4	0.19710
5	0.30741
6	0.26233
7	0.12248
8	0.03009
9	0.00350
10	0.00015

For these data $T = n_{111} + n_{112} = 1 + 2 = 3$. For a one-tail test against the alternative $\theta < 1$ the relevant p is the sum of the probabilities associated with the observed T and all lesser T, i.e. $p = 0.06607 + 0.01032 + 0.00053 = 0.07692$ so we would not usually reject the hypothesis of independence. StatXact confirms this p value using a different but equivalent statistic.

A practical demonstration of the test is its application to the data in Table 12.7.

Example 12.16 For the data in Table 12.7 I showed in Example 12.14 that there was insufficient evidence to reject the hypothesis of a common odds ratio across the age strata. The Gart test provides an exact test of the hypothesis that that common odds ratio is unity (implying independence) between drugs and side effects. The StatXact program also provides a confidence interval for the true θ. As indicated above the procedure for obtaining this requires theory beyond the scope of this book, involving noncentral hypergeometric distributions for determining the relevant t. However, I indicate the practical implications for these data.

StatXact gives the Mantel-Haenszel point estimator (12.9) for these data as $\theta = 3.216$ and also a maximum likelihood estimator obtained in a way described in the StatXact manual but not discussed here of $\theta = 3.237$ in close agreement. The Gart test for $\theta = 1$ gives a one-sided p-value relevant to the alternative hypothesis $H_1 : \theta > 1$, of $p = 0.0059$. The exact 95% confidence interval for θ is $(1.253, 9.480)$ whilst that based on the asymptotic result using the Breslow–Day statistic (12.10) is $(1.306, 7.921)$. The asymptotic two-sided p given by the Manzel–Haenszel test is 0.0080 compared to an exact two-sided p obtained by combining probabilities less than or equal to that observed from each tail of 0.0095, or 0.0118 by doubling the one-tail probability. The asymptotic results here are reasonably in line with the exact theory outcomes. In passing, note that for technical reasons StatXact uses a test statistic that is not the sum of the frequencies in cell (1.1) but the sum of those in my cell (1,2). This leads to the same probabilities but reciprocals of point estimators of θ and some differences to confidence intervals. This gives no difficulty in practice as one may use StatXact to give the values here by reversing columns 1 and 2 in the data.

12.6 Multiway tables

Section 12.5 covered inferences specific to sets of k 2×2 tables, which may be combined as one three dimensional k×2×2 table. The methods developed there were however fairly specific to 2×2 tables arranged in k strata where typically the strata may be specified by what is usually called a covariate (e.g. age in a situation like that in Table 12.7). In this section I look briefly at how some of the ideas relating to inference in 2-way r×c contingency tables generalize to 3-way r×c×s tables. The simplest case is a 2×2×2 table.

Although I did not draw attention there to it being so, I introduced a simple 2×2×2 table in Example 12.1. In that example I put the cart before the horse, introducing first a 2×2 table before admitting that it could be broken into two separate 2×2 tables, one for urban, the other for rural areas. That breakdown was given in Table 12.2.

This is the best way to present a three way table in two dimensions. The three dimensions are **locality** (urban and rural) **treatment** (standard or new drug) **outcome** (no effect or cure). The first two dimensions represent explanatory variables and outcomes is a response variable. In this representation I have taken cross sections of the three dimensional table, each cross section referring to a different locality, often referred to as stratification by locality. Each of the cross-sectional sub-tables is sometimes referred to as a **partial** table. The original 2×2 table that I gave as Table 12.1 was obtained by superimposing the above two partial tables by adding corresponding cell entries and is often referred to as a **marginal** table; it ignores the effect of localities.

Clearly by taking cross sections based on the other dimensions we get different sets of partial and marginal tables. If we stratify by treatment the partial tables are the two in Table 12.11 and the associated marginal table ignoring treatments is that in Table 12.12. In Exercise 12.9 I ask you to write down the corresponding tables when we stratify by outcomes. Interpretationally, this final marginal table ignores outcomes and simply tell us whether there is an association between treatments and locality, reflecting features of the way the experiment was conducted. Clearly the concept of two-way partial and marginal tables extends to $r \times c \times s$ tables, where, for example, if I stratify by r I will get r partial tables each $c \times s$ and superimposing these on each other and adding the counts in each superimposed cell gives one $c \times s$ marginal table.

Concepts such as odds ratios and loglinear models extend to three and higher way tables and often lead to informative analyses. In section 12.1 I showed that although chi-squared or Fisher exact tests all demonstrated association for each of the 2×2 tables in Tables 2.1 and 2.2 inspection of the data indicated that the direction of association was in the opposite direction for the table in Table 12.1 (the marginal table in the notation of this section) to that for the subtables in Table

Table 12.11 Responses in urban and rural areas to each of two drugs

| | Standard drug | | New Drug | |
	Urban	Rural	Urban	Rural
No effect	500	350	1050	120
Cure	100	350	350	180

Table 12.12 Numbers cured or not cured in urban and rural areas, irrespective of drug

	Urban areas	Rural areas
No effect	1550	470
Cure	450	530

12.2 (the partial tables). If we base our tests on odds ratios this difference in the nature of associations immediately become apparent.

Example 12.17   For the two tables in Table 12.2 the odds ratios are easily calculated as 1.67 and 1.50. For that in Table 12.1 the odds ratio is 0.8556. Recalling that an odds ratio $\theta = 1$ corresponds to independence and that odds ratios less than 1 indicate association in the opposite direction to odds ratios greater than 1, Simpson's paradox is immediately evident. For the two tables in Table 12.11 the odds ratios are 5 and 4.5 respectively and for Table 12.12 the odds ratio is 3.88. There is no paradox in this case, but ignoring which treatment is given has diluted the evidence of association. Can you see why this should be?

Agresti (1990, section 5.2.2) gives an excellent real-data example of Simpson's paradox. The data is used to explore associations between death penalty verdicts, defendant's race and victim's race.

A full discussion of the use of the loglinear model in 3-way tables is given by Agresti (1990, sections 5.2 and 5.3). Many of the ideas carry over from 2-way tables but the variety of interaction terms of interest is greater. These take into consideration the natures of association. For example, denoting the category distribution variates by X, Y, Z the model appropriate when Y is independent of both X and Z is different from that when Y is independent of X only when Z is fixed (conditional independence). The reader should refer to Agresti for details.

Exercises

12.1   Confirm that for a 2×2 table that independence implies that if m_{ij} are the expected numbers in cell (i, j) then $m_{11}m_{22} = m_{21}m_{12}$.

12.2   Show that under independence in a 2×2 table the conditional probabilities satisfy the relationship $\pi_{(1)1}\pi_{(2)2} = \pi_{(2)1}\pi_{(1)2}$.

12.3   Verify the relationship between odds ratios $\theta_{(st)(st)} \times \theta_{(st)(tu)} = \theta_{(st)(su)}$ given in section 12.3

12.4   Verify the column midrank scores given in Example 12.7.

12.5   Using an approach similar to that in Example 12.3, confirm an association between cure rate and type of drug for urban patients using the Table 12.2 data.

12.6   Confirm the values of θ_u, ϕ_u and σ_u quoted in Example 12.8 for the data in Table 12.2.

12.7   Confirm the value of M^2 quoted in Example 12.11.

12.8   Confirm the probabilities given in Table 12.10 calculated for the data in Example 12.15.

12.9   Compile subtables analogous to those in Tables 12.11 and 12.12 for stratification by outcomes for the data given originally in Table 12.2.

12.10   Hand *et al* (1994, Set 77) give data from a report in the *Daily Telegraph* (04.07.67) on voting at an Anglican Church Assembly on a motion to allow the ordination of women as priests . How do the voting patterns for the three groups, bishops, clergy and laity differ? How widely do you consider any inferences apply; e.g. does it make sense to think of any or all of these groups to be random samples from some hypothetical population?

	Vote		
	Yes	No	Abstain
House of bishops	1	8	8
House of clergy	14	96	20
House of laity	45	207	52

*12.11 Hand et al (1994) give the following data published by an American Insurance company [Metropolitan Life Insurance Company, *Statistical Bulletin*, **60**, (1979)] for deaths in sporting parachuting in each of 3 years graded by experience measured by number of jumps. Does the pattern of experience in relation to deaths change over time and if so how do you interpret it? Does it conform with any *a priori* intuitive ideas you may have?

Number of jumps	Year		
	1973	1974	1975
1–24	14	15	14
25–74	7	4	7
25–199	8	2	10
200+	15	9	10

The number of jumps was not reported for two deaths in 1974.

*12.12 Neyzi, Alp and Othon (1975) and Hand et al (1994, Set 286) give data categorizing 13–14 year old Turkish girls by socio-economic class of parent on a scale 1 to 4 and breast-development on a scale 1 (no development) to 5 (full development). Is there evidence of association between class and development? If so, explore the nature of that association.

Socio-economic class	Breast development				
	1	2	3	4	5
1	2	14	28	40	18
2	1	21	25	25	9
3	1	12	12	12	2
4	6	17	34	33	6

*12.14 The data below are part of a set given by Hand et al (1994, Set 374) for numbers of male and female candidates standing for the Labour and Conservative parties in the 1992 UK general election in different regions of the UK. Is there evidence of an association between parties and sex and if so does the nature of this association differ between regions?

Region	Political party			
	Conservative		Labour	
	M	F	M	F
Southwest	101	8	84	25
Southwest	45	3	36	12
Greater London	26	8	57	27
West Midland	50	8	43	15
Northwest	65	8	57	16
Wales	36	2	34	4
Scotland	63	9	67	5

13

Further categorical data analysis

13.1 Partitioning of chi-squared statistics

In this chapter I cover several topics on analysis of contingency tables where there is often a need to modify the approach in accord with the nature of the data and its relevance to answering specific questions.

It sometimes helps an analysis if we partition a chi-squared or a likelihood ratio statistic with k degrees of freedom into independent additive components in much the way the total sum of squares is partitioned in an analysis of variance. More than one mode of partitioning is usually possible.

Some basic rules for partitioning into subtables giving statistics with the desired additive property are listed by Agresti (1990, sections 3.3.5–7). He gives the following necessary conditions:

- Subtable degrees of freedom must sum to the degrees of freedom of the original table.
- Each cell count in the original table must appear in one and only one subtable.
- Each marginal total of the original table must be a marginal total for one and only one subtable.

To meet these conditions Lancaster (1949) suggested partitioning an $r \times c$ table into $(r-1)(c-1)$ tables each 2×2 with 1 degree of freedom where the 2×2 tables have the form:

$$\begin{array}{cc} \sum_{s<i}\sum_{t<j} n_{st} & \sum_{s<i} n_{sj} \\ \sum_{t<j} n_{it} & n_{ij} \end{array} \qquad (13.1)$$

For example, for a 3×3 table

$$\begin{array}{ccc} n_{11} & n_{12} & n_{13} \\ n_{21} & n_{22} & n_{23} \\ n_{31} & n_{32} & n_{33} \end{array}$$

this partitioning gives the 4 tables

(i)
$$\begin{array}{cc} n_{11} & n_{12} \\ n_{21} & n_{22} \end{array}$$

(ii)
$$\begin{array}{cc} n_{11}+n_{21} & n_{12}+n_{22} \\ n_{31} & n_{32} \end{array}$$

(iii)
$$\begin{array}{cc} n_{11}+n_{12} & n_{13} \\ n_{21}+n_{22} & n_{23} \end{array}$$

(iv)
$$\begin{array}{cc} n_{11}+n_{12}+n_{21}+n_{22} & n_{13}+n_{23} \\ n_{31}+n_{32} & n_{33} \end{array}$$

It is easily verified (Exercise 13.1) that this partitioning satisfies the above necessary conditions. The partitioning is not unique. For example, if rows and columns of the original table are permuted the chi-squared statistic for the whole table is unchanged but clearly the partitioning above will be altered. Although under the null hypothesis of independence asymptotically the chi-squared statistics for each sub-table are independent and additive the Pearson chi-squared statistic computed for each of the 2×2 tables (i) – (iv) above do not in general sum to the corresponding statistic for the original table. However, if the likelihood ratio statistic (12.6) is used the components are additive.

Partial partitioning (i.e. not necessarily to single degrees of freedom) is also possible. In practice one should choose partitionings that help to answer relevant questions about association.

Example 13.1 This example is from Sprent (1993). Drugs used in chemotherapy often cause side effects of varying severity. Two such drugs were tested on 100 patients, 60 receiving drug A and 40 drug B. Numbers showing none, one, or both of two side effects – hair loss (HL) and visual impairment (VI) – were recorded. The results are given in Table 13.1.

A meaningful partitioning is into single degree of freedom components that examine possible association between drugs and each single side effect, between drugs and either single or double side effects, and between over-all side effect status and drugs.

For Table 13.1 the overall value of the chi-squared statistic is $X^2 = 20.80$ and $G^2 = 22.13$ each with 3 degrees of freedom. One degree of freedom is associated with the table

	HL	VI
Drug A	9	4
Drug B	3	16

and for this $X^2 = 9.41$ and $G^2 = 9.72$. A further degree of freedom is associated with the table

	HL or VI	HL and VI
Drug A	13	16
Drug B	19	2

and for this $X^2 = 11.02$ and $G^2 = 12.24$. The remaining degree of freedom is associated with the table

	One or two side effects	None
Drug A	29	31
Drug B	21	19

Table 13.1 Incidence of side effects, hair loss (HL) and visual impairment (VI) for two drugs

	Side effect status			
	HL	*VI*	*HL and VI*	*None*
Drug A	9	4	16	31
Drug B	3	16	2	19

for which $X^2 = 0.17$ and $G^2 = 0.17$. The component X^2 sum to 20.60, slightly less than the value $X^2 = 20.80$ obtained for Table 13.1, but the component G^2 sum to 22.13, the value for the original table. You should verify (Exercise 13.2) that this partitioning satisfies the necessary conditions for additivity quoted above.

Interpretation is straightforward. The high values of G^2 (or X^2) for Table 13.1 indicates (using either an asymptotic result or an exact test) an association between drugs and side effects. The first component indicates this association is strong among those exhibiting only one side effect, drug A being more likely to result in hair loss, and drug B in visual impairment. The second component indicates that among those who suffer side effects drug A is much more likely than drug B to give rise to a double side effect. The third component clearly indicates no evidence of association between drugs and presence of some (unspecified) side effect or effects, i.e. the drugs are equally likely to produce at least one side effect. What action is appropriate in the light of these findings would depend upon the importance of each side effect. If visual impairment were slight and temporary it might be rated less serious than hair loss if that were permanent and that would indicate a clear preference for drug B. If visual impairment were severe and permanent it would be more serious than hair loss and the better drug to use is less clear; a decision might rest with factors such as whether two side effects combined are appreciably worse than only one.

An important aspect of being able to choose components when partitioning is that in a certain sense this introduces ordering. This feature is not present in an over-all chi-squared test because the test is unaltered by permutation of rows and columns. However, if we permute rows and columns before carrying out the partitioning in (13.1) it gives a choice of partitioned 2×2 component tables that depends upon the ordering of rows and columns in the chosen permutation. This is important in applications where rows or columns have some natural ordering and the somewhat different partitioning described later in sections 13.2 and 13.3 are invoked.

13.2 A new look at location and dispersion

Best (1994) uses a different partitioning of chi-squared into components, one analogous to that into linear and quadratic components in a regression-type analysis of variance. He was interested in attitudes of consumers to *sweetness* in chocolate, but he points out that the partitioning is generally valuable in looking for both location and dispersion differences and can be extended to examining differences in other features such as skewness.   The approach utilises a partitioning of the chi-squared statistic X^2, rather than G^2 as was the case in the previous section. It is based on a proposal by Lancaster (1953). I illustrate it here using data given by Best.

Example 13.2. The data in Table 13.2 are from a study comparing responses of Japanese and Australian consumers to various sweet foods. The data are ratings for sweetness of Japanese chocolate on a scale 1 to 7 by each class of consumer, a score of 1 indicates the consumer did not like the sweetness and 7 that he/she thought it very good.

The mean sweetness scores for Australian consumers is $\bar{x} = (2 \times 1 + 1 \times 2 + 6 \times 3 + \ldots +$ $6 \times 7)/33 = 4.91$ and for Japanese consumers the mean is $\bar{y} = 4.87$. Not surprisingly neither the t-test (two sided $p = 0.92$) nor the WMW test (exact two-sided $p = 0.45$) indicate a significant difference between means. On the other hand treating the data as counts arranged in a 2×7 contingency table the chi-squared statistic is $X^2 = 10.70$ and for the asymptotic test $p = 0.10$ and for the exact test $p = 0.08$, closer to values suggesting significance. Use of these tests is open to criticism. In the light of the spread of the data and the fact that the sample mean scores agree if rounded to one decimal place any test that **did** indicate a location difference should be suspect! Further, inspection of the data suggests that any difference in preference is more like one in dispersion, the scores for Japanese consumers being more concentrated around the score 5, almost half giving a score of 5, whereas there is more diversity in Australian scoring, less than one-quarter giving a score of 5 and almost one half giving a score outside the set {4, 5, 6}, whereas less than one-sixth of the Japanese give scores outside that set. The chi-squared test takes no account of the order implicit in the column scores.

The indication of dispersion differences suggest that if a test for location differences were required then something like a Behrens–Fisher test or one of its nonparametric or bootstrap analogues (section 6. 6) would be more appropriate, but the closeness of the sample means makes any such test somewhat pointless.

With a strong indication that there is no location difference, interest shifts to a possible dispersion difference hinted at above. An obvious approach is to use one or more of the nonparametric dispersion tests given in section 6.5. For instance, StatXact gives the following exact p-values:

Seigel–Tukey	$p = 0.0032$
FABDB	$p = 0.0033$
Mood	$p = 0.0021$

Table 13.2 Numbers of Australian and Japanese consumers awarding each score on a scale 1 (dislike) – 7 (approve strongly) for sweetness of Japanese chocolate

Consumer nationality	1	2	3	4	5	6	7
Australian	2	1	6	1	8	9	6
Japanese	0	1	3	4	15	7	1

Sweetness liking score

The procedure for choosing components used in Example 13.1 could be applied to Table 13.2 and you should verify (Exercise 13.4) that this gives the components

2	1		3	6		9	1		10	8		18	9		27	6
0	1		1	3		4	4		8	15		23	7		30	1

An interpretation is that the first column of the kth component ($1 \le k \le 6$) is the cumulative count of scores not exceeding k and the second column gives counts of those scoring $k + 1$. Each component is therefore appropriate to tests of association between (i) nationality and (ii) current and lower scores. This partitioning would pick up a sudden shift in national preferences at a particular score. Visual examination of Table 13.2 and the separate components given above suggests that such switches may occur between scores 3 or lower and 4 and also between scores 6 or lower and 7. However, neither these, or any other component gives a $p < 0.10$ in the exact likelihood ratio test using G^2. The testing procedure here might in broad terms be regarded as one of looking for large differences between the cumulative distribution functions and in that sense has analogies with the Smirnov test. It would seem to be of limited interest.

Best suggested partitioning chi-squared into components as proposed by Lancaster (1953) giving independent linear, quadratic and high order components. The technicalities of computing test statistics for such components is more complex than that for the simple cases considered so far in this chapter. Best gives details of the procedure and if software to carry out the necessary computations is not available he provides sufficient detail to construct an algorithm. His paper should be consulted for the technical but important practical details.

The linear component test (effectively a location test) is shown by Best to be equivalent to the WMW location test if the *scores* used are not those assigned by the tasters but are mid-rank scores based on these. The quadratic component, which is a measure of spread or dispersion, based on mid-rank scores has an associated $p = 0.005$, close to the p-values quoted above for the Siegel–Tukey, FABDB and Mood tests.

Each of the linear and quadratic components has 1 degree of freedom. The remaining 4 degrees of freedom give a value of the statistic X^2 well below that needed for significance.

Best and Rayner (1996) give other examples of partitioning of chi-squared into linear, quadratic and higher order components in various situations with categorical data. The method is not confined to mid-rank scores, and for more general monotone scores the linear component statistic and the corresponding tests are equivalent to using a linear-by-linear association analysis with those scores.

A key advantage of the partitionings advocated by Best and Rayner is that they let us not only test for location and dispersion differences but also, by using higher order components, to examine features like differences in skewness or kurtosis. These authors developed the methods for use primarily in investigations into consumer responses to food products and here one is clearly interested not only in whether the average score, usually on some arbitrary scale, has the same mean (or median) for different groups of consumers, but whether there is a different spread of scores. In particular, does one group of consumers break strongly into two sub-classes one of whom find a product highly acceptable, and the other who find the product obnoxious, while those in another consumer group nearly all find the product moderately acceptable? In a test each group may then well give very similar mean scores, but any sensible dispersion measure for the scores will show differences between groups.

13.3 More thought for food

To compare preferences for competing food varieties such as different brands of tomato soup or different varieties of apple a number, N, of potential consumers may be asked to rank each of k brands or varieties from 1 to k in order of preference. The idea clearly extends to consumer goods other than food. If tied rankings are not allowed the results of such tests may be expressed in a two-way square $k \times k$ table, rows corresponding to each of k brands or varieties and columns to each of the k ranks. The entries in the cells (i, j) of the table are the numbers among the N testers giving rank j to brand i. This model was considered by Anderson (1959) who gave the data in Table 13.3 for a study in which 123 consumers were asked to rank three varieties of snap beans in order of preference. The basic difference between this and most contingency tables considered so far in this and the previous chapter is that the cell entries are not sampled from 369 individual units. Instead each of the 123 testers separately allocates the ranks 1, 2 and 3 to each of the 3 varieties. Since each consumer uses each rank he or she is represented once in each column so one could validly carry out a chi-squared test on any column as a goodness of fit test for some hypothesis. For column 1, for example, the hypothesis of no preference implies that the expected allocation of ranks for each variety is 41. In other words, no preference implies a uniform distribution of ranks over varieties and the chi-squared test described in section 6.7 may be used with, here, 2 degrees of freedom. While such tests may be carried out on any column the tests are clearly not independent, for if they were we could add the components and get a chi-squared test with 6 degrees of freedom, whereas clearly with fixed marginal totals the total chi-squared only has 4 degrees of freedom for once the entries in the first two rows and columns are given those in the third row and column are fixed.

In the general case with r varieties and N testers (or tasters) the results may be put in the $r \times r$ form in Table 13.4

Applying a chi-squared test for uniform allocation of rank 1 as in section 6.7 the expected value for each variety is N/r, and thus the relevant test statistic under H_0 is $X_1^2 = \sum_i (n_{i1} - N/r)^2/(N/r)$. Under H_0 standard theory tells us that asymptotically X_1^2 has a chi-squared distribution with $r - 1$ degrees of freedom. The expectations for each cell take the same form as they do for independent sampling from Nr units when marginal totals are fixed. If we compute the analogue of X_1^2

Table 13.3 Number of testers giving various rankings to three varieties of snap beans

Variety	Rank 1	2	3	Total
I	42	64	17	123
II	31	16	76	123
III	50	43	30	123
Total	123	123	123	369

Table 13.4 Format for N testers giving ranks 1 to r to r varieties or brands

Variety	Rank 1	2	.	.	r	Total
1	n_{11}	n_{12}	.	.	n_{1r}	N
2	n_{21}	n_{22}	.	.	n_{2r}	N
.						
.	.	.	.	.	.	.
r	n_{r1}	n_{r2}	.	.	n_{rr}	N
Total	N	N	.	.	N	rN

for each column, as already pointed out their sum will not have a chi-squared distribution with $r(r-1)$ degrees of freedom since the X_r^2 are not independent, However, writing $X^2 = \sum X_r^2$, using the standard result for the mean of a sum of r variables, not necessarily independent, it follows that under H_0, $E(X^2) = r(r-1)$. The analogous statistic, X'^2, say, for multinomial sampling from a total of rN units has for the square table $(r-1)^2$ degrees of freedom and $E(X'^2) = (r-1)^2$. Anderson noted that $E[(r-1)X^2/r] = (r-1)^2$ and conjectured that if X^2 where adjusted by multiplication by $(r-1)/r$ that the resulting statistic might have a chi-squared distribution with $(r-1)^2$ degrees of freedom under H_0, and proved this to be so in Anderson (1959). The proof requires an understanding of the general theory of multivariate distributions for correlated variables and is not given here. The practical implication is that in this situation the usual chi-squared statistic must be reduced by a factor $(r-1)/r$ to obtain the appropriate statistic having asymptotically a chi-squared distribution with $(r-1)^2$ degrees of freedom under H_0. For the data in Table 13.3 Anderson found $X^2 = 79.56$ whence, for $r = 3$, the appropriate test statistic value is $X'^2 = 2\times79.56/3 = 53.04$ with 4 degrees of freedom and $p < 0.0001$.

Anderson suggests several alternative analyses for data of this type. In particular he discusses a partitioning of X'^2 into 4 additive components each with a single degree of freedom, pointing out that, as I showed above in other contexts, such partitionings are not unique and the important thing is to choose those that have an informative interpretation. In particular Anderson considers one partitioning that leads to a test equivalent to the Friedman test in section 8.2.

He considers for a 3×3 table a partitioning into 4 orthogonal contrasts. These and their interpretations are:

C_1:	$n_{21} - n_{23}$	Linear ranking effect for variety II.
C_2:	$n_{11} - n_{13} - (n_{31} - n_{33})$	Difference between linear rankings for I and III.
C_3:	$n_{21} - 2n_{22} + n_{23}$	Quadratic ranking effect for II.
C_4:	$n_{32} - n_{12}$	Difference between quadratic rankings for I, III.

The interpretation of C_4 is clearer if we write it in an alternative form

$$[(n_{11} - 2n_{12} + n_{13}) - (n_{31} - 2n_{32} + n_{33})]/3$$

This equivalence is easily established by noting that $n_{i1} + n_{i2} + n_{i3} = N$, $(i = 1, 3)$

Each single degree of freedom component of X'^2 take the form $C_i^2/\text{var}(C_i)$. For this 3-variety case we obtain the variances of each n_{ij} and the covariances between pairs by noting that for each observer there are only 6 possible allocation of ranks to the three varieties. In order of varieties these may be written $(1, 2, 3)$, $(1, 3, 2)$, $(2, 1, 3)$, $(2, 3, 1)$, $(3, 2, 1)$, $(3, 1, 2)$. Under the hypothesis that there is no variety preference all these rankings are equally likely. Thus the probability that variety 1 will be ranked 1 is 1/3 and the probability it will not be ranked 1 is 2/3, and if we denote by X_{ij} the random variable that takes the value 1 if variety i is allocated rank j and the value zero otherwise, i.e., the X_{ij} are Bernouilli random variables and so in particular $E(X_{11}) = 1/3$ and $\text{var}(X_{11}) = 2/9$. Since all N tasters rank independently it follows that $E(n_{11}) = N/3$ and $\text{var}(n_{11}) = 2N/9$. A moment's reflection shows that these results hold for all n_{ij}. However, as I have emphasized the n_{ij} are not independent of one another so we need the covariances between them. For determining $\text{var}(C_1)$, for example, we need to know the covariance of n_{21} and n_{23}. We first find the covariance between X_{21} and X_{23}. For the six possible allocation of ranks given above the values of X_{21} and X_{23} are:

Preference order	(1, 2, 3)	(1, 3, 2)	(2, 1, 3)	(2, 3, 1)	(3, 2, 1)	(3, 1, 2)
X_{21}	0	0	1	1	0	0
X_{23}	0	1	0	0	0	1

All outcomes have equal probability so it follows that $E(X_{21}) = E(X_{23}) = 1/3$ and $E(X_{21}X_{23}) = 0$, whence $\text{cov}(X_{21}, X_{23}) = -E(X_{21}).E(X_{23}) = -1/9$. Since all N observers rank independently it follows that $\text{cov}(n_{21}, n_{23}) = -N/9$. It is not difficult to see that this result holds for any two cells in the same row or in the same column of the relevant 3×3 table. However, (Exercise 13.5) for cells in different rows and columns similar arguments to those above establish that $\text{cov}(n_{ij}, n_{kl}) = N/18$. Using the above variances and covariances it is easily established (Exercise 13.6) that $\text{var}(C_1) = \text{var}(C_4) = 2N/3$ and $\text{var}(C_2) = \text{var}(C_3) = 2N$.

The choice of single degree of freedom contrasts above is arbitrary, but they have the interesting feature that by combining the chi-squared components corresponding to the first two contrasts, i.e. $(3C_1^2 + C_2^2)/(2N)$, it is not difficult (Exercise 13.7) to show that this may be put in the form $\sum_i (n_{i3} - n_{i1})^2/N$. If the statistic is written in an alternative form $\sum (r_i - \bar{r})^2/N$, it can fairly easily be shown to be equivalent to the Friedman statistic T in (8.2), where r_i is the rank total for the ith variety. Clearly this is equivalent to Kendall's coefficient of concordance.

Example 13.3 For the data in Table 13.3 it is easily confirmed that $C_1 = 31 - 76 = -45$ and that the corresponding component of X'^2 is $3 \times (45)^2/(2 \times 123) = 24.70$ and in like manner that the remaining components are 0.10, 22.86 and 5.38, summing to $X'^2 = 53.04$. To calculate the Friedman statistic for these data we note that $r_1 = 42 + 2 \times 64 + 3 \times 17 = 221$, and similarly $r_2 = 291$ and $r_3 = 226$ and $\bar{r} = 246$, whence the Friedman statistic is $T = 24.80$. This is easily verified (Exercise 13.8) if we note that here, in the notation used in (8.2) that $t = 3$ and $b = 123$. The remaining 2 degrees of freedom in X'^2 are associated with departures from linearity (differences in location) and may be regarded as indicators of differences in dispersion. For the data in Table 13.3 it is fairly clear that there is a monotonic falling off in preference for variety III (row 3), but the pattern is different for the other two varieties. Variety I is most commonly ranked 2, a lesser number prefer it to each of the other varieties and only a few rank it 3, whereas for variety II more than half the testers rank it 3, but amongst the remainder there are more who rank it 1 than there are who rank it 2, giving a larger dispersion of rankings for this variety than for either of the others.

Anderson suggests that if the experimenter is more interested in varietal contrasts at the more favourable ranks (1 and 2) a better set of contrasts would be

$$D_1 = n_{21} - n_{11}, \quad D_2 = n_{31} - 2n_{21} + n_{11}, \quad D_3 = n_{12} - n_{13} - (n_{32} - n_{33}), \quad D_4 = n_{23} - n_{22}.$$

The components D_1 and D_2 are linear and quadratic comparisons for the first rankings (of main interest in this approach) and the remaining two examine differences in the second and third rankings. A number of writers have considered problems of this type including Scholz and Stephens (1987), Best (1994) and Best and Rayner (1996).

13.4 Matched pair data and repeated responses

Entries in a square contingency table may arise from matched pairs. This and closely related situations are considered by Agresti (1990, Chapter 10). Many of the concepts extend from paired to general repeated measurement situations such as some arising in medicine where groups of patients are examined and diagnosed as fitting one of several categories and then are recategorized on a later occasion and interest centres on whether patients change categories between examinations.

A matched pair situation arises in, for example, studies of the relationship between occupations of parents and children when both parents and children are categorized by the same set of occupations, where again samples are dependent.

For such matched pair studies responses may be summarized in a two-way table in which rows represent the classification for, in this case parents' occupations and columns represent children's occupations. The table is necessarily square. Here is an example in another context.

Example 13.4 When workers are exposed to potential health hazards medical examinations are frequently held at regular intervals and any change in health status of employees is monitored to detect the onset of new symptoms or a worsening of existing ones. Abnormalities may or may not be connected with working in a particular environment. Interest centres on changes in the incidence or extent of specific abnormalities with time. The following data were obtained for workers in an industrial plant who were exposed to fumes from which there was a potential risk of respiratory impairment. In each of the years 1974 and 1977 employees were subjected to a set of tests to measure symptoms of breathlessness and allocated to one of four categories according to increasing evidence of a tendency to display breathlessness. Category A implied no abnormality and categories B, C and D increasing levels of breathlessness. For 114 employees taking part in both surveys the relationship between categories in 1974 (rows) and 1977 (columns) are given in Table 13.5. For reasons of commercial confidentiality further details not needed for the present discussion are not given.

An individual who is placed in the same category in 1974 and in 1977 is included in the count in one and only one diagonal cell. The 10 individuals recorded in cell (2,1) are people who showed some slight breathlessness (grade B) in the 1974 survey but no indication of breathlessness (grade A) in 1977. In broad terms all those below the diagonal from top left to bottom right exhibited less severe breathlessness symptoms in 1977 than they did in 1974, while counts in cells above the diagonal indicate more severe symptoms in 1977 than those observed in 1974. The null hypothesis is that changes in symptom incidence are equally likely to be in either direction between the surveys against the alternatives (i) that there has been an improvement in health status between surveys or (ii) a deterioration between surveys, or (iii) a change in one direction or the other. The first two alternative lead naturally to a one-tail test and the third to a two-tail test.

Do you recognise this problem as a generalization of one met earlier? The McNemar test introduced in section 6.2 and illustrated in Example 6.3 is a special case with only 2 rows and 2 columns. We based the test there on a comparison of off-diagonal cells (2, 1) and (1, 2) for which under the null hypothesis the expected counts are equal, or equivalently that the probability π_{21}, π_{12} of an observation falling into each of these cells was the same.

Table 13.5 Relation between respiratory grades A, B, C, D for 114 workers in 1974 and 1977

		1977 grading			
		A	B	C	D
	A	74	7	4	1
1974 grading	B	10	5	1	1
	C	3	4	2	1
	D	0	0	1	0

In the 2×2 table the hypothesis $H_0 : \pi_{12} = \pi_{21}$ is one of symmetry and if we assume multinomial sampling with the cell probabilities $\pi_{11}, \pi_{12}, \pi_{21}, \pi_{22}$ it is easy to show (Exercise 13.9) that $\pi_{12} = \pi_{21}$ implies $\pi_{1+} = \pi_{+1}$ which in turn implies $\pi_{2+} = \pi_{+2}$. The hypothesis H_0 defines symmetry in a 2×2 table and the equalities $\pi_{i+} = \pi_{+i}$, $i = 1,2$ define marginal homogeneity. For 2×2 tables either condition implies the other. For a general $r \times r$ table symmetry requires $\pi_{ij} = \pi_{ji}$ for all i, j and marginal homogeneity that $\pi_{i+} = \pi_{+i}$ for $1 \le i \le r$. It is easily verified (Exercise 13.10) when $r > 2$ that symmetry implies marginal homogeneity but that the converse is not necessarily true.

In section 6.2 I pointed out that McNemar's test was effectively a sign test where under the null hypothesis the off-diagonal element of a 2×2 matrix, n_{12} was under H_0 distributed binomially with parameters $m = n_{12} + n_{21}$ and $\pi = \frac{1}{2}$, implying that asymptotically $Z = (n_{12} - \frac{1}{2}m)/[m(\frac{1}{2})(\frac{1}{2})]^{\frac{1}{2}} = (n_{12} - n_{21})/(n_{12} + n_{21})^{\frac{1}{2}}$ has a standard normal distribution under H_0. Its square has a chi-squared distribution with 1 degree of freedom. It is easy to show Agresti (1990, section 10.2.1) that for a general matched pair table with r rows and columns this test for symmetry extends to give a statistic $X^2 = \sum_j \sum_{i<j} \{[(n_{ij} - n_{ji})/(n_{ij} + n_{ji})^{\frac{1}{2}}]^2\}$ which under the hypothesis of symmetry has a chi-squared distribution with $\frac{1}{2}r(r-1)$ degrees of freedom. Clearly this reduces to the McNemar statistic when $r = 2$. This test was first proposed by Bowker (1948).

Example 13.4 (continued) The hypothesis $H_0 : \pi_{ij} = \pi_{ji}$ is clearly relevant in this problem. Any evidence that it does not hold would indicate a change in health status for some employees at least (either for better or worse) between 1974 and 1977. The asymptotic result must be suspect for so small a data sample, suggesting perhaps that a bootstrap approach or one based on randomization theory would be more appropriate unless the X^2 statistic clearly indicates nonsignificance. It is easily verified that $X^2 = (10 - 7)^2/7 + (4 - 3)^2/7 + 1 + 0 + 1 + 0 = 2.67$. With 6 degrees of freedom this gives no indication of significance. I do not know of any programs giving an exact test. Care is needed to specify the appropriate permutation set under the restrictions implied by H_0.

The Bowker test takes no account of order in the response categories in this example from none to severe symptoms. Tests that take ordering into account are discussed in Agresti (1990, section 10.4) using loglinear models. Agresti also discusses models for testing marginal homogeneity which is sometimes a more realistic model than one of symmetry.

13.5 Trends in contingency tables

In a $2 \times c$ table the two rows may represent binomial responses and the columns counts corresponding to first row binomial probabilities $\pi_{11}, \pi_{12}, \pi_{13}, \ldots, \pi_{1c}$ where one may want to test the hypothesis $H_0 : \pi_{11} = \pi_{12} = \pi_{13} = \ldots = \pi_{1c}$ against a one-sided alternative $H_1 : \pi_{11} \leq \pi_{12} \leq \pi_{13} \leq \ldots \leq \pi_{1c}$ (or a similar hypothesis with all inequalities reversed). An appropriate test was proposed by Cochran (1954) and Armitage (1955) and is often referred to as the Cochran–Armitage test. A large number of smokers might be divided into sets classified by smoking status, e.g. nonsmokers, < 10 cigarettes per day, 10 to 20 cigarettes per day, more than 20 per day, and the numbers from each category classified as normal or abnormal on some criteria such as chronic cough, excessive blood-pressure, evident throat damage.

Exercise 13.5 StatXact demonstrates the test with the data in Table 13.6 given by Graubard and Korn (1987).

The maximum likelihood estimates of p, the probability of no deformation, assuming independent binomial distributions in each column are respectively $p_1 = 17066/(17066+48) = 0.9972$ and similarly $p_2 = 0.9974$, $p_3 = 0.9937$, $p_4 = 0.9921$, $p_5 = 0.9737$. There is a slight indication that the probability of no malformation decreases (or that the probability of malformation increases) with alcohol intake but doubt about this being meaningful may arise because of the small numbers of mothers who are heavy drinkers and also in any one drinking category the relatively small proportion of babies with abnormalities.

Clearly any test of monotonically increasing or decreasing π_i, where the actual values of the π_i are not specified, is potentially a test of linear-by-linear association. Cochran and Armitage proposed equally spaced scores 1, 2, 3, 4, 5. StatXact provides a program that uses these scores, designating it the Cochran–Armitage test, but the test may also be carried out using the linear-by-linear association program (with row scores 1, 2). The one tail p-value is $p = 0.1046$.

Other scores are possible, as in Example 12.7. Indeed, the assumption of a binomial model within each column is unnecessarily restrictive; all that is required to justify a linear-by-linear association model is a monotone trend in proportions. Ideally the choice of scores should be one that imparts a near-linearity to any such trend. In practice scores usually reflects our perception of the likely relationship between some characteristic of a population and dose level. For these data Graubard and Korn suggest the mean alcohol intake as scores, i.e. 0, 0.5, 1.5, 4 and 7 where the choice of 7 is arbitrary but not unreasonable. With these scores the linear-by-linear association test gives an exact one-tail $p = 0.0168$, which is an appreciable reduction on the p-value using Cochran–Armitage scores.

Table 13.6 Congenital sex organ malformation in children characterized by mother's alcohol consumption (From Gruabard and Korn, 1987)

| | Alcohol consumption (number of drinks per day) | | | | |
	0	<1	1–2	3–5	≥6
Malformation					
Absent	17066	14464	788	126	37
Present	48	38	5	1	1

Another scoring system sometimes chosen is mid-ranks, but even more than in Example 12.7, these are here strongly influenced by high numbers in the lower drinking groups relative to those for heavier drinkers. The scores here are 8557.5, 24.365.5, 32013, 32473, 32,555.5. The scores for the last three columns (where most of the changes in malformation probability take place) are given little different weight. For these scores the one-tail $p = 0.2861$.

The temptation to experiment with many scoring systems should be avoided and when no other system is obviously appropriate a default scoring system 1, 2, 3, 4, . . . has much to commend it.

In Example 13.5 a scoring system that might find support on clinical grounds would be to combine the first two columns on the reasoning that low intake of alcohol may be known not to be dangerous. There might also be a suspicion (backed by evidence of past experience) that heavier drinkers tend to under-report alcohol intake, so it might be argued that while those reporting 1–2 drinks per day may average 1.5, those reporting 3-5 may drink rather more with an average about 5 and those reporting 6 or more perhaps average 10. Thus one might combine the first two columns and give new column scores 0, 1.5, 5, 10. For a linear-by-linear test with these scores StatXact gives an exact one-tail $p = 0.0099$, slightly but not dramatically lower than for the value $p = 0.0168$ with scores proportional to alcoholic consumption. That the result is not too different for these two scoring schemes is comforting in that it suggests the analysis is not too sensitive to differences between scoring system that may be regarded as realistic under what most people would regard as rational assumptions.

Incidentally, in this section for convenience in layout of tables I have departed from the usual convention of allocating explanatory categories to rows and responses to columns. InTable 13.6 clearly rows represent the responses and columns are explanatory variable categories.

13.6 Statistics and the law

In Example 1.2 I presented data from Gastwirth and Greenhouse (1995) on promotions among 100 employees in a firm where the employees are classified as belonging to either a majority group or a minority group where I supposed the minority group were some specific ethnic group.   In firm A the situation was:

	Promoted	Not Promoted	Total
Minority group	1	31	32
Majority group	10	58	68

Gastwirth and Greehouse point out that if the data were presented in a court case alleging discrimination on ethnic grounds, then courts will often require a further explanation from a company to justify that its policy is not discriminatory on those grounds if the statistical evidence indicates rejection at the 5% significance level of the hypothesis that promotion and ethnic grouping are independent. If it were significant the employer would be called upon to justify the discrepancy on grounds other than ethnic considerations. For example, they might then argue that promotion required either agreement or willingness to work unsocial hours or spend long periods away from home and that relatively few in the minority group met that requirement. I mentioned in Example 1.2 and it should now be clear that an appropriate test for association between ethnic group and promotion policy would be Fisher's exact test. I pointed out that although the proportion of the minority group promoted was only 1/32, or 3.125% compared to the overall rate of 11 per cent the data did not provide evidence of association significant at the 5% level in the Fisher test; indeed the exact one-tail $p = 0.0765$

I also gave the following data from Gastwirth and Greenhouse for a comparable group of employees recruited from the same population of potential and eligible employees for a second firm B. For this firm the data were

	Promoted	Not Promoted	Total
Minority group	2	46	48
Majority group	9	43	52

Here the proportion of the majority group promoted was 4.167%, greater than the corresponding proportion for firm A, so clearly, I argued, tongue-in-cheek, there is not going to be significant evidence of association.

However, the one-tail $p = 0.0351$ in the Fisher exact test indicates association between promotion and ethnic grouping at the 5% level required by many courts.

Here is the explanation I promised in Example 1.2 for the anomaly.   It may partly be accounted for by the fact that there are relatively fewer (32) of the minority group in the pool eligible for promotion for employer A compared to the 48 in that group for employer B, because sample sizes do influence the behaviour of the Fisher exact test as I pointed out in section 6.6.   This in turn is a consequence, as Gastwirth and Greenhouse indicate, of differences between hiring policies for the two firms.   These are reflected in the row totals for the above tables. In 100 employees hired, firm A took on only 32 from the minority group and firm B took on 48. Now if, as Gastwirth and Greenhouse supposed in this hypothetical example in the workforce available for hire there were 40 in every 100 belonging to the minority group these figures do not show significant discrimination in either firm's hiring policy. This is easily confirmed because the number of minority hirings has a binomial distribution with $p=0.4$ and $n = 100$ (Exercise 13.11). However in the same exercise you will find that the *hiring policies* differ significantly between the two firms. The smaller number of minority hirings reduces the power of the Fisher exact test applied to firm A relative to the power for the test where there is a surplus of minority hirings (firm B). This extra power is enough to produce a significant result despite the higher percentage promotions among minority hirings (or the higher odds ratio if one makes the comparison on that basis). There is a timely warning here about the need for care when using conditional tests when the marginal totals used for conditioning may themselves be conditional upon a further factor, in this case hiring policy. Similar difficulties arise in other fields, especially in medical statistics, if one conditions on events in the causal path, e.g. in a study of lung cancer and smoking it would be unwise to condition upon patient status for other disorders such as emphysema or hypertension which are often related to smoking.

Use of statistical arguments in Courts and also in forensic medicine is a fast growing field. Courts on the whole prefer straightforward and not too complex analyses providing the logic is sound. A definitive work on statistical methods in a legal context is Gastwirth (1988) and other books dealing with statistics in legal contexts include Finkelstein and Levin (1990) and Aitken (1995).

Many statisticians complain of a reluctance by lawyers and the judicial system to consider statistical arguments, finding this strange in view of the common requirement of proof beyond reasonable doubt in criminal cases and the tendency to decide many civil disputes on a balance of probabilities. The reluctance to accept statistical arguments may be partly an inbred hostility to attempts to quantify uncertainty coupled to the non-unique nature of such quantification. This becomes apparent in adversarial systems where prosecution and defence experts may often present plausible analyses of the same data that lead to different conclusions. This situation is particularly likely to arise when each side presents different but valid analyses that differ in power, or that test subtly different hypotheses. The above discussion of Example 1.2 indicates the sort of anomaly that may arise because of power differences when the same test is applied to two data sets.

I conclude this section with simple examples that illustrate some approaches and a few potential problems that may arise in statistical analysis of employment data.

Example 13.6. When age discrimination is alleged in compulsory lay-off of employees courts often compare lay-off rates among those aged less than 40 and those aged over 40. Gastwirth (1997) gives for a case settled out of court before trial the following data for lay-offs:

	Age < 40	*Age 40+*
Not laid-off	787	1113
Laid-odd	18	50

Here the Fisher exact test is appropriate for association between age and lay-off policy and if the alternative hypothesis specifies discrimination against older employees the exact test one-tail $p = 0.0085$ indicates association. Courts often have a trusting belief in the conventional 5% significance level. Even if a two-tailed test were deemed appropriate on the grounds that there may well be discrimination against younger (rather than older) employees, the result would still be significant. However, as Gastwirth points out if more data are available there may be more powerful tests that give a lower p-value. Indeed for this case Gastwirth gives the following more detailed breakdown.

Age	*<40*	*40–49*	*50–59*	*60+*
Not laid-off	787	632	374	107
Laid-off	18	14	18	18

For these data the Cochran–Armitage or one of the related tests described in section 13.5 is appropriate. Without further information there is no reason to use scores other than the Cochran–Armitage 1, 2, 3, 4 for columns and 1, 2 for rows. With these scores the exact one-tail $p = 0.000\ 000\ 07$ giving even stronger evidence that the layoff policy adversely affected older workers and called for careful judicial scrutiny.

Example 13.7 The following data set was brought to my attention by Joseph Gastwirth and concerns lay-offs in a case that came before a US court in 1994 [Maidenbaum *versus* Bally's Park Place Inc., 68 FEP Cases 1245 (D.C. N.J., 1994), *affirmed* (3rd. Cir. 1995)].

Age	*<40*	*40–44*	*45–49*	*50–54*	*55–59*	*60+*
Not laid off	122	44	22	8	8	3
Laid-off	5	3	4	2	0	0

In this case the judge ruled that in his view there was no discrimination against older employees because no employees in the two oldest groups were laid-off. The Cochran–Armitage test with the default scores 1, 2, 3, 4, 5, 6 for columns here gives a one-tail exact $p=0.1108$, supporting the judge's view. However if the Fisher exact test is applied to a dichotomy of those aged under 40 and those aged 40 or more the one-tail $p = 0.078$. This is still above any conventional significance level, but less than p for the Cochran–Armitage test. This is indicative of a loss of power with the latter test when there are small numbers in some categories, particularly the over-sixties. In these circumstances tests are sometimes sensitive to small data changes. If, for example, there had been the same total number laid off, but one less person laid off in the *45–49* age group and one person laid off in the *60+* group the exact one-tail $p = 0.0343$ (Exercise 13.12) in the Cochran–Armitage test. The statistical explanation for this reduction in the p-value is clear. The lay-off rate in the highest age group has increased from zero to 33.3% with only a less dramatic change in one other lay-off rate. This result is statistically interesting but courts are often rightly wary about entertaining arguments introducing hypothetical data. Gastwirth (1996) illustrates the dangers of this and discusses in nontechnical terms other difficulties that may arise with statistical evidence in equal employment opportunity litigation.

Example 13.8 The following data are derived from some given in Gastwirth (1988) Table 6.9. In essence the data arose in a case where employees were grouped into two categories. Those in the high-ranking category received merit pay awards and those in the low-ranking category did not. If a higher proportion of blacks than whites were allocated to the lower-ranking category it could be argued that there was a salary discrepancy accounted for by race. The data below are for 5 different districts:

Race *District*	*Black* *High rank*	*Low rank*	*White* *High rank*	*Low rank*
NC	24	9	47	12
NE	10	3	45	8
NW	5	4	57	9
SE	16	7	54	10
SW	7	4	59	12

In all regions the proportion of Blacks in the low rank category is greater than the proportion of whites, yet only in the NW region does the Fisher exact test indicate significant association. The situation is rather like that with the data in Table 12.7 considered in Examples 12.14 and 12.16. For the data here the Mantel-Haenszel test or the Zelen exact test for a common odds ratio followed by a Gart test to see if that common ratio differs significantly from unity would be appropriate. For these data, for example, StatXact gives an exact $p = 0.709$ for the Zelen test indicating no evidence against a common odds ratio. The Mantel–Haenszel estimate of the common odds ratio is 2.166 with a 95% confidence interval (1.244, 2.770).

Gastwirth discussed these data giving more background information and data and also discusses the desirability or otherwise of incorporating data from one further district where there were no blacks employed.

The above examples give no more than a flavour of the potential role of statistics in a legal framework.

Exercises

13.1 Verify that the partitioning of a 3×3 table into four 2×2 tables given on p.364 satisfies the necessary conditions for additivity of the chi-squared components given on p.363.

13.2 Confirm that the partitioning in Example 13.1 satisfies the necessary conditions for additivity of the chi-squared components given on p. 363.

13.3 For the data in Table 13.2 plot the sample cumulative distribution functions on the same graph. Does it surprise you that the Smirnov test for these data does not indicate a population distributional difference?

13.4 Check whether the partitioning of the 2×7 Table 13.2 into 2×2 tables given in Example 13.2 satisfies the necessary conditions given on p.363.

13.5 For data as in Table 13.4 when $r = 3$ show that if $i \neq k$ and $j \neq l$ then $\text{cov}(n_{ij}, n_{kl}) = N/18$.

13.6 For conditions C_1, C_2 (p. 370) show that $\text{var}(C_1) = 2N/3$ and that $\text{var}(C_2) = 2N$.

13.7 For the contrasts C_1 to C_4 on p.370 show that the sum of the chi-squared components corresponding to C_1 and C_2, i.e. $(3C_1^2 + C_2^2)/2N$, is equivalent to the Friedman statistic (8.2).

13.8 Verify all calculations in Example 13.3

13.9 Establish that for the McNemar test that if $\pi_{12} = \pi_{21}$ this implies marginal homogeneity, i.e. that $\pi_{i+} = \pi_{+i}$, $i = 1, 2$.

13.10 Prove that for an $r \times r$ table with paired data responses considered in section 13.4 that symmetry implies marginal homogeneity but that the converse is not necessarily true.

13.11 For the employment statistics example discussed at the start of section 13.6 confirm that the hiring policy of neither firm indicates ethnic discrimination using test at the 5% significance level, but that there is evidence that the firms may operate different policies.

*13.12 For the revised data in Example 13.7 confirm the computed one-tail $p = 0.0343$ for a Cochran–Armitage test.

*13.13 Norton and Dunn (1985) and Hand et al (1994, Set 24) give snoring and heart disease status for 2484 patients. Is there evidence that snoring and heart disease are related?.

Heart disease	Snoring status			
	Never	Occasional	Most nights	Every night
Yes	24	35	21	30
No	1335	603	192	224

*13.14 The UK Department of Education and Science gave the following data in *Statistics of Education, 1980, Vol.2, School leavers.* The data are also given by Hand et al (1994, Set 91) The four achievement categories represent increasing levels in the English examination system, a being the lowest and d the highest. The data are numbers achieving each category in Mathematics and French for 383 boys and 366 girls. Explore the relationship, if any, between sex and achievement in the two subjects:

Subject	Mathematics				French			
Achievement	a	b	c	d	a	b	c	d
Males	82	13	176	112	289	10	44	40
Females	74	23	184	85	221	8	74	63

*13.15 Dunn (1989) and Hand et al (1994, Set 406) give data on grades from 0 (poor) to 4 (excellent) to each of 29 candidates by two examiners. Is there evidence that one examiner tends to be more lenient than the other?

Candidate	1	2	3	4	5	6	7	8	9	10	11	12	13	14	15	16	17	18
Examiner A	1	0	0	2	0	4	0	0	0	2	1	2	0	4	4	1	0	1
Examiner B	2	0	0	2	0	3	0	0	0	3	2	3	1	3	3	2	2	2

Candidate	19	20	21	22	23	24	25	26	27	28	29
Examiner A	2	0	2	4	0	0	4	0	1	3	2
Examiner B	3	0	3	4	0	0	3	2	2	4	3

14

Data-driven or model-driven

14.1 A merging of approaches

Expanding the range of data-driven analyses leads to a closer relationship between these and what I have called model-driven methods. This is consistent with the comment in section 1.2 that a data-driven approach almost inevitably leads to the concept of a model even if that model is only vaguely specified. The difference is that in a data-driven approach models are chosen to satisfy conditions imposed by or suggested by the data or our knowledge about what they represent, whereas in the model-driven approach the data are either assumed to fit a pre-specified model, or if they do not, they are adapted to fit it by procedures like transformations. Experienced statisticians use both approaches, making a choice of which, and of which technique in each, is appropriate for every new problem and at each stage of an analysis.

This chapter is a brief introduction to a broad class of model-driven methods where the models are so flexible that they adapt to fit many kinds of data and represent something of a synthesis of data- and model-driven approaches.

I consider first binary responses; e.g. the outcome of a treatment may be success or failure, manufactured items may be classed satisfactory or defective, the issue at a single birth may be male or female. We may represent such outcomes for an individual unit by a random variable Y which takes only the values 0 and 1. We arbitrarily denote one outcome as a *success* and the other as a *failure* and associate a success with $Y = 1$ and a failure with $Y = 0$. If we have independent observations Y and for each $\Pr(Y = 1) = \pi$ and $\Pr(Y = 0) = 1 - \pi$, then the number of successes r in a sequence of n such observations is a random variable X with a binomial distribution $B(n, \pi)$.

When binary responses generate a binomial variable we often want to know how changes in some other variable or variables affect the value of π. For example, if batches of insects are subjected each to a different dose of a pesticide the proportion of insects dying in each batch is likely to increase as the dose level increases. The proportion in a sample dying when subjected to a given dose provides an estimate of the relevant π for that concentration.

Similarly, if patients with a skin disorder are given a range of doses, e.g. 50. 75, 100 150, 200, 250 mg of a drug to treat that disorder and all observations are on different patients a clinician might anticipate that over that range the proportion of cures may increase with increasing dose. If we denote the ith dose level by x_i and r_i among n_i patients who receive this dose are cured , then $p_i = r_i/n_i$ is an estimate of the binomial parameter π_i for that group and we often use data from such an experiment to estimate a relationship between dose, x, regarded as an explanatory variable, and response as reflected in the variable π. The 'binomial' assumption is often an over-simplification and there are a number of alternative models that allow, for example, flexibility in the probabilities of response between individuals receiving the same dose, but I do not consider these.

Situations where the probability associated with a binary response is related to some explanatory variable such as dose have a superficial resemblances to classic regression where, in a simple model we are interested in a response Y which is distributed $N(\mu, \sigma^2)$ where σ is a constant but μ is an unknown linear function of one or more explanatory variables. Here a model of the form

$$\mu_i = E(Y_i|x_i) = \alpha + \beta x_i \qquad (14.1)$$

is fitted to data by least squares and might be used for estimation of, or tests of hypotheses about, the parameters α, β or for prediction. In the latter case we almost certainly want some estimate of how good the predictions are. We might also carry out diagnostic tests like those in section 11.3 to see how well the model does its job.

The simplicity of inferences associated with linear regression on the mean in the normal case hinges on the mathematical simplicity of the least squares procedure which is itself only appropriate if the nuisance parameter σ^2 is constant for all x, even though σ^2 may be unknown. The method is often aided by the kindness of nature in generating approximate linear relationships between explanatory and response variables that are close to straight lines in the bivariate case or relations easily put in a linear regression format for two or more explanatory variables.

In the spirit of (14.1) in toxicity studies one might try to fit a least squares linear regression of p_i on x_i to estimate $\pi_i = E(p_i|x)$. There are two good reasons for not doing this. Since π is a probability its values are restricted to the interval $0 \le \pi \le 1$, yet, by its nature a linear relationship will for some values of x give values outside this range. Further, least squares is no longer an optimum method of fitting in this situation for the variance is not constant, but is itself a function of π having its maximum when $\pi = \frac{1}{2}$ and declining to zero when $\pi = 0$ or 1. These limitations were recognised more than 60 years ago and alternative methods of estimating $E(p|x)$ were introduced, particularly in connection with problems about toxicity of insecticides and more generally about reactions of living material to chemicals.

In experiments to measure toxicity of an insecticide separate batches of insects are subjected to different concentrations of that insecticide over a range of values expected to give a wide spread of proportions showing a given response, usually death. Records are then made of the proportions dying at each concentration based upon the binary responses death, scored as 1, or survival, scored as 0. If the proportion dying is plotted against dose a variety of curves may arise. Sometimes a transformation of x gives a relationship that is easier to interpret. I illustrate a common situation involving binary response data in Examples 14.1 which shows the diversity of problems of this type by taking a classic example not related to toxicity or responses to a drug.

Example 14.1 The data in Table 14.1 given by Milicer and Szczotka (1966) and also in Hand *et al* (1994, Set 476) was analysed by Finney (1971) and again by Morgan (1992).

In Figure 14.1 the proportion (p) having menstruated is plotted against age (x). A curve to fit these data would be more or less symmetric and you probably recognise an appropriate one as not unlike that for the cumulative distribution function for a normal or perhaps some other symmetric distribution that has a longer or shorter tail.

Clearly the plot in Figure 14.1 does not suggest even an approximate straight line relationship between p and x except perhaps for values of x between about 12 and 14 and if such a relationship were fitted it would suggest values of $p < 0$ for x below about 11.5 and values of $p > 1$ for x above about 14.5! One gets a better fit to the data over the observed range of x by fitting a cubic in x but a moment's reflection indicates that extrapolation to values of x outside the range 9 to 17 may result in statistically nonsensical values of p. Further, nonhomogeneity of variance of the p_i reduces the efficiency of least squares as a method of fitting.

The appeal of a straight line relationship for simplicity of interpretation of experimental data stimulated a search for transformations that would produce approximate straight-line relationships between some function of p and x in situations like that considered in Example 14.1 An early approach was based on the similarity of response curves like that in figure 14.1 to the cumulative distribution function for a normal distribution. A technique called probit analysis is based on this similarity and provides a method to determine a straight-line relationship between a function of the response called a probit and an explanatory variable. A full account of the method and its applications is given by Finney (1971). In general, when either seems appropriate, probit analysis gives similar result to the method described in the next section, but the latter has wider applicability and the transformation involved has a physical interpretation in terms of odds, so this method has now largely replaced probit analysis.

Table 14.1 Age of menarche in 3918 Warsaw girls in 1965

Mean age of group	Number in group	Number having menstruated	Proportion having menstruated
9.21	376	0	0
10.21	200	0	0
10.58	93	0	0
10.83	120	2	0.02
11.08	90	2	0.02
11.33	88	5	0.06
11.58	105	10	0.10
11.83	111	17	0.15
12.08	100	16	0.16
12.33	93	29	0.32
12.58	100	39	0.39
12.83	108	51	0.47
13.08	99	47	0.47
13.33	106	67	0.63
13.58	105	81	0.77
13.83	117	88	0.75
14.08	98	79	0.81
14.33	97	90	0.93
14.58	120	113	0.94
14.83	102	95	0.93
15.08	122	117	0.96
15.33	111	107	0.96
15.58	94	92	0.98
15.83	114	112	0.98
17.58	1049	1049	1.00

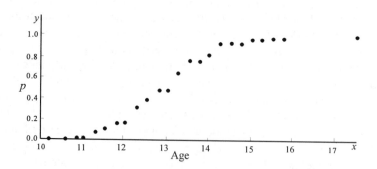

Figure 14.1 Proportions of Warsaw girls having menstruated in different mean age groups.

14.2 Logistic regression

A more versatile transformation than the probit transformation is associated with a family of distributions known as logistic distributions. I show in section 14.3 how analyses based on this transformation are applicable to some simple procedures given in earlier chapters.

For a single explanatory variable x suppose that for given $x = x_i$ the probability of a success is π_i. I further assume that

$$\pi_i = \frac{\exp(\alpha + \beta x_i)}{1 + \exp(\alpha + \beta x_i)} \tag{14.2}$$

where α and β are constants.

It is easily shown (Exercise 14.1) that

$$\lambda_i = \log[\pi_i/(1-\pi_i)] = \alpha + \beta x_i. \tag{14.3}$$

This means that the logarithm of the odds ratio, often called the logit, is linearly related to x_i. The relationship (14.3) is often referred to as a **logistic regression**. The idea extends immediately to any number of explanatory variables and for binary data logistic regression plays a role analogous to that of linear regression for continuous response variables. The model (14.2) may look restrictive but is no more so than the assumptions of normality and homogeneity of variance for the usual least squares regression model. It will not fit every set of data and indeed regression diagnostics for binary data and logistic regression are as relevant as they are when using a least squares model. The model may well break down, for example, if some explanatory variables are not included or if there is asymmetry in the curve relating the π_i and the x_i.

*14.3 Estimation of logistic regression parameters

Maximum likelihood is widely used for estimating the parameters α, β in (14.3) and the asymptotic theory for this model-based approach is described in many books including Cox and Snell (1989), who also describe a permutation based exact inference procedure valuable for small data sets. Modern software makes possible exact inferences based on this permutation theory. One such package, LogXact, is specifically for logistic regression and some idea of the range of problems it tackles is indicated in the LogXact manual and excellent examples of problems amenable to logistic analysis are given in Chapter 1 of Cox and Snell.

*This section gives a slightly more advanced treatment of a specialist topic and may be omitted at first reading.

I discuss briefly inference for the one explanatory variable case and give simple examples of its application. More complex problems involving several explanatory variables are analysed in specialist texts like Cox and Snell.

Readers not familiar with the concept of the likelihood function and maximum likelihood estimation may wish to consult a standard statistical text on these subjects for more detail but the treatment below is reasonably self contained.

Suppose we have a series of n observation (x_i, y_i) where the x_i are explanatory variables and the y_i are observed values 0 or 1 of independent binary variable Y_i taking the value 1 with probability π_i given by (14.2). Under the independence assumption the joint probability of the observed outcome is

$$\Pr(Y_1 = y_1, Y_2 = y_2, \ldots, Y_n = y_n) = \frac{\exp(\alpha \sum_i y_i + \beta \sum_i x_i y_i)}{\prod_i [1 + \exp(\alpha + \beta x_i)]} \qquad (14.4)$$

The expression (14.4) regarded as a function of the unknowns α, β is the likelihood function. Readers familiar with maximum likelihood estimation will recognise $t_0 = \sum_i y_i$ and $t_1 = \sum_i x_i y_i$ as sufficient statistics for determining α, β. If you are not familiar with the concept of sufficient statistics what you need to know for present purposes is that these statistics contain all the information in the data that is relevant to inferences about α and β. The physical interpretation of t_0 is that it is the number of Y_i that take the value 1 while t_1 is the sum of the x_i associated with those Y_i that take the value 1 in our sample.

The maximum likelihood point estimates of α, β are the values that maximize (14.4). Taking logarithms in (14.4) gives a function called the loglikelihood and it is easily established that the same values maximize both the likelihood and the loglikelihood. In practice it is generally easier to determine the maximum of the loglikelihood. If a maximum exists it is a solution of the equation obtained by equating the derivatives of the loglikehood with respect to each parameter to zero. In this, like many situations where the principle is invoked, iterative methods are needed to obtain estimates. Computer software to obtain solutions is widely available and the same programs usually compute confidence intervals and appropriate p-values for use in hypothesis testing. Asymptotic results often suffice for all but small samples. What inferences one wishes to make depend very much on the problem being tackled. One may want to estimate α and β or some function of these. The latter is common in toxicity studies where one is often interested, especially when comparing, say, different insecticides, in the dose x that will kill half the population, often called the median lethal dose. This is the value of x in (14.3) corresponding to $\lambda = 0$ since clearly $\lambda = 0$ implies $\pi = \frac{1}{2}$. If the estimates of α, β are $\hat{\alpha}$, $\hat{\beta}$ an appropriate estimate of median lethal dose obtained by setting $\lambda = 0$ is $x = -\hat{\alpha}/\hat{\beta}$. In other cases, as in classic least squares regression, we are interested in estimating, or testing hypotheses about, β only.

Exact conditional inference based on permutation theory is a data-driven approach described by Cox and Snell (1989, sections 2.1.3 and 2.1.4) in the more general context of any number, p, of explanatory variables. I only consider here inferences about β with one explanatory variable. In this situation α is regarded as a nuisance parameter which is fixed, even if unknown. To make inferences about β we consider the distribution of a statistic T_1 corresponding to t_1 conditional upon a statistic T_0 being fixed at its observed value t_0. An argument for using this approach due to R A Fisher is that when we are given only t_0 that statistic alone tells us nothing about β. However the value of t_0 is relevant to the precision with which we draw conclusions about β, so it is therefore appropriate to argue conditionally upon its observed value.

From the definition of conditional probabilities it follows that

$$\Pr(T_1 = t_1 \mid T_0 = t_0) = \Pr(T_0 = t_0, T_1 = t_1)/\Pr(T_0 = t_0). \qquad (14.5)$$

Using (14.4) and remembering the definitions of t_0, t_1 it follows that the numerator of (14.5) is

$$\Pr(T_0 = t_0, T_1 = t_1) = C(t_0, t_1) \; \frac{\exp(\alpha t_0 + \beta t_1)}{\prod_i (1 + \exp(\alpha + \beta x_i))} \qquad (14.6)$$

where $C(t_0, t_1)$ is the number of distinct binary sequences that yield the observed t_0, t_1. The denominator in (14.5) is obtained by summing an expression of the form (14.6) over all possible $t_1 = u$. Remembering that we keep t_0 fixed for each value of u, the number of permutations is $C(t_0, u)$. In (14.5) then the denominator terms in (14.6) cancel as do the numerator terms that involve t_0, whence, putting $t_1 = t$, (14.5) may be written

$$p_T(t, \beta) = \frac{C(t_0, t)e^{\beta t}}{\sum_u C(t_0, u)e^{\beta u}} \qquad (14.7)$$

In particular when $\beta = 0$ this reduces to

$$p_T(t, 0) = \frac{C(t_0, t)}{\sum_u C(t_0, u)} \qquad (14.8)$$

so that in that case the distribution of T_1 given t_0 is determined by the combinatorial coefficients.

Example 14.2. A value (0 or 1) of a binary variable Y was recorded at each of 8 values of an explanatory variable X. The results were:

X	0	3.5	4.0	4.2	4.6	7.0	9	10.5
Y	1	1	0	1	0	0	0	0

Test the hypothesis $H_0: \beta = 0$ against the alternative $H_1: \beta \neq 0$.

For these data $t_0 = \sum_i y_i = 3$ and $t_1 = \sum_i x_i y_i = 0 + 3.5 + 4.2 = 7.9$. Conditional upon $t_0 = 3$ (i.e. three successes), inspecting the data we see there is no other permutation of the observed Y giving $t_1 = 7.9$. (Remember t_1 is the sum of the x_i corresponding to the $y_i = 1$.) Thus the numerator of (14.8) is $C(t_0, t_1) = 1$. There are $^8C_3 = 56$ ways to allocate 3 successes ($y = 1$) to the 8 values of x_i, so the denominator of (14.8) is 56 and thus under H_0 we have $p_T(t, 0) = 1/56$. The strongest evidence against H_0 would be a bunching of the 3 successes at either low or high values of the x_i. These correspond to extreme values of the statistic T_1. If the statistic T_1 does not look familiar turn to Example 1.3. It is identical with and has the same distribution as the statistic S_1 in that example. Thus the test here of zero association between X and the logit, λ, and thus by implication between X and π, reduces to the Fisher–Pitman test. Indeed the data in this example is equivalent to the Group 1 allocation of measurements in Example 1.3. The procedure here is a complementary way of viewing the one used there. Our attitude here is that if there is no treatment effect allocation to groups may be looked upon as independent of X, whereas in Example 3.1 we looked upon the X as being independent of allocation to groups.

It is easily verified that the observed T_1 is the second lowest of all possible values and thus falls in a lower-tail critical region of size 2/56 and the corresponding two-tail $p = 4/56 = 0.0714$. If the X are replaced by ranks the test is equivalent to a WMW test.

Even with specialist programs like LogXact exact inference is only practical for relatively small sample sizes.

The more common method of solving problems involving logistic regression is by maximum likelihood. For a simple problem it is relatively easy to write down the loglikelihood derived from (14.4).

Example 14.3. The data in Example 13.5 on a possible relationship between maternal alcohol consumption and certain malformations in offspring provides suitable material for a logistic regression analysis if appropriate X scores are assigned. For illustrative purposes I use mean alcohol consumptions, i.e. $x = 0, 0.5, 1.5, 4, 7$. From Table 13.6 we easily find:

Mean Consumption	0	0.5	1.5	4	7
Number with abnormalities	48	38	5	1	1
Numbers in study	17114	14502	793	127	38

Counting (ironically?) a malformation as a success, we find $t_0 = 48 + 38 + 5 + 1 + 1 = 93$ and $t_1 = 0 \times 48 + 38 \times 0.5 + \ldots + 1 \times 7 = 37.5$. Using (14.4) the loglikelihhood is easily shown to be

$$93\alpha + 37.5\beta - [17114 \log(1 + e^{\alpha}) + 14502 \log(1 + e^{\alpha + 0.5\beta}) + 793 \log(1 + e^{\alpha + 1.5\beta}) + 127 \log(1 + e^{\alpha + 4\beta}) + 38 \log(1 + e^{\alpha + 7\beta})]$$

and maximization calls for an iterative process to estimate α, β that is available in many standard statistical packages. I omit details but the fitted logistic regression is

$$\lambda = -5.961 + 0.316x$$

and an asymptotic 95% confidence interval for β given by LogXact is $(0.0707, 0.5624)$ and the two-sided p-value for testing $H_0: \beta = 0$ is $p = 0.0116$, not surprising in view of the confidence interval for β indicating rejection of H_0. These results are broadly in line with those in

Example 13.5. Unfortunately exact conditional test results were not available from LogXact because of the large sample sizes in some groups.

Example 14.4. For the data in Table 14.1 LogXact gives the logistic regression equation

$$\lambda = -21.228 + 1.632x \qquad (14.9)$$

with asymptotic 95% confidence interval for β of (1.5166, 1.7476). For given x one computes the expected value of λ from (14.9) and hence estimates the probability π for that x and expected responses for that x. Finney (1970, Table 5.1) gives these expected numbers using both a probit and a logit fit. There is good agreement between the two methods and each fits the data well. Setting $\lambda = 0$ gives the estimated median age of menarche to be 13.01.

While the median value corresponding to $\pi = 0.5$ is often of interest in studies of toxicity as a yardstick for comparing toxicities it is not a useful statistic for the data in Example 14.3 because it would involve extrapolation beyond the data range. From that example the estimated median toxic intake is $x = 18.86$ drinks per day. It is debatable whether a mother with that alcohol consumption would conceive or if she did whether she would survive to give birth.

14.4 Generalized linear models

The logistic regression model is important and applicable to a wide range of problems as indicated in Cox and Snell (1989), Collett (1991) and Agresti (1996). It is a special case of the **generalized linear models** proposed by Nelder and Wedderburn (1972) in a versatile model-driven approach that acts as an ideal catalyst between data- and model-driven inference. A definitive treatment of theory and application is given by McCullagh and Nelder (1989) and Dobson (1990) is an excellent introductory text.

The models are concerned with relationships between a response variable Y and one or more explanatory variables x, with subscripts if necessary, to distinguish between them, or they may be written as a vector x In the classic regression model the linear relationship is between the mean $E(Y|x)$ and x, i.e.

$$E(Y|x) = \beta_0 + \beta_1 x_1 + \beta_2 x_2 + \ldots + \beta_p x_p. \qquad (14.10)$$

In logistic regression (section 14.2) a linear relation between $E(P|x) = \pi$ and x is inappropriate but one may exist between a function of π (the logit) and x.

Generalized linear models extend this concept of linear relationships between a function of the mean response and explanatory variables. Nelder and Wedderburn recognized that there is a broad class of distributions for which some function of the responses has a mean that could be expressed as a linear function of explanatory variables. The appropriate function, known as the link function, depends on the distribution of responses.

The responses must belong to the **exponential family of distributions**. The exponential distribution itself is only one member of the family.

A distribution with a parameter θ of interest belongs to the exponential family if its probability mass function (discrete distribution) or its probability density function (continuous distribution) can be written in the form:

$$f(y, \theta) = s(y)t(\theta)e^{a(y)\,b(\theta)}$$

or equivalently writing $s(y) = e^{d(y)}$ and $t(\theta) = e^{c(\theta)}$

$$f(y, \theta) = \exp[a(y)b(\theta) + c(\theta) + d(y)]. \qquad (14.11)$$

Parameters other than θ are nuisance parameters and are treated as known and included in the functions a, b, c, d. Often $a(y) = y$ and then the distribution is said to be in canonical form with natural parameter $b(\theta)$.

Example 14.5. Many common distributions are expressible in the form (14.11). For a $N(\mu, \sigma^2)$ distribution if we are interested in μ we may write the probability density function

$$f(y, \mu) = \exp[-\tfrac{1}{2}(y/\sigma)^2 + (y\mu/\sigma^2) - \tfrac{1}{2}(\mu/\sigma)^2 - \tfrac{1}{2}\log(2\pi\sigma^2)]$$

which is in canonical form with $a(y) = y$, and natural parameter $b(\mu) = \mu/\sigma^2$.

For the binomial $B(m, \pi)$ distribution if we are interested in π then if Y is the number of successes the usual form of the probability mass function is

$$f(y, \pi) = {}^mC_y\pi^y(1 - \pi)^{m-y}$$

which is equivalent to

$$f(y, \pi) = \exp\{y\log[\pi/(1 - \pi)] + m\log(1 - \pi) + \log({}^mC_y)\}$$

which again is in canonical form with natural parameter the logit, i.e. $\log[\pi/(1 - \pi)]$.

For the Poisson distribution

$$f(y, \lambda) = \lambda^y e^{-\lambda}/(y!)$$

which may be written

$$f(y, \lambda) = \exp[y\log\lambda - \lambda - \log(y!)]$$

in canonical form with natural parameter $\log\lambda$.

Nelder and Wedderburn established under very general conditions that if responses Y_i, which may be related to a set of explanatory variables x_i, each having distributions from the same member of the exponential family, then if μ_i is the mean of Y_i, there exists some function $g(\mu)$ such that

$$g(\mu_i) = x_i^T\beta.$$

The function g is called the link function and is often closely related to the natural parameter. For the normal distribution it is the mean μ itself. For the binomial distribution as we saw in section 14.2 it is the logit which is the natural parameter and for the Poisson distribution it is the natural parameter log λ.

The generalized linear model is one of the most widely used model-based tools in modern statistical inference. While most statistical software at all but the elemntary level has facilities for using it, the specialized package GLIM is strongly recommended for the serious user.

I have aimed in this book to give some indication of the interlocking threads in data-driven analyses. The generalized linear model extends these threads further from a model-driven base, as I indicated for the logistic regression model in section 14.3.

Exercises

14.1 Using (14.2), establish the relationship (14.3).

*14.2 Smith (1967) and also Hand et al (1994, Set 56) give these data on the levels of creatinine kinase (CK) measured on patients suspected of having had a heart attack. Further tests then indicated whether each patient had actually had a heart attack. As the level of CK may be determined fairly quickly the interest was in whether levels of CK are diagnostically useful. The levels in the data are given in ranges of an appropriate unit. Consider the desirability of, and any practical problems that may arise in using logistic regression in the analysis of these data, performing such an analysis if you consider it appropriate. Do you consider any other analysis equally or more appropriate? If so perform it. Make a careful summary of your conclusions.

CK range	<40	40–80	80–120	120–160	160–200	200–240	240–280	280–320
Heart attack	2	13	30	30	21	19	18	13
No attack	88	26	8	5	0	1	1	1

CK range	320–360	360–400	400–440	440–480	480 and over
Heart attack	19	15	7	8	35
No attack	0	0	0	0	0

References

Books highly recommended for more detailed treatment of specific topics covered in this text or related matters are marked with an asterisk*

Adichie, J.N. (1967) Estimates of regression parameters based on rank tests. *Ann. Math. Statist.*, **38**, 894–904.

Agresti, A. (1984) *Analysis of Ordinal Categorical Data.* New York: John Wiley & Sons.

*Agresti, A. (1990) *Categorical Data Analysis.* New York: John Wiley and Sons.

*Agresti, A. (1996) *An Introduction to Categorical Data Analysis.* New York: John Wiley & Sons.

*Aitken, C. G. G. (1995) *Statistics and the Evaluation of Evidence for Forensic Scientists.* Chichester: John Wiley and Sons.

*Altman, D.G. (1991) *Practical Statistics for Medical Research.* London: Chapman & Hall.

Altshuler, B. (1970) Theory for the measurement of competing risks in animal experiments. *Math. Biosciences*, **6**, 1–11.

Andersen, P.K., Borgan, O., Gill, R.D. and Keiding, N. (1982) Linear nonparametric tests for the comparison of counting processes with application to censored survival data. *International Statistical Review*, **50**, 219–58.

Anderson, R.L. (1959) Use of contingency tables in the analysis of consumer preference studies. *Biometrics*, **15**, 582–90.

Andrews, D.F. (1974) A robust method for multiple linear regression. *Technometrics*, **16**, 523–31

Andrews, D.F., Bickel, P.J., Hampel, F.R., Huber, P.S., Rogers, W.H. and Tukey, J.W. (1972) *Robust Estimates of Location. Survey and Advances.* Princeton: Princeton University Press.

Ansari, A.R. and Bradley, R.A. (1960) Rank sum tests for dispersion. *Ann. Math. Statist.*, **31**, 1174–89

Armitage, P. (1955) Tests for linear trends in proportions and frequencies. *Biometrics*, **11**, 375–86.

*Atkinson, A.C. (1985) *Plots, Transformations and Regression.* Oxford: Clarendon Press.

Azzalini, A. and Bowman, A.W. (1990) A look at some data on the Old Faithful geyser. *Applied Statistics*, **39**, 357–66.

Bailey, R.A. (1991) Strata for randomized experiments. *J. Roy. Statist. Soc. B.*, **53**, 27–78.

Bardsley, P. and Chalmers, R.L. (1984) Multipurpose estimation from unbalanced samples. *Applied Statistics*, **33**, 290–9.

*Barnett, V. and Lewis, T. (1994) *Outliers in Statistical Data.* 3rd. edn. Chichester: John Wiley & Sons.

Bartlett, M.S. (1936) Some notes on insecticide tests in the laboratory and in the field. *J. Roy. Statist. Soc., Suppl.*, **3**, 185–94.

Bartlett, M.S. (1949) Fitting a straight line when both variables are subject to error. *Biometrics*, **5**, 207–12.

Bassendine, M.F., Collins, J.D., Stephenson, J., Saunders, P. and James, O.W.F. (1985) Platelet associated immunoglobulins in primary biliary cirrhosis: a cause of Thrombocytopenia? *Gut*, **26**, 1074–9.

Belsey, D.A., Kuh, E., and Welsch. R.E. (1980) *Regression Diagnostics*. New York: John Wiley & Sons.

Bennett B.M. (1968) Rank order tests of linear hypotheses. *J. Roy. Statist. Soc.*, B, **30**, 483–9.

Berry, D.A. (1987) Logarithmic transformations in ANOVA. *Biometrics*, **43**, 439–56.

Berry, G. and Armitage, P. (1995) Mid-*P* confidence intervals: a brief review. *The Statistician*, **44**, 417–23.

Best, D.J. (1994) Nonparametric comparison of two histograms. *Biometrics*, **50**, 538–41.

Best, D.J. and Rayner, J.C.W. (1996) Nonparametric analysis for doubly ordered two-way contingency tables. *Biometrics*, **52**, 1153–6.

Birnbaum, Z.W. and Tingey, F.H. (1951) One-sided confidence contours for probability distribution functions. *Ann. Math. Statist.* **22**, 592–6.

Bishop, Y.M.M., Fienberg, S.E. and Holland, P. (1975). *Discrete Multivariate Analysis: Theory and Practice*. Cambridge, Mass: MIT Press.

Blomqvist, N. (1950) On a measure of dependence between two random variables. *Ann. Math. Statist.*, **21**, 593–600.

Blomqvist, N. (1951) Some tests for dichotomization. *Ann. Math. Statist.*, **22**, 362–71.

Bodmer, W.F. (1985) Understanding statistics. *J. Roy. Statist. Soc.*, A., **148**, 69–81

Bowker, A.H. (1948) A test for symmetry in contingency tables. *J. Amer. Statist. Assn.*, **43**, 572–4.

Bowman, K.O. and Shenton, I.R. (1975) Omnibus test contours for departure from normality based on $\sqrt{b_1}$ and b_2. *Biometrika*, **62**, 243–50.

*Bradley, J.V. (1968). *Distribution-free Statistical Tests*. Englewood Cliffs, NJ: Prentice Hall.

*Breslow, N.E. and Day, N.E.(1980) *Statistical Methods in Cancer Research. I. The Analysis of Case–Control Studies*. Lyons:WHO/IARC Scientific Publications No. 12.

Brown, G.W.and Mood, A.M. (1948) Homogeneity of several samples. *The American Statistician*, **2**, 22.

Brown, G.W. and Mood, A.M. (1951) On median tests for linear hypotheses. *Proc. 2nd. Berkeley Symp.*, 159–66.

Brownlee, K.A. (1965) *Statistical Theory and Methodology in Science and Engineering* 2nd edn. New York: John Wiley & Sons.

Burns, K.C. (1984) Motion sickness incidence: distribution of time to first emesis and comparision of some complex motion conditions. *Aviation Space and Environmental Medicine*, **56**, 521–7.

Capon, J. (1961) Asymptotic efficiency of certain locally most powerful rank tests. *Ann. Math. Statist.*, **32**, 88–100.

Carroll, R.J. (1980) A robust method for transformations to achieve approximate normality. *J. Roy. Statist. Soc.*, B, **42**, 71–8.

Carroll, R.J. (1982) Two examples of transformations with possible outliers. *App. Statist.*, **31**, 149–52.

Carroll, R.J. and Rupert, D. (1988) *Transformation and Weighting in Regression*. New York: Chapman & Hall.

Carter, E.M. and Hubert J.J. (1985) Analysis of parallel line assays with multivariate responses. *Biometrics*, **41**, 703–10.

*Chatfield, C. (1983) *Statistics for Technology*. London: Chapman & Hall.

Chatfield, C (1988) *Problem solving: A Statistician's Guide*. London: Chapman & Hall.

Christensen, R.A. (1990) *Log Linear Models*. New York: Springer–Verlag.

Cochran, W.G. (1950) The comparison of percentages in matched samples. *Biometrika*, **37**, 246–66.

Cochran, W.G. (1953) *Sampling Techniques*. New York: John Wiley & Sons.

Cochran, W.G. (1954) Some methods for strengthening the common χ^2 test. *Biometrics*, **10**, 417–54.

Cochran, W.G. and Cox, G.M. (1950) *Experimental Designs*. New York: John Wiley & Sons.

Cohen, A. (1983) Seasonal daily effect on the number of births in Israel. *Applied Statistics*, **32**, 328–35.

*Collett, D. (1991) *Modelling Binary Data*. London: Chapman & Hall.

*Conover, W.J. (1980) *Practical Nonparametric Statistics*. New York: John Wiley & Sons.

Cook, R.D. (1977) Detection of influential observations in linear regression. *Technometrics*, **19**, 15–18.

*Cook, R.D. and Weisberg, S. (1982) *Residuals and Influence in Regression*. New York: Chapman & Hall,

Cox, D.R. (1964) Some applications of exponential ordered scores. *J. Roy. Statist. Soc.*, B, **26**, 103–10.

Cox, D.R. (1984) Interaction. *International Statistical Review*, **52**, 1–31.

*Cox, D.R. and Snell, E.J. (1981) *Applied Statistics – Principles and Examples*. London: Chapman & Hall.

*Cox, D.R. and Snell, E.J. (1989) *Analysis of Binary Data* 2nd edn. London: Chapman & Hall.

Cox, D.R., and Stuart, A. (1955) Some quick tests for trend in location and dispersion. *Biometrika*, **42**, 80–95.

Cramér, H. (1928) On the composition of elementary errors. *Skandinavisk Aktuarietidskrift*, **11**, 13–74, 141–180.

Dale, G., Fleetwood, J.A., Weddell, A., Ellis, R.D. and Sainsbury, J.R.C. (1987) Beta endorpin: a factor in 'fun run' collapse? *British Medical Journal*, **294**, 1004.

David, F.N. and Barton, D.E. (1958) A test for birth-order effects. *Ann. Hum. Eugenics*, **22**, 250–7.

Dinse, G.E. (1982) Nonparametric estimation for partially-complete time and type of failure data. *Biometrics*, **38**, 417–31.

*Dobson, A. J. (1990) *An Introduction to Generalized Linear Models*. London: Chapman & Hall.

Dolkart, R.E., Halperin, B. and Perlman, J. (1971). Comparison of antibody responses in normal and alloxan diabetic mice. *Diabetes*, **20**, 162–7.

Draper, D., Hodges, J.S., Mallows, C.L. and Pregibon, D. (1993) Exchangeability and data analysis. *J. Roy. Statist. Soc.*, A., **156**, 9–37.

Draper, N.R. and Smith, H. (1980) *Applied Regression Analysis*, 2nd edn. New York: John Wiley & Sons.

Dunn, G. (1989) *Design and Analysis of Reliability Studies*. London: Edward Arnold.

Durbin, J. (1987) Statistics and statistical science. *J. Roy. Statist. Soc.*, A., **150**, 177–91.

*Edgington, E.S. (1995) *Randomization tests*, 3rd edn. New York: Marcel Dekker, Inc.

Efron, B. (1979) Bootstrap methods: another look at the jackknife. *Ann. Statist.* 7, 1–26.

*Efron, B. and Tibshirani, R.J. (1993) *An introduction to the Bootstrap*. London: Chapman & Hall.

*Everitt, B.S. (1992) *The Analysis of Contingency Tables*. 2nd edn. London: Chapman & Hall.

Fiegl, P. and Zelen, M. (1965) Estimation of exponential survival probabilities with concomitant information. *Biometrics*, **21**, 826–38.

Fieller, E.C. (1940) The biological standardization of insulin. *J. Roy. Statist. Soc., Suppt.*, **7**, 1–64.

Fieller, E.C. (1944) A fundamental formula in the statistics of biological assay, and some applications. *Quart. J. Pharm.*, **17**, 117–23.

Finkelstein, M.O. and Levin, B. (1990) *Statistics for Lawyers*. New York: Springer.

Finney, D.J. (1971) *Probit Analysis* 3rd edn. Cambridge: Cambridge University Press.

Fisher, N.I. (1993) *Statistical Analysis of Circular Data*. Cambridge: Cambridge University Press.

Fisher, R.A. (1935) *The Design of Experiments*. Edinburgh: Oliver & Boyd.

Fisher, R.A. (1946) *Statistical Methods for Research Workers*, 10th edn. Edinburgh: Oliver & Boyd.

Fisher, R.A., Corbet, A.S. and Williams, C.B. (1943) The relation between the number of species and the number of individuals in a random sample of an animal population. *J. Animal Ecol.*, **12**, 42–57.

Fisher, R.A. and Yates, F. (1957) *Statistical Tables for Biological, Agricultural and Medical Research* 5th edn. Edinburgh: Oliver & Boyd.

Fligner, M.A. and Rust, S.W. (1982) A modification of Mood's median test for the generalized Behrens–Fisher problem. *Biometrika*, **69**, 221–6.

Fligner, M.A. and Verducci, J.S. (1988) A nonparametric test for judges' bias in an athletics competition. *Applied Statistics*, **37**, 101–10.

Freeman, G.H. and Halton, J.H. (1951) Note on an exact treatment of contingency, goodness of fit and other problems of significance. *Biometrika*, **38**, 141–9.

Freund, J.E. and Ansari, A.R. (1957) *Two Way Rank Sum Tests for Variance*. Blacksburg, VA: Virginia Polytechnic Institute Report to Ordnance Research and NSF, 34.

Friedman, M. (1937) The use of ranks to avoid the assumption of normality implicit in the analysis of variance. *J. Amer. Statist. Assn.*, **32**, 675–701.

Gart, J. (1970) Point and interval estimation of the common odds ratio in the combination of 2×2 tables with fixed marginals. *Biometrika*, **57**, 471–5.

Gastwirth, J.L. (1965) Percentile modifications of two sample rank sum tests. *J. Amer. Statist. Assn.*, **60**, 1127–41.

Gastwirth, J.L. (1968) The first median test: A two-sided version of the control median test. *J. Amer. Statist. Assn.*, **63**, 692–706.

Gastwirth, J. L. (1985) The use of maximin efficiency robust tests in combining contingency tables and survival analysis. *J. Amer. Statist. Assn.*, **80**, 380–4.

*Gastwirth, J.L. (1988) *Statistical Reasoning in Law and Public Policy*. Orlando:Academic Press.

Gastwirth, J.L. (1996) Statistical issues arising in equal employment litigation. *Jurimetrics Journal*, **36**, 353–70.

Gastwirth, J.L. (1997) Statistical evidence in discrimination cases. *J.Roy. Statist. Soc. A*, **160**, 289–303.

Gastwirth, J.L. and Greenhouse, S.W. (1995) Biostatistical concepts and methods in the legal setting. *Statistics in Medicine*, **14**, 1641–53.

Gastwirth, J.L. and Wang, J.L. (1987) Nonparametric tests in small unbalanced samples: application in employment discrimination cases. *Can. J. Statist.*,**15**, 339–48.

Gat, J.R. and Nissenbaum, A. (1976) Limnology and ecology of the Dead Sea. *Nat. Geog. Soc. Research Reports – 1976 Projects*, 413–18.

Gehan, E.A. (1965*a*) A generalized Wilcoxon test for comparing arbitrarily singly censored samples. *Biometrika*, **52**, 203–23.

Gehan, E.A. (1965*b*) A generalized two-sample Wilcoxon test for doubly censored data. *Biometrika*, **52**, 650–3.

*Gibbons, J.D. and Chakraborti, S. (1992) *Nonparametric Statistical Inference*. 3rd. edn. New York: Marcel Dekker, Inc.

Gideon, R.A. and Hollister, R.A. (1987). A rank correlation coefficient resistant to ouliers. *J. Amer. Statist. Assn.*, **82**, 656–66.

Gilchrist, W. (1984) *Statistical Modelling*. Chichester: John Wiley & Sons.

*Good, P. (1994) *Permutation Tests. A Practical Guide to Resampling Methods for Testing Hypotheses*. New York: Springer–Verlag.

Goodman, L.A. and Kruskal, W.H. (1954) Measures of association for cross classification. *J. Amer. Statist, Assn.*, **49**, 732–64.

Graubard, B.I, and Korn, E.L. (1987) Choice of column scores for testing independence in ordered $2 \times k$ contingency tables. *Biometrics*, **43**, 471–6.

*Green, P.J. and Silverman, B.W. (1993) *Nonparametric Regression and Generalized Linear Models: A Roughness Penalty Approach*. London: Chapman & Hall.

Grizzle, J.E. (1965) The two-period change-over design and its use in clinical trials. *Biometrics*, **21**, 467–80.

Grizzle, J.E., Starmer, C.F., and Koch, G.G. (1969) Analysis of categorical data by linear models. *Biometrics*, **25**, 489–504.

Groggel, D.J. and Skillings, J.H. (1986) Distribution-free tests for main effects in multifactor designs. *American Statistician*, **40**, 99–102.

Hajék. J. and Sidak, Z. (1967) *Theory of Rank Tests*. New York: Academic Press.

Hall, P. (1992) *The Bootstrap and Edgeworth Expansion*. New York: Springer–Verlag.

Hall, P. and Wilson, S.R. (1991) Two guidelines for bootstrap hypothesis testing. *Biometrics*, **47**, 757–62.

Hampel, F.R. (1971) A generalized quantitative definition of robustness. *Ann. Math. Statist.* **42**, 1887–96.

Hampel, F.R. (1974) The influence curve and its role in robust estimation. *J. Amer. Statist. Assn.*, **69**, 383–93.

*Hand, D.J., Daly, F., Lunn, A.D., McConway, K.J. and Ostrowski, E. (1994) *Small Data Sets* London: Chapman & Hall.

Hodges, J. L. (1967) Efficiency in normal samples and tolerance of extreme values for some estimates of location. *Proceedings of the 5th Berkeley Symposium on Statistics and Probability*. Berkeley: University of California.

Hodges, J.L. and Lehmann, E.L. (1956) The efficiency of some nonparametric competitors of the *t*-test. *Ann. Math. Statist.*, **27**, 324–35.

Hodges, J.L. and Lehmann. E.L. (1963). Estimates of location based on rank tests. *Ann. Math. Statist.*, **34**, 598–611.

Huber, P.J. (1972) Robust statistics: a review. *Ann. Math. Statist.* **43**, 1041–67.

Huber, P.J. (1977) *Robust Statistical Procedures.* Philadelphia: Society for Industrial and Applied Mathematics.

*Huber, P.J. (1981) *Robust Statistics.* New York: John Wiley & Sons.

Huber, B. J. (1983) Minimax influence of bounded-influence regression. *J. Amer. Statist. Assn.,* **78**, 66–80.

Hussain, S.S. and Sprent, P. (1983) Nonparametric regression. *J. Roy. Statist. Soc., A.,* **146**, 182–91.

Iman R.L. and Davenport, J.M. (1980) Approximations of the critical region of the Friedman statistic. *Communications in Statistics,* **A9**, 571–95.

Iman, R.L., Hora, S.C. and Conover, W.J. (1984) Comparison of asymptotically distribution-free procedures for the analysis of complete blocks. *J. Amer. Statist. Assn.,* **79**, 674–85.

Jaeckel, L.A. (1972) Estimating regression equations by minimizing the dispersion of residuals. *Ann. Math. Statist.,* **43**, 1449–58.

Jarque, C.M. and Bera, A.K. (1987). A test for normality of observations and regression residuals. *Internat. Statist, Rev.,* **55**, 163–72.

Jonckheere, A.R. (1954) A distribution-free *k*-sample test against ordered alternatives. *Biometrika,* **41**, 133–45.

Kalbfleish, J.D. and Prentice, R.L. (1980). *The Statistical Analysis of Failure Time Data.* New York: John Wiley & Sons.

Kaplan, E.L. and Meier, P. (1958) Nonparametric estimation from incomplete observations. *J. Amer. Statist. Assn.,* **53**, 457–81.

Kauman, W.G., Gottstein, J.W. and Lantican, D. (1956) Quality evaluation by numerical and subjective methods with application to dried veneer. *Biometrics,* **12**, 127–53.

Kempthorne, O. (1952) *Design and Analysis of Experiments.* New York: John Wiley & Sons.

Kempthorne, O. (1955) The randomization theory of experimental inference. *J. Am. Statist. Assn.,* **50**, 946–67.

Kendall, M.G. (1938) A new measure of rank correlation. *Biometrika,* **30**, 81–93.

*Kendall, M.G. and Gibbons, J.D. (1990) *Rank correlation Methods,* 5th edn. London: Edward Arnold.

Kimber, A.C. (1987). When is a χ^2 not a χ^2? *Teaching Statistics,* **9**, 74–7.

Kimura, D.K. and Chikuni, S. (1987) Mixtures of empirical distributions: an iterative application of the age–length key. *Biometrics,* **43**, 23–35.

Klotz, J.H. (1962) Nonparametric tests for scale. *Ann. Math. Statist.,* **33**, 495–512.

Koch. G.G. (1972) The use of nonparametric methods in the statistical analysis of the two-period change-over design. *Biometrics,* **28**, 577–84.

Kolmogorov, A. N. (1933) Sulla detterminazione empirica di una legge di distribuzione. *G. Inst. Ital. Attuari,* **4**, 83–91.

Kolmogorov, A.N. (1941) Confidence limits for an unknown distribution function. *Ann. Math. Statist.* **12**, 461–3.

Kruskal W.H. and Wallis, W.A. (1952) Use of ranks in one-criterion variance analysis. *J. Amer. Statist. Assn.,* **47**, 583–621

Lancaster, H.O. (1949) The derivation and partitioning of χ^2 in certain discrete distributions. *Biometrika,* **36**, 117–29.

Lancaster, H.O. (1953) A reconciliation of χ^2, considered metrical and enumerative aspects. *Sankhya,* **13**, 1–9.

Lee, A.J. (1965) New observations on distribution of neoplasms of female breast in certain European countries. *Brit. Med. J.*, **1**, 488–90.

*Lehmann, E.L. (1975) *Nonparametric Statistical Methods Based on Ranks.* San Francisco: Holden Day, Inc.

Lilliefors, H.W. (1967) On the Kolmogorov–Smirnov test for normality with mean and variance unknown. *J. Amer. Statist. Assn.*, **62**, 399–402.

Lindley, D.V. and Scott, W.F. (1995) *New Cambridge Elementary Statistical Tables.* 2nd edn. Cambridge: Cambridge University Press.

*McCullagh, P. and Nelder, J.A. (1989) *Generalized Linear Models.* 2nd edn. London: Chapman & Hall.

Mack, G.A. and Skillings, J.H. (1980) A Friedman type test for main effects in a two-factor ANOVA. *J. Amer. Statist. Assn.* **75**, 947–51

McNemar, Q. (1947) Note on the sampling error of the difference between correlated proportions or percentages. *Psychometrika*, **12**, 153–7.

Mann, H.B. and Whitney, D.R. (1947) On a test of whether one of two random variables is stochastically larger than the other. *Ann. Math. Statist.*, **18**, 50–60

*Manly, B.F. J. (1997) *Randomization, Bootstrapping and Monte Carlo Methods in Biology.* 2nd. edn. London: Chapman & Hall.

Mantel, N. (1966) Evaluation of survival data and two new rank order statistics arising in its consideration. *Cancer Chemotherapy Reports*, **50**, 163–170.

Mantel, N. and Haenszel, W. (1959). Statistical aspects of the analysis of data from retrospective studies of disease. *J. Nat. Cancer. Res. Inst.*, **22**, 719–48.

Mapes, R. and Dajda, R. (1976) Children and general practitioners, the general household survey as a source of information. *Sociological Review Monographs*, **M22**, 99–110.

Maritz, J. S. (1979) On Theil's method in distribution free regression. *Austral. J. Statist.*, **21**, 30–5.

*Maritz, J.S. (1995) *Distribution-free Statistical Methods*, 2nd edn. London: Chapman & Hall.

Mathisen, H.C. (1943) A method of testing the hypothesis that two samples are from the same population. *Ann. Math. Statist.*, **14**, 188–94.

Mazess, R.R., Peppler, W.W. and Gibbons, M. (1984) Total body composition by dual photon (^{153}Gd) absorptiometry. *Amer. J. Clin. Nutr.*, **40**, 834–9.

Milicer, H. and Szczotka, F. (1966) Age at menarche in Warsaw girls in 1965. *Human Biol.*, **38**, 199–203.

Mood, A.M. (1954) On the asymptotic efficiency of certain nonparametric two-sample tests. *Ann. Math. Statist.*, **25**, 514–22.

Morgan, B.J.T. (1992) *Analysis of Quantal Response Data.* London: Chapman & Hall.

Nanji, A.A. and French, S.W. (1985) Relationship between pork consumption and cirrhosis. *The Lancet*, **i**, 681–3.

Neave, H.R. (1981) *Elementary Statistical Tables.* London: George Allen & Unwin, Ltd.

Neave, H.R. and Worthington, P.L. (1988) *Distribution-free Tests.* London: Unwin Hyman.

Nelder, J.A. and Wedderburn, R.W.N. (1972) Generalized linear models. *J. Roy. Statist. Soc.,A*, **135**, 370–84.

Newman, H.H., Freeman, F.N. and Holzinger, K.J. (1937) *Twins.* Chicago: University of Chicago Press.

Neyzi, O., Alp, H. and Orhon, A. (1975) Breast development of 318 12–13 year old Turkish girls by socio-economic class of parents. *Ann. Human Biol*, 2, **1**,49–59.

Norton, P.G. and Dunn, E.V. (1985) Snoring as a risk factor in disease: an epidemiological survey. *British Medical Journal*, **291**, 630–2.

Page, E.B. (1963) Ordered hypotheses for multiple treatments: a significance test for linear ranks. *J. Amer. Statist. Assn.*, **58**, 216–30.

Peto, R. and Peto, J. (1972) Asymptotically efficient rank-invariant test procedures. *J. Roy. Statist. Soc., A.*, **135**, 185–206.

Peto, R, Pike, M.C., Armitage, P., Breslow, N.E., Cox, D.R., Howard. S.V., Mantel, N., McPherson, K., Peto, J. and Smith, P.G. (1977) Design and analysis of randomized clinical trials requiring prolonged observation of each patient. *British J. Cancer*, **35**, 1–39.

Pitman, E.J.G. (1937a) Significance tests that may be applied to samples from any population. *J. Roy. Statist. Soc., Suppt.*, **4**, 119–130.

Pitman, E.J.G. (1937b) Significance tests that may be applied to samples from any population, II. The correlation coefficient test. *J. Roy. Statist. Soc., Suppt.*, **4**, 225–32.

Pitman, E.J.G. (1938) Significance tests that may be applied to samples from any population, III. The analysis of variance test. *Biometrika*, **29**, 322–35

Pitman, E.J.G. (1948) Mimeographed lecture notes on nonparametric statistics. Columbia University.

Pocock, S.J. (1983) *Clinical Trials: A Practical Approach.* Chichester: John Wiley & Sons

Pothoff, R.F. (1963) Use of the Wilcoxon statistic for a generalized Behrens–Fisher problem. *Ann. Math. Statist.*, **34**, 1596–99.

Prentice, R.L. and Marek, P. (1979) A qualitative discrepancy between censored data rank tests. *Biometrics*, **35**, 861–7.

Quade, D. (1979) Using weighted rankings in the analysis of complete blocks with additive block effects. *J. Amer. Statist. Assn.*, **74**, 680–3.

Quenouille, M. H. (1949) Approximate tests for correlation in time series. *J. Roy. Statist. Soc., B.*, **47**, 18–44.

*Rees, D.G. (1994) *Essential Statistics*. 3rd edn. London: Chapman & Hall.

Robins, J., Breslow, N. and Greenland, S. (1986). Estimation of the Mantel–Haenszel variance consistent in both sparse data and large-strata limiting models. *Biometrics*, **42**, 311–23.

Ross, G.J.S. (1990) *Nonlinear estimation.* New York: Springer-Verlag.

Rutherford, E and Geiger, M. (1910) The probability variations in the distribution of alpha-particles. *Phil. Mag. Series 6*, **20**, 698–704.

Savage, I.R. (1956) Contributions to the theory of rank order statistics I. *Ann. Math. Statist.*, **27**, 590–615.

Scholz, F.W. and Stephens, M.A. (1987) K-sample Anderson–Darling tests. *J. Amer. Statist. Assn.*, **82**, 918–24.

Schrader, R.M. and Hettmansperger, T.P. (1980). Robust analysis of variance based upon a likelihood ratio criterion. *Biometrika*, **67**, 93–101.

Sen, P.K. (1968) Estimation of the regression coefficient based on Kendall's tau. *J. Amer. Statist. Assn.*, **63**, 1379–89.

*Senn, S. J. (1992) *Cross-over Trials in Clinical Research.* Chichester: John Wiley & Sons.

Shapiro, S.S. and Wilk, M.B. (1965) An analysis of variance test for normality (complete samples). *Biometrika*, **52**, 591–611.

Shirley, E.A.C. (1977) A nonparametric equivalent of Williams' test for contrasting increasing dose levels of a treatment. *Biometrics*, **33**, 386–9.

Shirley, E.A.C. (1987) Application of ranking methods to multiple comparison procedures and factorial experiments. *Applied Statistics*, **36**, 205–13.

*Siegel, S. and Castellan, N.J. (1988) *Nonparametric Statistics for the Behavioral Sciences*. 2nd edn. New York: McGraw-Hill.

Siegel, S. and Tukey. J.W. (1960) A nonparametric sum of ranks procedure for relative spread in unpaired samples. *J. Amer. Statist. Assn.*, **56**, 1005.

Smirnov, N.V. (1939) On the calculation of discrepancy between empirical curves of distribution for two independent samples (in Russian). *Bulletin Moscow University*, **2**, 3–16.

Smith, A.F. (1967) Diagnostic value of serum-creatinine-kinase in a coronary care unit. *Lancet*, **2**, 178.

Snedecor G.W and Cochran, W.G. (1967) *Statistical Methods*, 6th edn. Ames: The Iowa State University Press.

Snee, R.D. (1985) Graphical display of the results of three treatment randomized block experiments. *Applied Statistics*, **34**, 71–7.

Spearman, C. (1904) The proof and measurement of association between two things. *Amer. J. Psychol.* **15**, 72–101.

*Sprent, P. (1993) *Applied Nonparametric Statistical Methods*, 2nd edn. London: Chapman & Hall.

Stablein, D.M. and Koutrouvelis, I.A. (1985) A two-sample test sensitive to crossing hazards in uncensored and singly censored data. *Biometrics*, **41**, 643–52.

Stuart, A. and Ord, J.K. (1991) *Kendall's Advanced Theory of Statistics. Vol 2*. 5th edn, London: Edward Arnold.

Sukhatme, B. V. (1957) On certain two sample nonparametric tests for variances. *Ann. Math. Statist.*, **28**, 188–94.

Terpstra, T. J. (1952) The asymptotic normality and consistency of Kendall's test against trend when ties are present in one ranking. *Indag. Math.*, **14**, 327–33.

Theil, H. (1950) A rank invariant method of linear and polynomial regression analysis. I, II, III. *Proc. Kon. Nederl. Akad. Wetensch. A*, **53**, 386–92, 521–5, 1397–1412.

Thomas. G.E. (1989) A note on correcting for ties with Spearman's ρ. *J. Statist. Comp. and Simulation*, **31**, 37–40.

Tukey, J.W. (1958) Bias and confidence in not quite large samples. Abstract, *Ann Math. Statist.*, **29**, 614.

van der Waerden, B.L. (1952). Order tests for the two-sample problem and their power. I. *Proc. Kon. Nederl. Akad. Wetensch. A.*, **55**, 453–8.

van der Waerden, B.L. (1953). Order tests for the two-sample problem and their power. II,III. *Proc. Kon. Nederl. Akad. Wetensch. A.*, **56**, 303–16.

von Mises, R. (1931) *Warscheinlichkeitsrechnung und Ihre Anwendung in der Statisk und Theoretischen Physik*. Leipzig: F. Deuticke.

Wald, A. and Wolfowitz, J. (1940) On a test whether two samples are from the same population. *Ann. Math. Statist.*, **11**, 147–62.

Walsh, J.E. (1949a) Application of some significance tests for the median which are valid under very general conditions. *J. Amer. Statist. Assn.*, **44**, 342–55.

Walsh, J.E. (1949b) Some significance tests for the median which are valid under very general conditions. *Ann. Math. Statist.*, **20**, 64–81.

Weiner, B. (1977) *Discovering Psychology*. Chicago: Science Research Association.

Weisberg, S. (1980) *Applied Linear Regression*. New York: John Wiley & Sons.

Westenberg, J. (1948) Significance tests for median and interquartile range in samples from continuous populations of any form. *Proc. Kon. Nederl. Akad. Wetensch., A.* **51**, 252.61.

Wilcoxon, F. (1945) Individual comparisons by ranking methods. *Biometrics*, **1**, 80–83.

Yates, F. (1984) Tests of significance for 2×2 contingency tables (with discussion). *J. Roy. Statist. Soc., A*, **147**, 426–463.

Yule, G.U. (1900) On the association of attributes in statistics. *Phil. Trans. Roy. Soc. Lond., A.*, **194**, 257–319.

Zelen, M. (1971) The analysis of several 2×2 contingency tables. *Biometrika*, **58**, 129–37.

Index

80% discount to students

StatXact -3 for Windows: Statistical Software for Exact Non parametric Inference

If you are currently a registered student you could buy StatXact 3.1 for Windows for only $195 plus shipping* (regular price $995). This offer includes the software and the full manual available on CD-ROM . Printed manuals are available for $95 plus shipping#.

The following exact procedures are available:
- goodness of fit tests
- one-sample tests
- two-sample tests for related or independent samples
- K-sample tests for related or independent samples
- tests for censored survival data
- tests on 2x2, and 2xc contingency with and without stratification
- tests on rxc contingency tables
- tests on measures of association and measures of agreement
- confidence intervals for categorical and continuous data
- permutation distributions for all test statistics.

For more information on StatXact visit the Web site http://www.cytel.com

To take advantage of this offer just fill in this card overleaf and send it together with a letter from your department or course instructor verifying your student status to:

Cytel Software Corp.
675 Massachusetts Avenue
Cambridge, MA02139-9571
USA

Phone: 617-661-2011
Fax: 617-661-4405
E-Mail: sales@cytel.com

☐ I wish to purchase StatXact 3.1 for windows for the special student price of $195 (plus shipping as applicable*) and enclose a letter verifying my student status.

☐ I wish to purchase a printed copy of the manual for the price of $95.00 (plus shipping as applicable#)

☐ Cheque enclosed for the sum of ……… (including shipping)

☐ Charge my order to my credit card checked below

☐ Visa   ☐ Mastercard

Card No. _____   Expiry date _____

Signature _____

Ship to:

_____
Name

_____
Address

_____

_____

_____

* Shipping for the CD-ROM costs $5 for US sales and $10 elsewhere
# Shipping of the printed manual costs $10 for US sales and $25 elsewhere

Please note to avail of these prices you must enclose this card when ordering.